Polymerization in Biological Systems

The Ciba Foundation for the promotion of international cooperation in medical and chemical research is a scientific and educational charity established by CIBA Limited – now CIBA-GEIGY Limited – of Basle. The Foundation operates independently in London under English trust law.

Ciba Foundation Symposia are published in collaboration with Associated Scientific Publishers (Elsevier Scientific Publishing Company, Excerpta Medica, North-Holland Publishing Company) in Amsterdam.

Associated Scientific Publishers, P.O. Box 1270, Amsterdam

Polymerization in Biological Systems

Ciba Foundation Symposium 7 (new series)

1972

Elsevier · Excerpta Medica · North-Holland
Associated Scientific Publishers · Amsterdam · London · New York

ISBN Excerpta Medica 90 219 4008 6
ISBN American Elsevier 0-444-10401-1

Library of Congress Catalog Card Number 72-86558

Published in 1972 by Associated Scientific Publishers, P.O. Box 1270, Amsterdam, and 52 Vanderbilt Avenue, New York, N. Y. 10017.
Suggested series entry for library catalogues: Ciba Foundation Symposia.

Ciba Foundation Symposium 7 (new series)

Printed in The Netherlands by Mouton & Co., The Hague

Contents

Participants

Symposium on Polymerization Reactions in Biological Systems, held at the Ciba Foundation, London, 14th–16th March 1972

Chairman: E. KATCHALSKI Department of Biophysics, Weizmann Institute of Science, Rehovoth, Israel

J. BADDILEY Microbiological Chemistry Research Laboratory, School of Chemistry, The University, Newcastle upon Tyne NE1 7RU

J. P. CHANGEUX Département de Biochimie Cellulaire, Institut Pasteur, 25 rue du Docteur Roux, Paris 15, France

D. CHAPMAN Department of Chemistry, The University, Sheffield S3 7HF

P. CORRADINI Università di Napoli, Istituto Chimico, Via Mezzocannone 4, I–80134 Naples, Italy

M. EIGEN Max-Planck-Institut für Physikalische Chemie, Bunsenstrasse 10, D–3400 Göttingen, Germany

A. ELLIOTT School of Biological Sciences, Department of Biophysics, King's College, 26–29 Drury Lane, London WC2

J. ENGEL Biozentrum der Universität Basel, Abteilung für Biophysikalische Chemie, Basel, Switzerland

P. J. FLORY Department of Chemistry, Stanford University, Stanford, California 94305, USA

W. GILBERT The Biological Laboratories, Harvard University, 16 Divinity Avenue, Cambridge, Massachusetts 02138, USA

H. E. HUXLEY MRC Laboratory of Molecular Biology, University Postgraduate Medical School, Hills Road, Cambridge CB2 2QH

E. KELLENBERGER Biozentrum der Universität Basel, Abteilung für Mikrobiologie, Klingelbergstrasse 70, CH–4056 Basel, Switzerland

A. KLUG MRC Laboratory of Molecular Biology, University Postgraduate Medical School, Hills Road, Cambridge CB2 2QH

M. LEVITT MRC Laboratory of Molecular Biology, University Postgraduate Medical School, Hills Road, Cambridge CB2 2QH

S. LIFSON Chemical Physics Department, Weizmann Institute of Science, Rehovoth, Israel

SUSAN LOWEY Children's Cancer Research Foundation Inc., 35 Binney Street, Boston, Massachusetts 02115, USA

V. LUZZATI Centre National de la Recherche Scientifique, Centre de Génétique Moléculaire, F–91 Gif-sur-Yvette, France

GERTRUDE E. PERLMANN The Rockefeller University, New York, NY 10021, USA

D. C. PHILLIPS Department of Zoology, Laboratory of Molecular Biophysics, South Parks Road, Oxford OX1 3PS

M. REVEL Department of Biochemistry, Weizmann Institute of Science, Rehovoth, Israel

R. SCHWYZER Institut für Molekularbiologie und Biophysik, Eidg. Technische-Hochschule, Zürich-Hönggerberg, CH–8049 Zürich, Switzerland

I. SILMAN Department of Biophysics, Weizmann Institute of Science, Rehovoth, Israel

M. SZWARC Polymer Research Center, Department of Chemistry, State University College of Forestry, Syracuse, NY 13210, USA

C. TANFORD Department of Biochemistry, Duke University Medical Center, Durham, North Carolina 27706, USA

A. TISSIÈRES Département de Biologie Moléculaire, Université de Genève, 30 quai de l'Ecole de Médecine, CH–1211 Genève, Switzerland

W. ZILLIG Max-Planck-Institut für Biochemie, Goethestrasse 31, Postschliessfach 64, 8 München 15, Germany

Editors: G. E. W. WOLSTENHOLME and MAEVE O'CONNOR

Introduction: the objectives

E. KATCHALSKI

Department of Biophysics, Weizmann Institute of Science, Rehovoth

At this symposium we have brought together polymer chemists, who are mainly interested in polymerization and the properties of synthetic polymers, and biologists, who are mainly interested in the biosynthesis, structure and biological function of biopolymers. Biologists realize by now that a thorough understanding of the synthesis and behaviour of biopolymers both *in vivo* and *in vitro* requires a correspondingly thorough knowledge of chemistry and physics. At the same time chemists have been developing new theoretical and experimental techniques which offer new lines of approach to those biologists who are primarily concerned with the formation of biopolymers and their modes of intra- and intermolecular interaction. I hope, therefore, that at this symposium we shall find a common language to cope with the problems to be discussed.

The layman often refers to this age as the age of plastics, in which nylon, polystyrene, polyvinylchloride and other synthetic polymers are in everyday use. Biopolymers, such as proteins, nucleic acids and polysaccharides, play a major role in determining the most essential processes in living organisms. They have, in one form or another, been utilized since time immemorial, and by now even the man in the street is familiar with some of their specific biological properties, yet the interests of the molecular biologists are still rather different from those of the synthetic polymer chemist. The polymer chemist is mainly interested in the mechanical, electrical and other physical properties of the polymers which he has produced. The molecular biologist is interested in nucleic acids, proteins and other biopolymers because they store information, because some of them act as catalysts, and others possess recognition patterns which allow them to recognize other molecules. There is much room for increased cooperation between the two groups of scientists with their somewhat divergent interests.

We are going to begin by discussing polymerization mechanisms and the techniques used in the preparation of synthetic polymers. In some cases, the mechanisms involved in the formation of synthetic polymers resemble those observed in the biosynthesis of natural polymers, e.g. condensation. However, the commonly used polymerizations, in which radicals and carbonium ions participate as reactive intermediates, have no counterpart in nature. I wonder why this is so?

Professor Fritz Lipmann, who most certainly is not a synthetic polymer chemist, has divided the reactions by which biopolymers are formed into two classes: the tail type of polymerization and the head type of polymerization (Lipmann 1968). The tail-type polymerizations are those in which the polymer has an inert tail with which activated monomers react successively. In the head type of polymerization the head of the polymeric chain is active; elongation of the polymeric chain occurs as a result of the reaction of the active head with the monomer generating a fresh active head. Why does nature choose the tail type of polymerization to produce polysaccharides, RNA and DNA, and why is the head type of polymerization used in the synthesis of proteins, lipids and terpenes? Certainly nature has had a long time to try out various types of polymerization before making its choice.

The molecular biologist studying the biosynthesis of natural macromolecules has gradually come to realize the complexity of the reactions involved. The classical biochemist is mainly familiar with homogeneous enzymic reactions. In fact he usually does his best to study enzymic reaction in dilute aqueous buffer solutions. Unfortunately, many of the biologically important polymerizations occur in heterogeneous media. Protein biosynthesis, for example, requires the participation of ribosomes, ribosome aggregates, the polysomes and numerous protein factors; moreover the endoplasmic reticulum seems to play a major role in organizing and regulating the process. This is somewhat annoying for the classical biochemist; it is also somewhat daunting for the classical synthetic polymer chemist who is aware of the fact that even 'simple' polymerizations catalysed by heterogeneous catalysts and leading to isotactic and syndiotactic polymers are still far from being thoroughly understood. Even the relatively simple biopolymerizations involved in cell wall formation present many intricate problems still to be solved.

In spite of the complexity of biopolymerizations, the molecular biologists have succeeded in recent years in reproducing *in vitro*, partially or completely, many such processes. It is thus possible to study in the test tube the various factors involved in the biosynthesis of proteins, polysaccharides and nucleic acids. The replication of nucleic acids *in vitro* has, moreover, allowed the study of evolution at the molecular level. The theoretical implications for molecular evolution of the beautiful experiments of Dr Spiegelman with Qβ-replicase will be discussed by Dr Eigen.

During the symposium we shall spend some time discussing the factors determining the secondary structure of synthetic polymers and biopolymers. Starting in the late 1940s, the polymer chemists attempted to derive many of the properties of synthetic linear polymers, both in the solid state and in solution, from a statistical treatment of the random coil (see, for example, Flory 1953). In recent years, however, it has become evident that intramolecular atomic interactions, bond lengths, bond angles and bond rotational potentials severely restrict the conformations which a linear polymer chain can assume. Furthermore, polymer–solvent interactions play an important role in determining the conformation of polymers in solution. In coping with these problems the polymer chemist has devised new theoretical approaches for predicting the macromolecular conformations of lowest free energy (Flory 1969). The new theoretical tools thus developed have been adopted by the theoretical protein, polysaccharide and nucleic acid chemists for evaluating some of the factors determining the highly specific conformations assumed by native biopolymers.

The above theoretical approach is essentially a thermodynamic one; the biologist, however, is interested not only in the relatively stable macromolecular conformations but also in the mechanisms and reaction rates involved in the transformation of a given polymer from one conformation to another. The biologist wants to know how fast a polypeptide chain coils into its active three-dimensional structure, how fast a double helix forms, and how fast it unwinds. Furthermore, rather subtle conformational changes seem to be directly involved in the activity of enzymes, antibodies and receptors, as illustrated by the recent findings of Perutz (1970) on haemoglobin and Lipscomb *et al.* (1969) on carboxypeptidase. Information as to the rates of such conformational transitions is thus crucial to the understanding of specific biological interactions. Unfortunately kinetic data on the rates of conformational transitions of both synthetic and biopolymers are still scanty.

In current statistical mechanical analyses of chain molecules it is still difficult to take into consideration interactions between residues far apart on the linear polymer chain. Such interactions are, however, of great importance in the determination of the tertiary and quaternary structures of proteins and other biopolymers.

Nature possesses a variety of means of attaining the appropriate conformations, associations and aggregations of biopolymers. These may be the direct result of the primary structure of the biopolymer. They may also result from the chemical modification or cross-linking of suitable residues, or from the reversible binding of metal ions, cofactors, or other ligands. In some cases polymers possessing a higher molecular weight than the final product are formed initially and are subsequently enzymically degraded to yield the final species. Phosphorylation of enzymes by the appropriate kinases can serve as an example of conformation and biological activity being modified as a result of chemical modification (for review see Holzer & Duntze 1971). Disulphide bridges in fibrillar and globular proteins are known to stabilize their secondary, tertiary and quaternary structures markedly. The transformation of zymogens to active enzymes as a result of partial proteolytic degradation has been shown in some cases to lead to a marked conformational change. Similar changes most likely occur when proinsulin is transformed into insulin and fibrinogen into fibrin. It has been shown recently that even formation of the triple helix of collagen is preceded by the formation of a procollagen molecule possessing a considerably higher molecular weight than collagen (Bellamy & Bernstein 1971). The two ribosomal RNA fractions also stem from a single polynucleotide chain which is cleaved enzymically to yield the desired fragments (Grierson *et al.* 1970).

In the latter stages of the symposium we shall concentrate on assembly processes, i.e. on interactions between the subunits of a given protein, and on protein–protein, protein–nucleic acid and protein–lipid interactions. All these specific interactions between macromolecules depend largely on the ability of one molecule to recognize another. This feature, which is so characteristic of biopolymers, has not been utilized by synthetic polymer chemists but it strikes me as being one of the most important characteristics of the living organism. This is how hormones recognize receptors, how an antibody recognizes an antigen, and maybe how one cell recognizes another. In this connection it is worth mentioning, for the sake of our polymer chemists, that the recently developed technique of affinity chromatography is also based on the ability of a synthetic macromolecule, suitably modified, to recognize a corresponding biopolymer (Porath 1967; Cuatrecasas & Anfinsen 1971). The affinity chromatography technique is currently being extended to the isolation of cell receptors, and even to the separation of specific classes of cells.

We shall obviously not be able to cover thoroughly all the topics I have outlined. I hope, however, that those who discuss a particular topic will state clearly what has been achieved, what are the focal problems to be solved, and what theoretical and experimental difficulties may be encountered in so doing.

References

BELLAMY, G. & BERNSTEIN, P. (1971) *Proc. Natl Acad. Sci. U.S.A.* **68**, 1138–1142

CUATRECASAS, P. & ANFINSEN, C. B. (1971) *Annu. Rev. Biochem.* **40**, 259–278

FLORY, P. J. (1953) *Principles of Polymer Chemistry*, Cornell University Press, Ithaca

FLORY, P. J. (1969) *Statistical Mechanics of Chain Molecules*, Interscience, New York

GRIERSON, D., ROGERS, M. E., SARTIRANA, M. L. & LOENING, U. E. (1970) *Cold Spring Harbor Symp. Quant. Biol.* **35**, 589–598

HOLZER, H. & DUNTZE, W. (1971) *Annu. Rev. Biochem.* **40**, 345–374

LIPMANN, F. (1968) In *Essays in Biochemistry* (CAMPBELL, P. N. & GREVILLE, G. D., eds.) Vol. 4, pp. 1–23, Academic Press, London

LIPSCOMB, W. N., HARTSUCK, J. A., REEKE, G. N., JR., QUIOCHO, F. A., BETHGA, P. H., LUDWIG, M. L., STEITZ, T. A., MUIRHEAD, H. & COPPOLA, J. C. (1969) *Brookhaven Symp. Biol.* **21**, 24–90

PERUTZ, M. F. (1970) *Nature (Lond.)*, **228**, 726–739

PORATH, J. (1967) In *Gamma Globulins: Structure and Control of Biosynthesis* (KILLANDER, J., ed.), (Nobel Symposium 3), pp. 287–294, Interscience, New York

Synthetic polymers, biopolymers and block polymers

MICHAEL SZWARC and CONRAD SCHUERCH

Polymer Research Center, Department of Chemistry, State University College of Forestry, Syracuse, New York

Abstract During the last two decades two important factors have become controllable in polyaddition, the technique yielding synthetic polymers of high molecular weight. Coordination or ionic propagation provided means to direct in a highly specific way the stereochemical course of the reaction, and the utilization of living polymers allowed us to regulate the molecular weight of the resulting polymers and to produce them in uniform size. Moreover, the use of living polymers permits the preparation of regular block polymers—long-chain molecules composed of sequences of monomers of desired length which follow each other according to a predetermined pattern.

Examples of stereoregular polymerization will be reviewed and the factors responsible for the stereospecificity discussed. The cationic polymerization of some anhydro sugars, developed by one of us, will be considered in greater detail. This reaction yielded, for the first time, synthetic linear polysaccharides of uniform structure, such as dextran, which exhibit optical and biological activity.

The incompatibility of polymers built from different units leads to the formation of mesomorphic structures in block-polymer systems. The principles governing the mesomorphic structures will be discussed and examples will be given of new types of materials obtained by this route.

Two distinct techniques, polycondensation and polyaddition, are used in the synthesis of high molecular weight polymers, but only the latter is discussed in this review. Three basic steps are involved in a typical polyaddition: the active centres from which polymeric molecules evolve are produced in the initiation step, monomeric units are added consecutively to growing polymers during propagation, and finally the ability to grow possessed by each active end of a growing polymer is eventually lost in some termination step. The propagation steps determine the stereochemical structure of the resulting polymer and the termination steps govern their molecular weight and molecular-weight distribution.

STEREOSPECIFIC POLYMERIZATION

In the past, synthesis of polymers led to random structures, that is—to apply modern terminology—the resulting product was atactic. Similarly, copolymerization of two or more monomers led to their random distribution along the polymer chain, alternating copolymers being an exception. The advent of ionic and coordination techniques of polymerization radically changed the situation. Natta made the striking discovery (Natta 1955; Natta *et al.* 1955) that stereospecific polymerization is both feasible and profitable on an industrial scale. His preparation of isotactic polypropylene was followed by numerous syntheses of other isotactic and syndiotactic vinyl or vinylidene polymers, that is polymers with identical or alternating pseudo-asymmetrical carbon atoms along their chains:

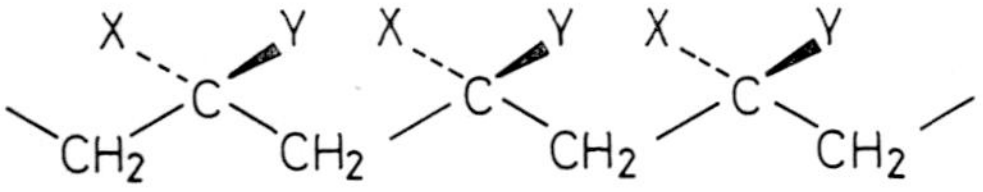

Isotactic chain

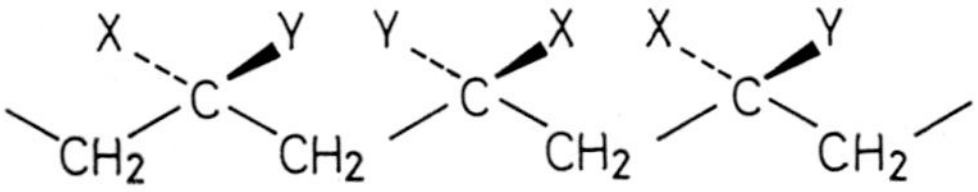

Syndiotactic chain

The anionic polymerization of isoprene performed in hydrocarbon medium with Li^+ acting as the cation led to all *cis*-1,4-polyisoprene,

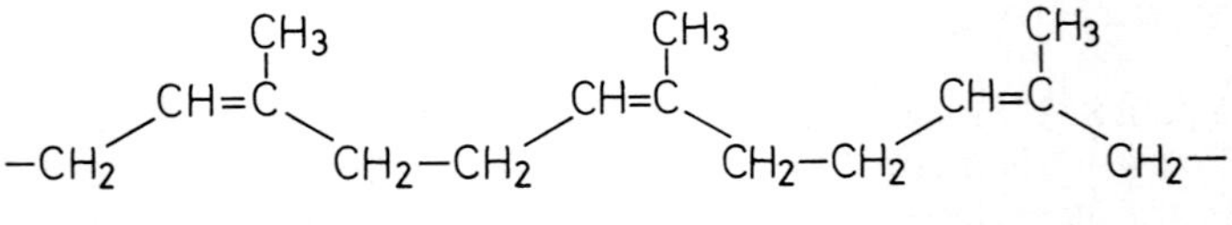

a synthetic material closely resembling the natural rubber. Soon after, catalysts were discovered which allow us to synthesize all *cis*-1,4 or all *trans*-1,4 polydienes, e.g. the synthetic all *trans*-1,4 polyisoprene represents a man-made analogue of another natural product—guttapercha. The power of stereoregular polymerization is more appreciated if one recalls that four distinct stereospecific polymers, namely, all *cis*-1,4-polybutadiene, all *trans*-1,4-polybutadiene, isotactic all 1,2-polybutadiene and syndiotactic all 1,2-polybutadiene, were produced from the same monomer—butadiene.

The techniques of stereospecific polymerization have been extended to other classes of monomers, for example to cyclic ethers. Price & Spector (1965), using ion-complex catalysts, synthesized a stereoregular polypropylene oxide from optically active propylene oxide although racemization occurs when the polymerization is performed under conventional conditions. This polymer possesses truly asymmetrical carbon atoms in its backbone:

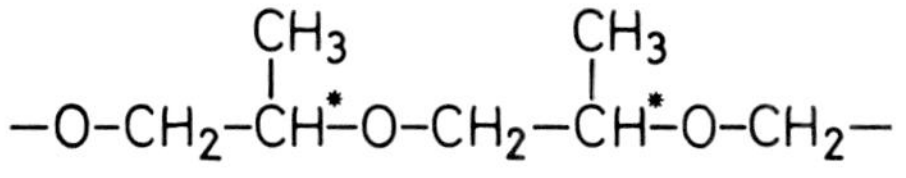

that is, the product is optically active if produced with the aid of a proper ionic (or coordination) catalyst from pure enantiomorph. This observation leads to a question: can a stereospecific system be devised which polymerizes only one enantiomorph in a racemic mixture? This problem has been examined by Pino (1965), who designed stereoelective catalysts and discussed their applications. Further development of stereospecific polymerization led to the synthesis of polymers in which each pair of two consecutive carbon atoms has a specific structure; thus, all *threo* or *erythro* chains have been prepared.

It is not possible to discuss in this brief review all the details of stereospecific polymerization. Suffice to say that the stereoregular polymers often show properties much different from those of the atactic (random) polymers and, hence, they have considerable practical as well as academic value. It would be advantageous, however, to consider the reasons why the stereospecificity operates in ionic or coordination polymerization and not in a radical polymerization.

The addition of a radical to a C=C bond is not guided by any preferential direction, although steric hindrance, usually not a very compelling reason, may favour one mode of the addition rather than another. In ionic or coordination addition a specific direction is introduced by the cation-anion axis of an ion-pair or coordination complex. This differentiates between the various modes of the addition and favours a stereospecific behaviour of the system. We therefore do not expect to find high stereospecificity in a process involving free ions.

A second and even more important factor favouring stereospecificity of ionic or coordination polymerization arises from the monomer–counter-ion interaction. The monomer, as well as the penultimate units of the growing polymer, may solvate the counter-ion and form an organized complex. The structure of such a complex compels the subsequent addition of the monomer to the growing polymer to proceed in a well-defined fashion. Moreover, the simultaneous coordination of the penultimate unit and the monomer with the counter-ion 'informs' the monomer about the configuration of the growing chain. Indeed, Fowells *et al.* (1967) demonstrated that the addition of selectively deuterated acrylate or methacrylate, 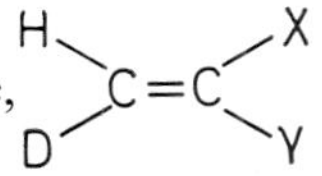, leads to an orderly array of the 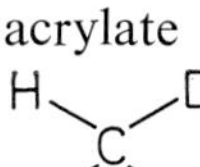 groups in the resulting isotactic polymer when the monomer, before its addition, acts as a solvating agent of the counter-ion, i.e. when the polymerization takes place in a poorly solvating medium the molecules of which do not compete efficiently with the monomer for the solvating sites. However, in a better solvating medium this preferential orientation is lost because then only solvent molecules participate in the solvation of the counter-ion.

The mode of coordination of the monomer to the counter-ion is even more specific in coordination polymerization in which the monomer becomes a true ligand before its addition to the chain. In such systems the nature of other ligands is of paramount importance in determining the stereospecificity of polymerization and, indeed, the stereospecificity may be altered by replacing one set of ligands by another.

LIVING POLYMERS

The anionic mode of polymerization allows us to control another feature of the reaction, namely its termination step. Termination is virtually unavoidable in radical polymerization because two radicals combine or disproportionate on encountering each other and either reaction leads to their annihilation, and hence to the loss of their reactivity. Such a bimolecular termination does not take place in ionic polymerization and, in fact, in proper conditions termination may be entirely eliminated in anionic polymerization (Szwarc 1956). The resulting polymer is known as living polymer; it ceases to grow when all the monomer is polymerized* but it retains its ability to continue its growth whenever additional monomer is supplied.

The concept of living polymers has numerous ramifications. Since polymerization proceeds without termination it becomes feasible to prepare polymers of nearly uniform molecular weight (having a Poisson distribution of molecular weights). Indeed, such polymers are produced now on a semi-industrial scale. Although spontaneous termination is eliminated a deliberate conversion of active ends of living polymers into inactive ends is still feasible. For example, the carbanions of living polystyrene or polydienes may be deprived of their ability to continue their growth if a proton-donating reagent is added; each of the resulting dead polymers then contains a C-H end-group. However, if CO_2 is added instead of a proton-donating reagent, an identical polymer is formed which possesses a COOH end-group in place of a C-H end-group. Similarly, the addition of ethylene oxide endows it with a hydroxylic end-group, and still

* Or, more correctly, when the monomer attains its equilibrium concentration characterizing the monomer-living polymer system (szwarc 1967).

other desired end-groups can be introduced by using other reagents. Hence, a series of polymers may be prepared, identical in every respect except the nature of their end-groups. The electron-transfer initiation (Szwarc 1956) leads to living polymers with both ends growing and thereafter their deliberate termination yields bi-functional polymers such as dicarboxylic acid, diols, etc. Extension of this work led to the preparation of star-shaped polymers—macromolecules with several chains protruding from a common centre. Each of their ends, if so desired, may be again endowed with any required functional end group.

BLOCK POLYMERS

The capacity of living polymers to start growing again when a suitable monomer is added leads to the synthesis of block polymers—macromolecules in which a long sequence of one monomer, A, is followed by a sequence of another monomer, B, and so on. Thus, a two-block polymer has the structure

AA......A.BB......B

A tri-block polymer may be of the type

AA....A.BB....B.AA....A

or

BB....B.AA....A.BB....B

or

A.A....A.BB....B.CC....C, etc.

The preparation of block polymers through the living-polymer technique is extremely simple. By 'feeding' a living polymer with monomer A one produces sequences of A's of the desired size and, if necessary, of uniform length. Thereafter, the feeding is continued with another monomer, say B, and a desired sequence of B is added then to the block of A. The only requirement is that a living poly-A is capable of reacting with B. If a living poly-B can initiate the polymerization of A, one can produce a series of polymers of identical size and identical composition, say 50:50 A to B, with any desired distribution of A's and B's along the chain. Hence, instead of random A-B copolymers, organized polymers with a predetermined pattern of monomers may be prepared.

Further simplification of preparation of tri-block polymers, A....A.B....B. A....A, is achieved if a living poly-B with both ends active is used in the synthesis. Feeding such a polymer with monomer A yields at once a tri-block polymer with equally long sequences of A flanking the centre sequence of poly-B.

The behaviour of tri-block polymers, polystyrene-polyisoprene-polystyrene, or polystyrene-polybutadiene-polystyrene, exemplifies some unusual properties of block polymers. Such polymers prepared in hydrocarbon medium with Li^+ counter-ions have the properties of vulcanized, i.e. cross-linked, rubber, and the tensile strength of the resulting material at room temperature is appreciable. However, at elevated temperatures the material melts, becomes fluid and can be recast in a new mould; hence, it is referred to as thermoplastic rubber. The high tensile strength is lacking, however, in both a tri-block polybutadiene-polystyrene-polybutadiene and a di-block polystyrene-polybutadiene, even when the composition and molecular weight of these polymers are identical to those of the thermoplastic rubber. We face here a striking manifestation of properties which are determined by the mode of distribution of monomers along the polymer chain.

What endows the thermoplastic rubber with its unusual properties? Its peculiar behaviour is determined by its morphology. As is well known, most of the polymers are incompatible with each other; hence the polystyrene segments of the thermoplastic rubber tend to aggregate, forming domains of polystyrene dispersed in a continuum of the rubbery polybutadiene or polyisoprene. These domains act as giant cross-linking centres when the material is kept below the glass temperature of polystyrene, and thus the properties of vulcanized rubber appear. At higher temperatures the polystyrene domains melt and the cross-linking behaviour is lost. The cross-linking characteristic requires each polybutadiene chain to possess two 'anchors' and therefore the polystyrene-polybutadiene-polystyrene tri-block, and not the polybutadiene-polystyrene-polybutadiene, exhibits the desired properties.

The incompatibility of polymers gives rise to various morphological structures, often referred to as mesomorphic structures, that spontaneously develop in concentrated, gel-like block-polymer solutions. The tendency for blocks of the same kind to become aggregated leads to domains the dimensions of which are determined by the molecular weight of the blocks, as well as by the nature and the amount of the added solvent. The length of the blocks limits the size of the domains. A domain of poly-A cannot be extended further away from its surface than the length of the poly-A block, otherwise a block of poly-B would be pulled into a domain of poly-A. This is precisely what the system tries to avoid.

What are the morphological forms of feasible domains? At least three forms are known: layers containing one kind of blocks alternating with layers built from the others, spheres of blocks A regularly dispersed in a continuum of blocks B, and parallel cylinders of A regularly protruding through the continuum of B. The morphology of the domains is governed by factors such as the nature of the segments and of the solvent, the composition and molecular weight of the block polymers, and the temperature and concentration of the gel.

The parallel layers of poly-A blocks alternating with layers of poly-B blocks seem to represent the simplest morphological structure. Since the layers are of uniform size they act as a diffraction grating, and for a sufficiently high molecular weight, i.e. for sufficiently thick layers, the colourless material acquires an interference colour which may be varied by varying the molecular weight of the

(a)

(b)

FIG. 1. Electron micrographs of polystyrene-polybutadiene block polymers in rigid medium. Sliced (*a*) perpendicularly and (*b*) parallel to the axis of the cylinder. (From Ruckel & Schuerch 1966.)

blocks. When shear takes place the colours change but the material relaxes thereafter and the original colour returns.

Studies of mesomorphic structures of block polymers were initiated by the pioneering investigations of Skoulios *et al.* (1960), Skoulios & Finaz (1961) and Sadron (1963). Di-blocks of polystyrene and polyethylene oxide were the first to be investigated by the low-angle X-ray scatter technique. The gels produced with nitromethane solvent resulted in cylindrical micelles whereas lamellae were observed when butyl phthalate was used as a solvent.

It is also interesting to note that the properties of the materials produced from block polymers after evaporation of the solvent depend on the nature of the solvent. For example, when cast from a solvent good for polystyrene but bad for polybutadiene, a block polymer of polystyrene and polybutadiene yields a film with properties resembling those of polystyrene film. However, a rubbery film is obtained when the material is cast from a solvent good for polybutadiene but poor for polystyrene.

Let us close this section by describing interesting and novel materials prepared by B. Gallot and C. Sadron. Various di-block polymers were prepared by the living-polymer technique and mesomorphic phases were developed using monomers such as styrene or methyl-methacrylate as the solvents. After the mesomorphic phase had been developed the monomeric solvent was polymerized, e.g. photochemically. The gel then becomes converted into a rigid material which may be sliced with a microtome and stained. Two different slices obtained from the same sample are shown in Fig. 1.

Inspection of these figures shows that the block polymers are forming long cylinders of polybutadiene (stained by OsO_4) in a matrix of polystyrene. The slicing is random; however if it happens to be in a plane perpendicular to the cylinder axes a photograph like that shown in Fig. 1*a* is obtained. On the other hand, if the slicing occurred in a plane parallel to the cylinder axes the resulting photograph looks like Fig. 1*b*.

SYNTHETIC STEREOSPECIFIC POLYSACCHARIDES

The most challenging field for the application of our knowledge of the mechanisms of stereoregular polymerization and of the nature of reaction intermediates is the synthesis of biopolymers. Of the three major polymer systems found widely in nature—nucleic acids, proteins and polysaccharides—only the polysaccharides require for their synthesis a chain-forming reaction under complete steric control. For this reason alone, their synthesis is perhaps the most difficult.

The chain-forming reaction is an acetal synthesis, formally between the hemiacetal function at $C_{(1)}$ and any one of several different hydroxyl groups on the substituent sugar units. The monomeric sugars are asymmetrical. The hexoses, for example, have four asymmetrical carbon atoms, one asymmetrical hemiacetal function, two possible ring sizes and four sites of linkage between the $C_{(1)}$ hemiacetal and a hydroxyl group. There are, therefore, 256 stereo-regular linear homopolysaccharides theoretically derivable from aldohexoses alone. To control the structure of the polymer produced, the ring size must be fixed and all but one hydroxyl group blocked before the chain-forming reaction is attempted.

By far the best results in terms of high stereospecificity and high molecular weights have been obtained in the cationic-ring-opening polymerization of some anhydro sugar derivatives. To date, three ring types have been investigated, namely 1,2-anhydro, 1,4-anhydro, and 1,6-anhydroglucopyranoses. For a variety of reasons ester groups are not satisfactory blocking groups in these syntheses and ether substituents must be used instead. Since the only known 1,2-anhydro sugar derivatives contain ester functions, their polymerization will not be discussed. The 1,4-anhydro and 1,6-anhydro sugar derivatives poly-merize in a manner reminiscent of the cationic polymerization of tetrahydro-furan. However, the asymmetry of these monomers gives more insight into the reaction and it is now apparent that there are well defined although rather narrow conditions within which stereospecific polymerizations can be achieved.

The mechanism usually proposed for tetrahydrofuran polymerization involves trialkyl oxonium ions as intermediates:

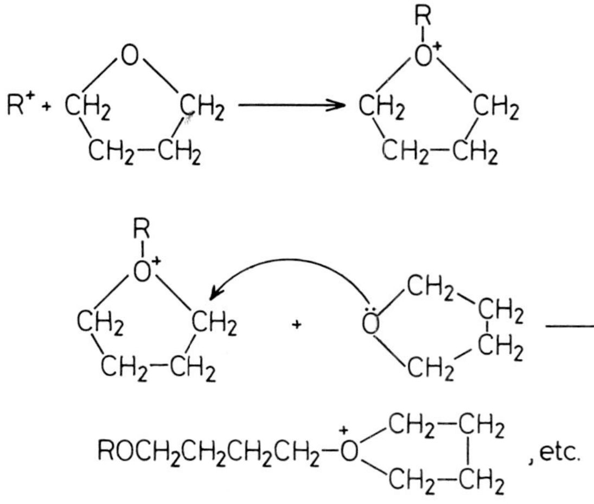

This reaction mechanism implies simultaneous bond breaking and bond formation and in the anhydro sugars it should result in stereospecific polymerization. In the 1,4-anhydro sugars two factors interfere. First there is little difference in basicity between the two ring oxygens and little difference in strain between the five- and six-membered rings. As a result, either ring oxygen may be attacked by the propagating ion with the corresponding ring-opening reaction. Therefore both pyranose and furanose rings are incorporated into the chain.

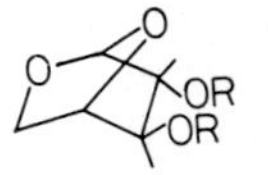

In spite of this complexity both reactions might still proceed stereospecifically. However both α and β furanosidic structures are in fact found in the polymer and their relative proportions vary with reaction conditions in ways which suggest that the interconversion of a close and a loose ion pair accompanies the polymerization. It is not possible to explain these and other results in terms of a trialkyl oxonium ion intermediate, and one must postulate, therefore, the presence of carbonium ions (see Fig. 2).

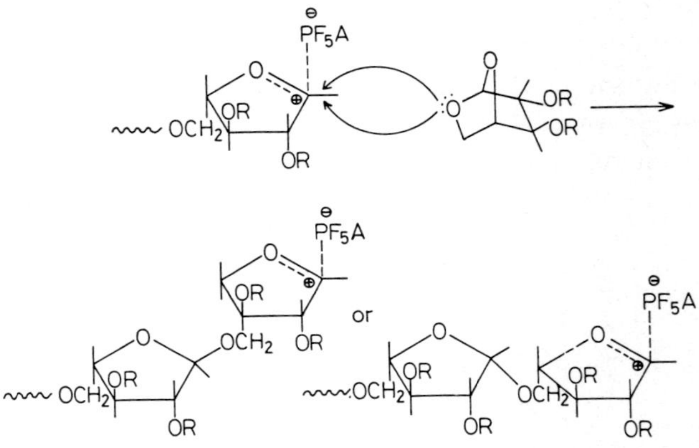

FIG. 2. Propagation by a carbonium ion mechanism.

The 1,6-anhydro sugar derivatives form a fused five- and six-membered ring. The ring-bridging by $C_{(6)}$ forces the substituents on $C_{(1)}$ and $C_{(5)}$ to be in axial conformation. In the anhydro gluco-, manno- and galacto-sugars (all of which have been polymerized), either two or three other alcoholic substituents are also axial. There is therefore substantial driving force from non-bonded interactions leading to the opening of the five-membered ring during polymerization, and the resulting polymers have the pyranose structure.

The stereoregular polymers are those expected from the attack of ring oxygen on $C_{(1)}$ of the trialkyl oxonium ion which proceeds simultaneously with the bond-breaking process (Fig. 3).

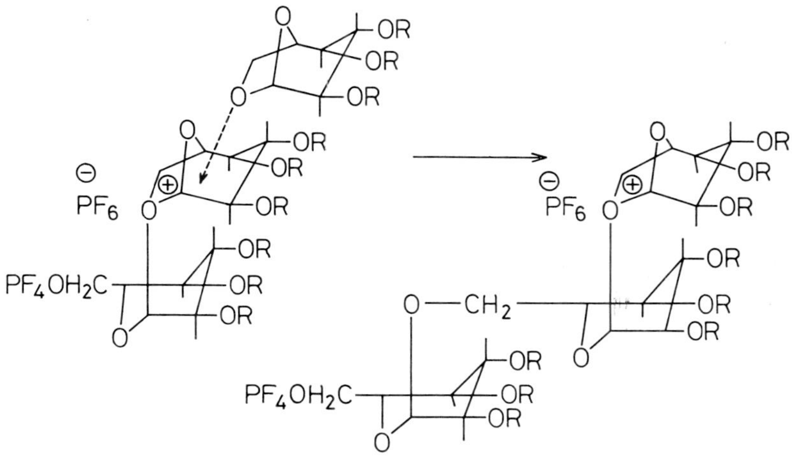

FIG. 3. 1,6-Anhydro ring-opening polymerization; stereospecific activated chain. (From Ruckel & Schuerch 1966.)

Although many of the main features of the polymerizations can be explained in terms of a propagating cation, the counter-ion also plays a role in the process. To date stereoregular polymers have been obtained only with Lewis acids that contain fluorine, i.e. with boron trifluoride, phosphorus pentafluoride and antimony pentafluoride, while antimony pentachloride or tin tetrachloride yield non-stereospecific polysaccharides. The temperature at which the polymerization proceeds is also a function of the counter-ion. Boron trifluoride etherate catalyses stereoregular polymerization at temperatures ranging from $+25°$ to $-20°C$; at lower temperatures the reaction does not occur. Phosphorus pentafluoride initiates stereospecific polymerization at lower temperatures, i.e. from about $-55°$ to $-78°C$ but above this temperature the reaction is no longer stereoregular. Apparently the stability and reactivity of the propagating ion is a function of the counter-ion.

Most of the work has been carried out with 1,6-anhydro-2,3,4,-tri-O-benzyl-β-D-gluco-, -manno-, and galactopyranose. These differ only in their configurations on $C_{(2)}$ and $C_{(4)}$ centres. However their rates of polymerization are quite different: manno>gluco>>galacto, and the maximum degrees of polymerization obtained follow the same order: manno>gluco>galacto. The polymers are amorphous because of the bulky aromatic side groups and readily soluble in a variety of solvents. However, the solubilities of the parent polysaccharides, obtained by debenzylation with sodium in liquid ammonia, are strikingly different. The glucan polymer is water-soluble with some tendency to crystallize out on standing. The mannan polymer is dispersible and partially soluble in water but requires dimethyl sulphoxide for its complete solution. The galactose polymer is insoluble in essentially all except some complex-forming solvents. This striking difference in solubility results solely from the difference in configuration around a single carbon atom attached to a hydroxyl group and not from a configurational difference at $C_{(1)}$ which differentiates starch from cellulose. Anhydrodisaccharides can also be polymerized to chains of 14–30 saccharide units, and their copolymerizations proceed normally.

The synthetic linear glucan has a backbone structure identical to that of clinical dextran, which is a slightly branched polysaccharide produced by bacteria and used as a blood volume expander. A similar dextran is implicated in caries formation by oral bacteria. The synthetic glucan has been tested enzymically by two independent laboratories and found to be 100% stereoregular and about 98% α1,6. Similar results have been obtained on the mannan. Failure to precipitate conconavalin A confirmed the linearity of both these synthetic polysaccharides. The mannan and galactan are unknown in nature except as fragments of more complex polysaccharides from yeast and fungi. A related α1,6-linked mannan with α1,2-linked mannose side chains is formed by a dermatophyte and causes allergic reactions. Similar allergenic effects are observed with the synthetic mannan and this unequivocally demonstrates that these reactions are induced by carbohydrate and not by protein or nucleic acid impurities. A similar mannan has also been reported to be active against viral infections by activating interferon production and to have caused regression of sarcoma 180 in experimental animals.

It is therefore obvious that these polysaccharides serve a useful function as model substances in immunological, serological, enzymic and pharmaceutical investigations. They also permit the systematic investigation of structure-property relationships.

References

Fowells, W., Schuerch, C., Bovey, F. A. & Hood, F. P. (1967) *J. Am. Chem. Soc.* **89**, 1396

Natta, G. (1955) *J. Polym. Sci.* **16**, 143

Natta, G., Pino, P., Corradini, P., Danusso, F., Mantica, E., Mazzanti, G. & Moraglio, G. (1955) *J. Am. Chem. Soc.* **77**, 1708

Pino, P. (1965) *Adv. Polym. Sci.* **4**, 393

Price, C. C. & Spector, R. (1965) *J. Am. Chem. Soc.* **63**, 2069

Sadron, C. (1963) *Angew. Chem. Int. Ed. Engl.* **75**, 472

Skoulios, A. & Finaz, G. (1961) *C.R. Hebd. Séances Acad. Sci. (Paris)* **252**, 3467
Skoulios, A., Finaz, G. & Parrod, J. (1960) *C.R. Hebd. Séances Acad. Sci. (Paris)* **251**, 739
Szwarc, M. (1956) *Nature (Lond.)* **178**, 1168
Szwarc, M. (1967) *Adv. Polym. Sci.* **4**, 457

A more detailed review of the subject may be found in:

Szwarc, M. (1968) *Carbanions, Living Polymers and Electron Transfer Processes*, Interscience Publishers, New York

The synthesis of stereospecific polysaccharides is described by:

Masura, V. & Schuerch, C. (1970) *Carbohydr. Res.* **15**, 65
Ruckel, E. R. & Schuerch, C. (1966) *J. Org. Chem.* **31**, 2233
Veruovic, B. & Schuerch, C. (1970) *Carbohydr. Res.* **14**, 199
Zachoval, J. & Schuerch, C. (1969) *J. Am. Chem. Soc.* **91**, 1165

Discussion

Katchalski: Why don't the various types of polymerization that you described, such as ionic polymerization and the coordination type, appear in biological systems? Is it because these types cannot occur in water?

Szwarc: That is not the only reason. Some of these polymerizations may proceed in aqueous solution. However, most of the monomers I discussed never appear in biological systems. For example, no isoprene is present in a rubber tree and biosynthesis of rubber proceeds through different intermediates than in its laboratory or industrial synthesis.

Katchalski: You indicated that some of your polymerization techniques lead to products possessing sharp molecular weight distributions of the Poisson type, yet the biologist is familiar with a sharper type of distribution where every molecule is of equal weight. Of course there are biopolymers such as polysaccharides which display polydispersity, but in most cases all the molecules of the biopolymer, such as those of a given protein or nucleic acid, are of the same size. Formation of molecules of the same size and structure is determined *in vivo* by a template mechanism. Can you suggest any other mechanisms for obtaining molecules of identical size?

Szwarc: The template mechanism usually operates whenever polymers of the type A–B–C–D– are produced, i.e. when there is a non-uniform pattern in the resulting polymer. The pattern as well as the size have to be determined by a template and hence the resulting polymer has a unique size. The biological way of achieving such a template polymerization is well known and it is unnecessary for me to elaborate this topic.

A synthetic template polymerization has been attempted in Germany (see

Kämmerer & Ozaki 1966). The idea was to produce a polyol, esterify it with acrylic acid and polymerize the vinyl monomer attached to such a 'template'.

Changeux: Could you make polymers of the compounds referred to as photochromes whose properties change when exposed to light?

Swarc: Yes. Prins (1971), for example, prepared gels which contain chromophoric groups, which upon irradiation isomerize and then soak up more water and therefore swell. In the dark the original isomer is re-formed and the gels let the water out. Smets (1968) has synthesized polymers with stilbene units in the chain. On irradiation the *trans* form isomerizes to the *cis*, and this substantially modifies the conformation of the polymer and therefore its structure and properties. Again, there is a relaxation which takes place in the dark.

Luzzati: The structural work on these systems originated from the study of lipids. The analogy is obvious. Lipid molecules consist of hydrocarbon chains and polar groups: the two moieties have different solubilities. This led us to try block copolymers in the presence of solvent, with the hope that the two regions would segregate according to their affinity with the solvent. Indeed several structures were obtained, which we had found previously in lipid-water systems.

The problem of monodispersity is quite interesting. In the field of proteins and nucleic acids we are now accustomed to dealing with monodisperse compounds, with a specific and unique chemical structure. But it is not always like that in biology: membranes are an interesting example where the lipids are highly polydisperse. In spite, or perhaps because, of that polydispersity, and a high disorder in the short-range conformation, lipids can display a large number of highly ordered phases.

Engel: Is the equilibrium in living polymers always on the side of the polymer, Dr Szwarc? Is there a difference in equilibrium constant between the binding of A to B and the binding of equal segments? That is, do you have to consider a redistribution of A and B in your polymers?

Szwarc: I presume that you are enquiring about the equilibrium between a living polymer and its monomer. A living polymer containing n units may add its monomer and yield an $n+1$-mer. According to the principle of microscopic reversibility the $n+1$-mer may degrade and yield an n-mer and monomer. Therefore, by virtue of the fact that the polymers are living, in my meaning of the term, they degrade as well as grow and therefore the conversion of monomer into polymer never goes to completion. Thus, the system attains a state of equilibrium at a concentration of monomer at which the rate of propagation is exactly balanced by the rate of depropagation. Living polymer is then in equilibrium with its monomer. One should ask, what is this equilibrium monomer concentration? That depends on the nature of the monomer, on the temperature and on the nature of the solvent, but not on the mechanism of polymerization. For living polystyrene at, say, room temperature the equilibrium concentration of the monomer in most solvents is of the order of 10^{-7} M, so for all practical purposes one can say polymerization proceeds to completion. For α-methylstyrene, the equilibrium concentration of the monomer at room temperature is about 1.5 M, hence an enormous amount of monomer is left in this system. But at −60 °C the equilibrium concentration decreases to an immeasurably low level and for all practical purposes the polymerization is then quantitative.

You ask whether I get a redistribution of monomers when a chain of A's is followed by a chain of B's. As B is added and then B grows, all the residual A becomes incorporated in the resulting polymer.

Engel: If the binding constant for A and B is very different from that between equal segments, then these mixed block copolymers of A and B will be eaten up, and you will end up with molecules which consist entirely of A or B segments.

Szwarc: Yes, but not always. Moreover the redistribution, if thermodynamically favoured, is an extremely slow process. If you make a long chain of B you will have to wait until one long chain of B is completely degraded and another long chain of B doubles its size. In practice, you don't need to worry about this happening.

Flory: If one disregards the structures of the anionic polymerization process, which is limited to addition or removal of monomers one at a time to the active chain terminus, and assumes instead that interchanges between two chains may occur—for example, as follows:

$$\left.\begin{matrix} \ldots\mathrm{AA\ AA}\ldots \\ + \\ \ldots\mathrm{BB\ BB}\ldots \end{matrix}\right\} \rightarrow \begin{matrix} \ldots\mathrm{AA\ AA}\ldots \\ |+| \\ \ldots\mathrm{BB\ BB}\ldots \end{matrix}$$

—then the rearrangements under discussion might readily occur. The interactions between A and B could then be accommodated, and if they are favourable for A to adjoin B, then the block copolymer, consisting of long segments of A and of B units, could readily be converted into the random state. In the opposite case, where the A,B contacts are highly disfavoured, the interchange processes could lead spontaneously to segregation, starting with the random case. Thus, the rearrangement of sequence may conceivably go either way, and this might be of some significance in biochemical systems.

Szwarc: That is correct but, as you said, it requires operation of another mechanism than the one in the polyadditions discussed in my paper.

Eigen: Whether you get a block or a mixed polymer depends on whether the cooperative interaction A...A A...B is preferred. The interaction constants must come in. If such a cooperative interaction exists, would you then find in the polymerization mechanism something like a cooperative length and could you calculate those from known interaction constants? Those cooperative regions are well known, for instance, in the helix formation of polypeptides and polynucleotides.

Szwarc: The point you mention is important if we start with a mixture of A and B. But in the preparation of block copolymers we start with A, complete its polymerization, then add B and again complete the polymerization. In that way, we produce block copolymers only.

Tanford: From what you said earlier, Dr Szwarc, I didn't think that this was an equilibrium situation but that the equilibrium was overwhelmingly on the side of polymer bond formation, and that what stops polymerization is lack of monomer. But you then implied in discussion that you are actually dealing with an equilibrium situation in which the monomer concentration is quite high—about 10^{-7} M. Does this not mean that if a polymer is dialysed against pure solvent, so that the monomer concentration falls below its equilibrium value, dissociation should occur? Similarly, if a block copolymer is allowed to stand in the presence of equilibrium concentrations of A and B, does it mean that randomization will slowly occur?

Szwarc: Randomization will occur if one waits 100 years or so. Dissociation after the monomer concentration has fallen below 10^{-7} M is not a practical proposition, although you are right in principle. For us 10^{-7} M is very low because we start with monomer concentrations of 0.1–1 M. Under these conditions, more than 99.9% of the monomer polymerizes. Surely one has to ask what is the rate of depropagation. To give an example—for styrene, the rate constant of depropagation at 25 °C is about 10^{-5}/s; so for all practical purposes, you are through with the preparation and you are through with the reaction and you do not worry about degradation. We have to keep in mind the differences in our yardsticks.

References

KÄMMERER, H. & OZAKI, S. (1966) *Makromol. Chem.* **91**, 1

PRINS, W. (1971) *Nat. Phys. Sci.* **230**, 70

SMETS, G. (1968) *J. Polym. Sci. Part C* **22**, 231

Mechanisms in the chemical syntheses of polypeptides and polynucleotides

R. SCHWYZER

Institute of Molecular Biology and Biophysics, Swiss Federal Institute of Technology (ETH), Zürich

Abstract Biological mechanisms for the synthesis of information-carrying biopolymers find their parallels in chemical synthetic reaction mechanisms. The following points are discussed:

Chemical polypeptide synthesis in general, with historical aspects; chemical mechanism of peptide bond formation; racemization mechanisms; chemical and strategic measures for avoiding racemization; parallels between chemical and biosynthetic peptide bond formation; chemical polynucleotide synthesis; chemical mechanisms of phosphate ester hydrolysis and the kinetic stability of cyclic 3′,5′-adenosine monophosphate.

Life's endeavours towards synthesis, in the living cell as in the laboratory, must remain within the limits set by the nature of the reacting molecules and follow the fundamental routes described more or less adequately by chemical reaction mechanisms. In the laboratory the pattern of synthesis of polypeptides and polynucleotides with defined sequences is similar to that used in the cell, in that amino acids and nucleotides are, at present, the most convenient building blocks and the peptide and the phosphodiester bonds are formed according to the standard chemical reactions (*a*), (*b*):

(*a*) $$Y\cdot NH\overset{R(Z)}{\overset{|}{C}}HCOOH + H_2N\overset{R'(Z)}{\overset{|}{C}}HCOOX \longrightarrow Y\cdot NH\overset{R(Z)}{\overset{|}{C}}HCO\text{–}NH\overset{R'(Z)}{\overset{|}{C}}HCOOX + H_2O$$

(*b*)

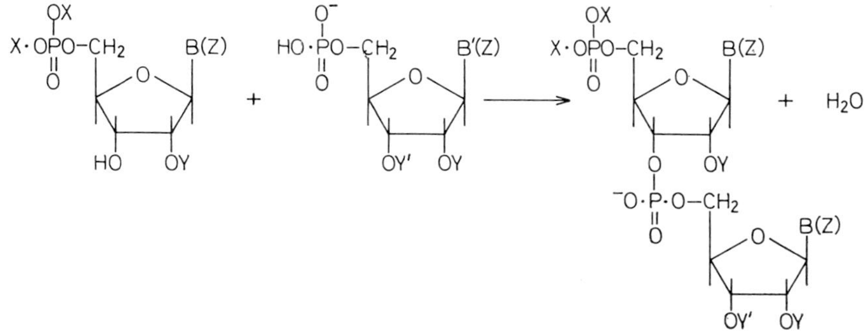

However, the chemist is still forced to introduce and selectively cleave complex protecting groups (X, Y, Y′ etc.) and to revert to a variety of reagents, stratagems and purification methods in order to produce even relatively short sequences. Thus, for example, the production of one of the longest synthetic peptide chains of high purity reported to date, that of an ACTH nonatriakonta peptide (Schwyzer & Sieber 1963, 1966; Schwyzer 1966), takes months, whereas the living cell synthesizes protein molecules with high precision in a matter of minutes. This applies to polynucleotides as well. It is the ambitious aim of synthetic chemistry to simulate the biosynthetic process to such a degree as to be able to use only one condensing reagent, to do away with protecting groups and to introduce automation, while achieving the same purity.

POLYPEPTIDE SYNTHESIS

About 20 years ago, the first synthesis of *S*-benzoyl and *S*-acetyl pantetheine in the laboratory was reported (Schwyzer 1952):

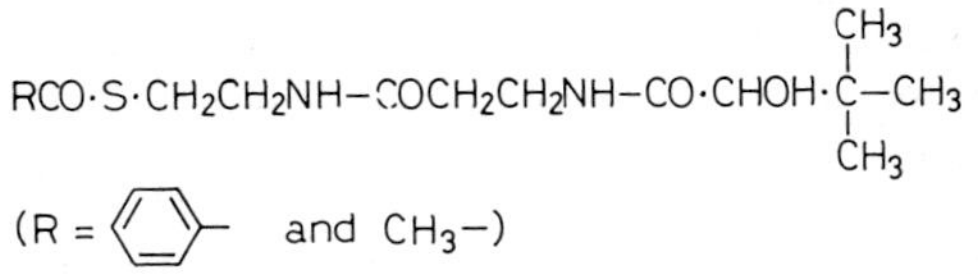

Studies on aminolytic reactions of these and other carboxylic acid thiol esters* with various substituents showed that two simple measures could be taken to enhance the reaction rate: (1) introduction of electronegative substituents into the thioalcohol moiety, and (2) addition of certain metal ions that are able to complex with sulphur atoms

(X = –I substituent, e.g. —COOH)

These observations were explained in terms of the reaction mechanism (—I substituent effects) (Schwyzer 1953; Schwyzer & Hürlimann 1954). Further considerations led to the preparation of amino acid cyanomethyl esters

$Y{\cdot}NH{\cdot}CHR{\cdot}CO{-}OCH_2CN$

as the first oxygen-ester type reagents of practical use for the synthesis of the peptide bond, and this resulted in the formulation of the concept of active

* Simple thiophenyl esters had been studied by others, for example by Wieland *et al.* (1951).

esters (or activated esters) (Schwyzer *et al.* 1954, 1955*a*, *b*, *c*, 1956). The more potent *p*-nitrophenyl esters (Bodánszky 1955; Bodánszky *et al.* 1955; Farrington *et al.* 1955; Wieland & Jaenicke 1956; Schwyzer & Sieber 1956*a*, *b*; Iselin *et al.* 1957) and other activated phenyl esters (Schwyzer & Sieber 1956*a*, *b*; Iselin *et al.* 1957; Pless & Boissonnas 1963) were soon prepared: their activity is explained by the comparatively strong −M effect of the appropriately substituted phenyl ester group:

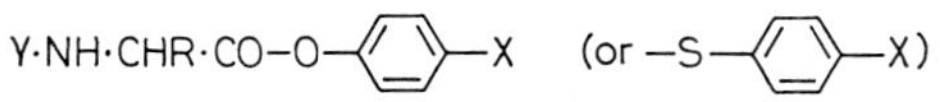

(X = $-NO_2$; $-SO_2CH_3$; $-COCH_3$; $-COOCH_3$)

Numerous ingenious modifications of the 'active ester' scheme have been elaborated since then and have produced some of the best and most widely applicable reagents for peptide synthesis (cf. Table 1 and many excellent reviews, e.g. Rudinger 1963; Schröder & Lübke 1965; Jakubke & Jeschkeit 1969; Law 1970; Jones 1970). Acid anhydrides and a few other types of reactive carboxyl compounds (Table 2) complete the list of chemical tools now available for the formation of the peptide bond (cf. the references just cited).

Chemical mechanism of peptide bond formation

The mechanism $B_{AC}2$ that is now accepted for peptide bond formation, which is identical with that formulated for aminolysis (Ingold 1953), explains the reactivity of (active) esters, thioesters, anhydrides and other active carboxyl derivatives (*c*):

(*c*)

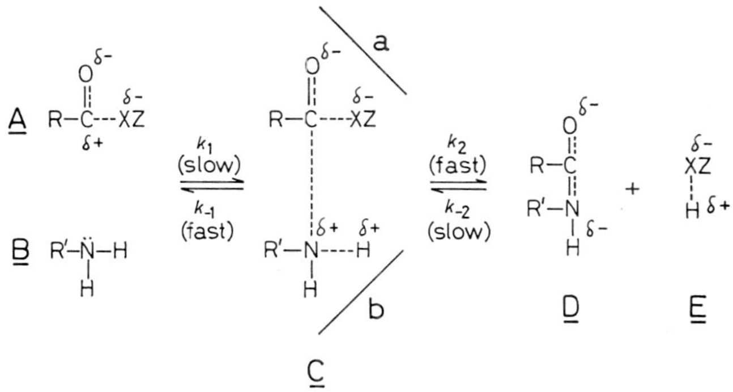

For active esters (X = O; Z = −I or −M group), the δ^+ character of the carboxyl carbon atom and hence its electrophilicity will be the more pronounced

TABLE 1

Typical active esters, RCO–XZ

Method	XZ	Educts	References
Carboxymethyl thiol ester	$-S.CH_2COOH$	RCOCl + HSZ	(1)
Cyanomethyl ester	$-O.CH_2CN$	$RCOO^-$ + Cl.Z	(2)
Carbodiimide (CI)	—O.C=NR' NHR''	RCOOH + R'N=C=NR''	(3)
Phenyl thiol ester	$-S.C_6H_5$	RCOCl + HSZ	(4)
p-Nitrophenyl thiol ester	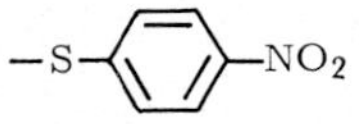	ZS–P(O)–SZ + RCOOH	(5)
p-Nitrophenyl ester (–ONp)	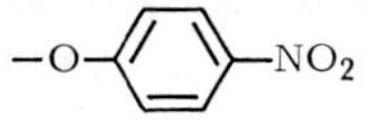	ZO–S(O)–OZ + RCOOH or RCOOH + HOZ + CI	(6)
p-Methanesulphonyl-phenyl ester	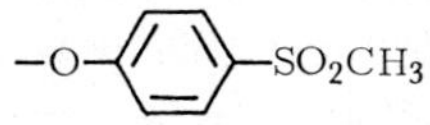	ZO–S(O)–OZ + RCOOH or RCOOH + HOZ + CI	(7)
2,4-Di-, 2,4,6-tri- and pentachloro-phenyl esters		RCOOH + HOZ + CI	(8)
Oxine esters	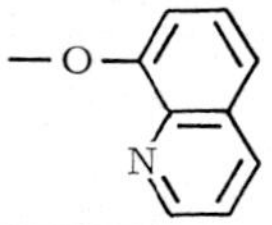	RCOOH + HOZ + CI	(9)
Thiopyridine ester	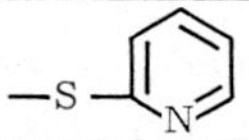	RCOOH + HSZ + CI	(10)
Dihydroxybenzamide esters (DHB)	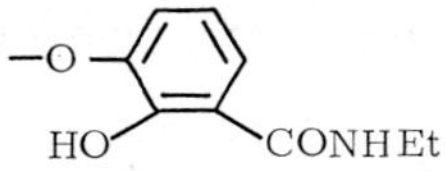		(11)

(1) Schwyzer 1953; Schwyzer & Hürlimann 1954.
(2) Schwyzer *et al.* 1954, 1955*a*, *b*, *c*, 1956.
(3) Sheehan & Hess 1955.
(4) Wieland *et al.* 1951.
(5) Farrington *et al.* 1955; Wieland & Jaenicke 1956.
(6) Bodánszky 1955; Bodánszky *et al.* 1955; Schwyzer & Sieber 1956*a*, *b*; Iselin *et al.* 1957.
(7) Iselin *et al.* 1957.
(8) Pless & Boissonnas 1963.
(9) Jakubke 1965; Jakubke & Voigt 1966; Jakubke *et al.* 1967.
(10) Lloyd & Young 1971.
(11) Kemp & Chien 1967.

TABLE 2

Selected acid anhydrides and other types of reactive carboxyl compounds*

Azides	$RCO—N_3$
Mixed anhydrides	RCO—O—COOEt RCO—O—COOBU[i]
Symmetrical anhydrides	RCO—O—COR
N-Carboxy anhydrides	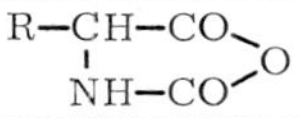
N-Carboxy thio-anhydrides	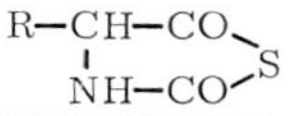
Imidazolides	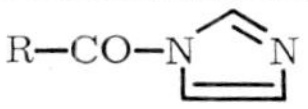
N-Hydroxypiperidine esters	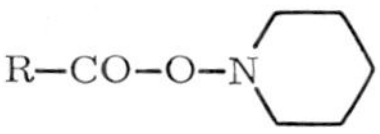
N-Hydroxyphthalimide esters	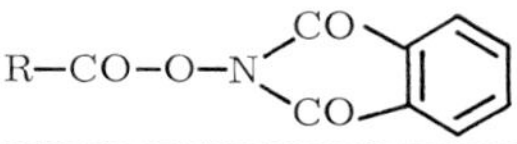
N-Hydroxysuccinimide esters	$R—CO—O—N(CO—CH_2)(CO—CH_2)$
p-Hydroxytris(dimethyl-amino)phosphonium esters	$R—CO—O—P^+(NMe_2)_3$ **

* See Rudinger 1963; Schröder & Lübke 1965; Jakubke & Jeschkeit 1969; Law 1970; Jones 1970.
** Gawne *et al.* 1969, 1971.

the stronger the —I and —M character of Z. (The reason for the better reactivity of thiol esters, X = S, compared to oxygen esters resides in the lower strength of the C—S bond* and its higher polarizability: approach of the nucleophil B can in itself induce an $OC^{\delta+}$------$^{\delta-}SZ$ dipole and thus be facilitated.) A bond is formed between the reacting species A and B to give C (which is assumed—in analogy to hydrolysis—to be an *intermediate* with relatively low energy barriers towards both A + B and D + E). The rate ratio of k_2/k_{-1} will become larger, the better the leaving group properties of $\overset{\delta-}{---}$-XZ are, compared to those of $R'NH_2\overset{\delta+}{---}$-. These leaving group properties roughly parallel the —I and —M influence of Z with given X, or, in other terms, the acidity of H—XZ (E).

It becomes clear from C (see (*c*), p. 25) that the surroundings will play a

* Sulphur p^3 electrons cannot interact as easily with the CO π system as oxygen p^2 electrons, leading to a decrease of bond order.

considerable role in regulating the thermodynamic stability of reactants, intermediate and products. Stabilization of C in solvents of high dielectric constant or with added salts results in an enhancement of the overall reaction rate A + B → D + E. General acid and general base catalysis will be expected to take place at the surfaces *a* and *b*, respectively (for an excellent discussion of catalysis and solvent effects, see Kemp 1972). Base catalysis expresses itself in the (generalized) rate law:

$$\frac{d[\text{peptide}]}{dt} = k[\text{ester}]\cdot[\text{amine}]\cdot[\text{base}]$$

where 'base' can be identical with 'amine' (particularly in non-basic solvents of low dielectric constant) or a (basic) solvent. (The role of the adjuvant 'base' or solvent molecules can merely be to stabilize the proton on its way to the leaving group function, X.)

In mechanism (*c*) with a probably tetragonal intermediate, the postulate of microscopic reversibility (i.e. ---XZ leaving and entering in a position that is geometrically equivalent to that of $R'NH_2$---) is fulfilled by the three-dimensional representation of Fig. 1, which also takes into account the necessity for maximum overlap between orbitals on $C^{\delta+}$ of A and N of B during the approach (pseudo inversion, somewhat like the Walden inversion in S_N2 reactions).

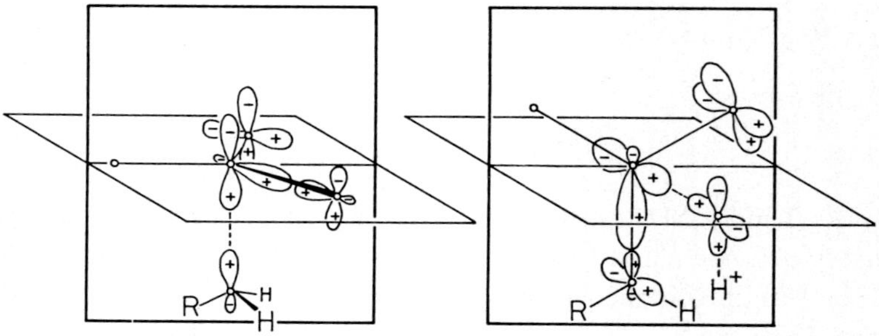

FIG. 1. Three-dimensional representation of reactants and products just before entering and just after leaving the transition intermediate. Plus and minus signs indicate possible phase relations of interacting atomic orbitals. Pseudo-rotation from horizontal to vertical plane.

Similar considerations hold for acid anhydrides (which are especially good leaving groups and produce a very strong δ^+ on the carbonyl carbon of A) and the other active carbonyl compounds.

Besides electronic effects, steric effects also operate. Strongly branched side-chains in R and R′ *C*-terminal and *N*-terminal amino acids respectively (e.g. valine, isoleucine) can drastically reduce reaction rates. This is also true for steric effects emanating from chain length and specific coiling of R and R′ where these residues are polypeptides.

Logically, therefore, stronger −I and −M substituents are needed to produce very highly reactive carboxyl derivatives. However, as Brenner (1967) and Young (1967) have clearly indicated, if one hopes to escape the Scylla of too small reaction rates, one will fall into the Charybdis of racemization.

Chemical mechanism of racemization during peptide bond formation

Racemization of carboxyl-activated amino acid derivatives occurs as a side reaction to peptide bond formation. It can proceed either through removal of the proton on the alpha carbon atom (*d*) or through an intermediate oxazolone (*e*) (Young 1967; Neuberger 1948; Goodman & McGahren 1965; Goodman & Glaser 1970; Liberek 1963):

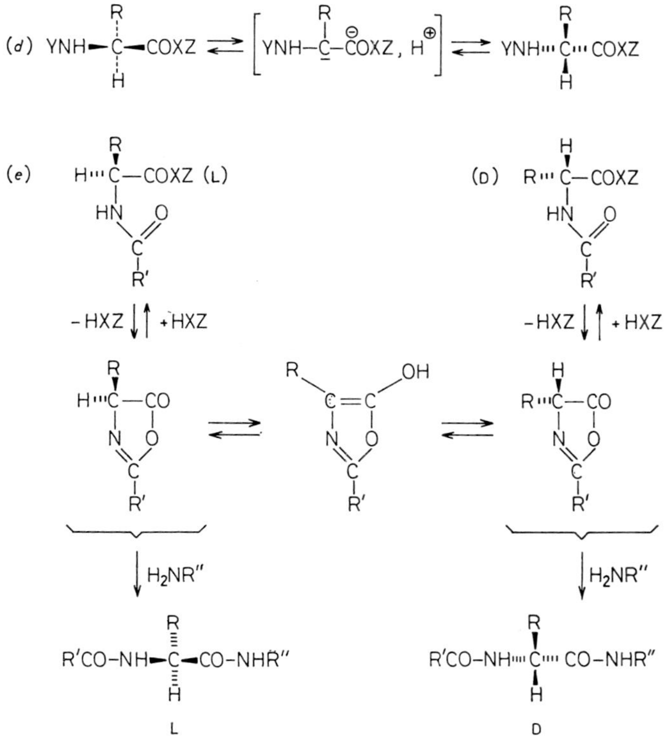

Both reactions will be favoured by the same factors as enhance the rate of aminolysis, in particular strong electronegativity of −XZ and Y− (R′CO−).

If the electronegativity of Y is reduced by replacement of R′CO = amino acyl or peptidyl by R′CO = aryloxy or alkyloxy carbonyl (e.g. benzyloxycarbonyl or *t*-butoxycarbonyl), the extent of racemization is also diminished. Reduction of the electronegativity of XZ will reduce the rate of both aminolysis and racemization and hardly produce a change in their ratio. However, a decrease in aminolysis by steric factors always increases the ratio of racemization to peptide bond formation.

Kemp (1972) has recently demonstrated that in the polar solvents favoured for the formation of peptide bonds, oxazolone formation is subject to specific base catalysis: the NH proton is removed in a first step. This can be achieved by the α-amino group of $R''NH_2$; as expected, addition of a salt of $R''NH_2$ to the reaction medium reduces specific catalysis. In solvents with low dielectric constant, general base catalysis would seem to prevail.

Two solutions to the problem have been attempted with more or less success:

(1) Design of active esters with high rates of aminolysis despite low electronegativity of —XZ.

(2) Strategies of peptide assembly designed to minimize racemization.

Condensation methods with low racemization levels

A number of very promising active esters and similar compounds have been found which reconcile the postulates of low electronegativity and high rates of aminolysis. They all provide anchimeric assistance (intramolecular stabilization)

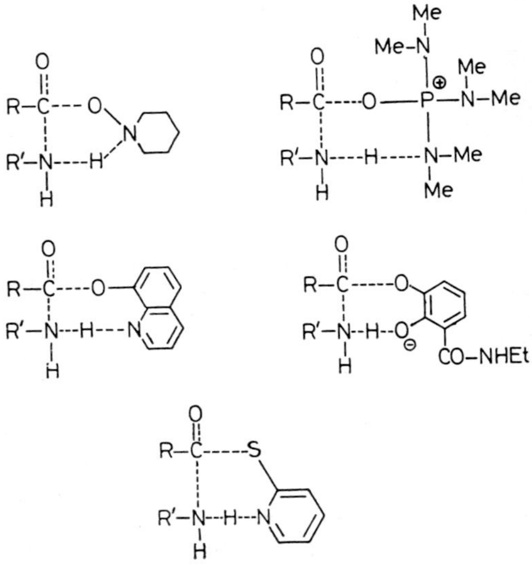

for the proton leaving the R′NH_2 group in (*e*), which appears to be the main factor of rate enhancement. This is exemplified by the reactions of 8-hydroxyquinoline (Jakubke 1965; Jakubke & Voigt 1966; Jakubke *et al.* 1967), 1-hydroxypiperidine (Young 1967; Hanford *et al.* 1965), 2,3-dihydroxy-*N*-ethylbenzamide (Kemp 1972; Kemp & Chien 1967), 2-thiolpyridine (Lloyd & Young 1971), and hydroxy-tris(-dimethylamino)-phosponium (Gawne *et al.* 1969, 1971) (see structure on p. 30).

*Strategic considerations (cf. the excellent review of Wünsch (*1971*))*

Difficulties arising from *C*-terminal-activated amino acid racemization can be avoided in two ways: (1) by stepwise synthesis and (2) by fragment condensation using peptide fragments with *C*-terminal glycine or proline residues (which are not prone to racemization).

Stepwise synthesis starts at the *C*-terminus and adds one N^{α}-protected amino acid at a time (followed, of course, after each condensation by removal of the protecting group):

(*f*)

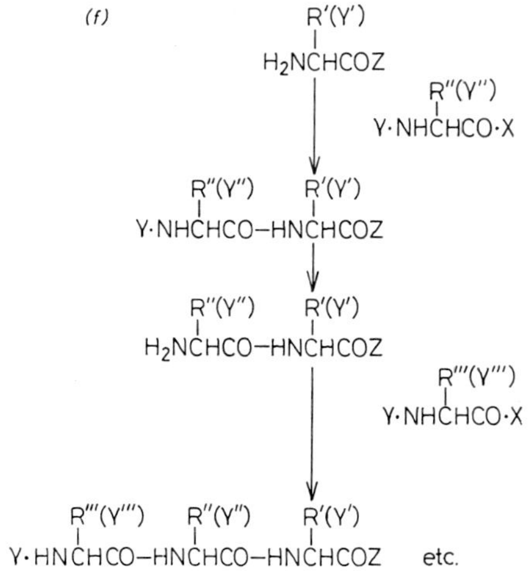

If Y has low electronegativity (for example if it is a urethane-type protecting group such as benzyloxycarbonyl or *t*-butoxycarbonyl) and X is properly selected, racemization at each step can be kept to a minimum ($<0.1\%$). This strategy was proposed some time ago (Boissonnas *et al.* 1955; Schwyzer & Li 1958; Schwarz & Bumpus 1959; Bodánszky & du Vigneaud 1959) and used on a number of occasions, e.g. the synthesis of an ACTH 25–39 pentadecapeptide with overall yields of $\simeq 8\%$ of highly purified product (Schwyzer & Sieber 1963, 1966; Schwyzer 1966). In order to avoid mechanical losses, Merrifield (Merrifield 1963; Merrifield *et al.* 1966; Stewart & Young 1969) introduced a solid-phase technique using an insoluble polymer support as the *C*-terminal carboxyl-protecting group **Z** in (*f*). The method appears to be very rapid, and amenable to automation, but the problem of side reactions and the accumulation of 'error sequences' is a serious drawback (cf. Wünsch 1971), because purification can only be attempted after completion of the synthesis. Intermediate purification (with mechanical losses!) is the advantage of the methyl ester technique (Z = $-OCH_3$) described by Schwyzer & Sieber (1963, 1966) and Schwyzer (1966). Reduction of mechanical losses and preservation of the possibility of intermediate purification are provided by the 'handle' techniques of Wieland & Racky (1968) and Camble, Garner & Young (1968), in which Z is a basic ester, allowing separation of products on a cation exchange column (*g*):

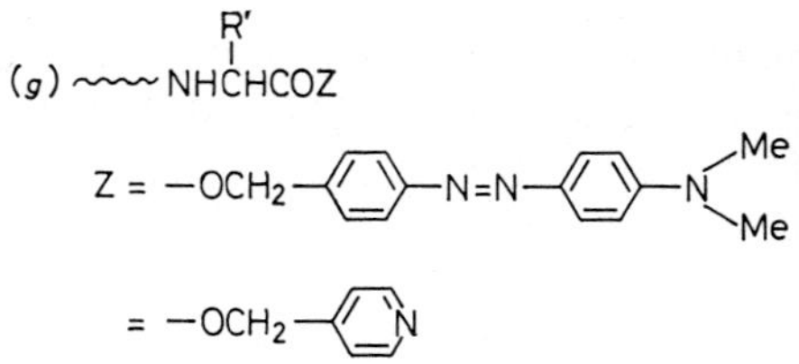

The 4-picolyl ester method has been applied very successfully to the preparation of angiotensin (Garner & Young 1971) and bradykinin (Schafer *et al.* 1971), and appears to be especially promising.

Parallels to biological synthesis (for a more detailed discussion see Schwyzer 1967)

Biological synthesis of polypeptides follows a stepwise scheme moving in the opposite direction: *N*- to *C*-terminus. An active ester of the growing peptide is used in the condensation step (*h*):

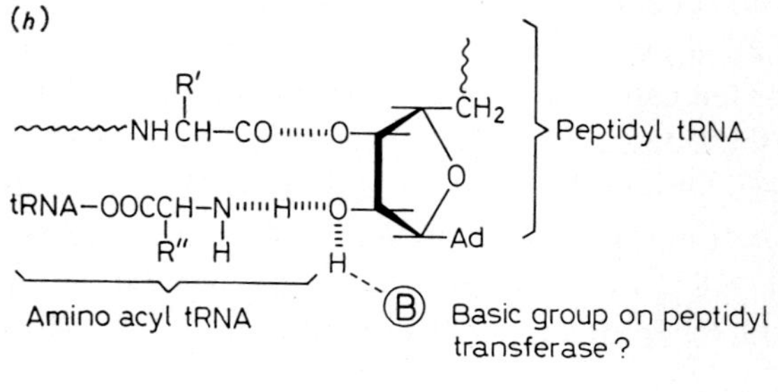

Perhaps anchimeric assistance in the ribose ester (which could be enhanced by the correct positioning of a proton acceptor group B) also comes into play (cf. Zachau 1960). It is, of course, essential that carboxyl activation other than that at the ribosome surface be kept to a minimum, in order to exclude polymerization reactions between dissolved amino acyl tRNA molecules. On the surface of the ribosome, all problems arising from a tendency to racemization or from high reactivity of other functional groups can be taken care of by an appropriate positioning in space of the reactive species. This could also provide further activation of the peptide-forming reaction by reducing the reaction entropy.

Recently, it has been shown that non-protein polypeptides such as certain peptide antibiotics can be assembled on enzyme surfaces in a non-classical manner. The synthesis of the cyclo-decapeptide gramicidin S, for example, uses thiol esters as active intermediates; one form is an active amino acyl pantetheine derivative (Lipmann 1971).

POLYNUCLEOTIDES

The most successful strategy for polynucleotide synthesis so far utilizes the stepwise procedure for the production of oligonucleotides, and 'block' (or 'fragment') condensation of oligo to polynucleotides (Khorana 1961; 1968). The stepwise procedure uses suitably protected deoxy-5′-mononucleotides (working in the 5′→3′ direction), and 3′-mononucleotides (working in the 3′→5′ direction) for deoxyribo- and ribo-oligonucleotides, respectively (*i*):

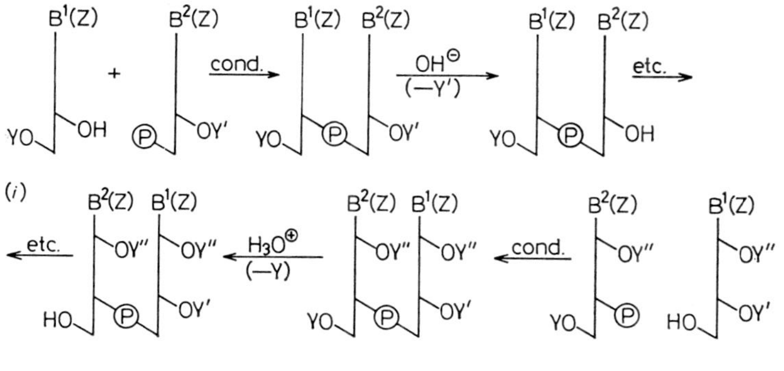

The required protecting groups (cf. Table 3) are: mild acid-labile Y, alkali-labile Y′ and Y″, and NH_3-labile Z (protection of amino groups on the base, B): this scheme takes into account the stabilities (or labilities) of deoxyribo- and ribo-polynucleotides.

TABLE 3

Protecting group combinations in chemical polynucleotide synthesis (Khorana 1961, 1968), cf. reactions (*b*), (*i*)

(1) Deoxyribonucleosides, deoxyribonucleotides

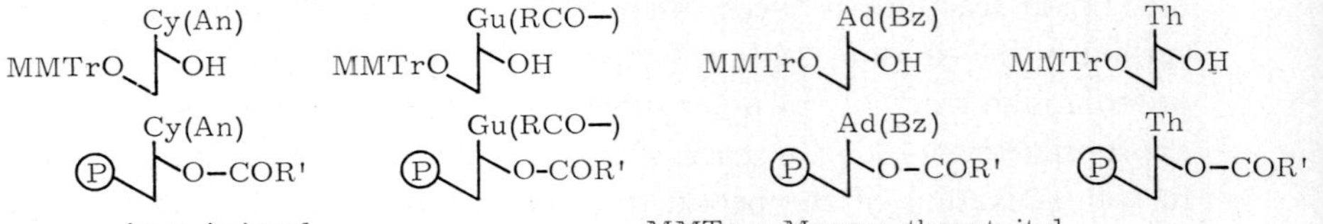

An = Anisoyl
Bz = Benzoyl
RCO = Isobutyryl or acetyl

MMTr = Monomethoxytrityl
R'CO = Acetyl

(2) 5'-Terminal phosphate protection as monocyanoethyl ester

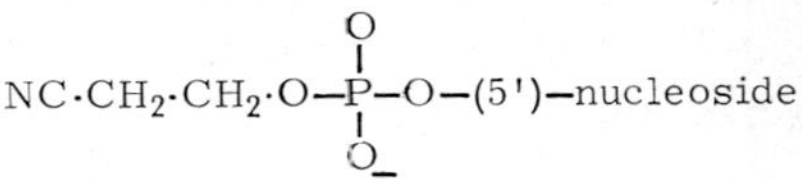

(3) Ribonucleosides, ribonucleotides

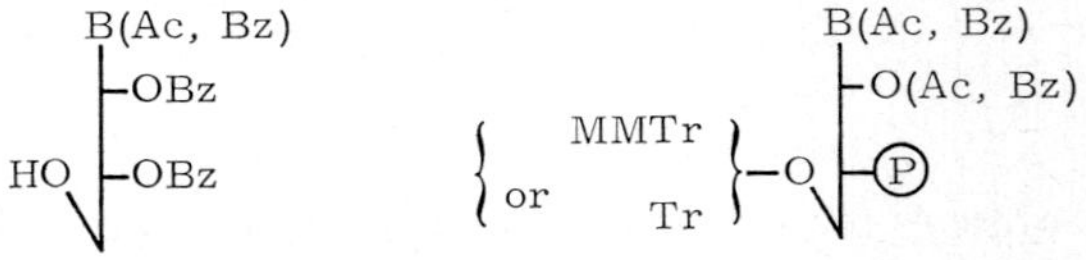

Heterocyclic base B protected, if Cy, Gu, Ad; unprotected if Ur .

Ac = Acetyl
Bz = Benzoyl

Tr = Trityl
MMTr = Monomethoxytrityl

Neither stepwise nor fragment condensations are endangered by racemization, but side reactions producing diphosphate are always possible and sometimes troublesome.

The most useful condensing agents are dicyclohexylcarbodiimide (DCCI) and sterically hindered benzene sulphonic acid chlorides:

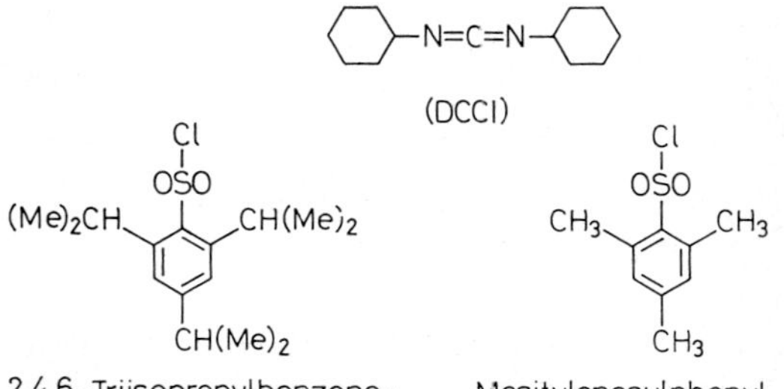

2,4,6-Triisopropylbenzene-sulphonyl chloride (TPS)

Mesitylenesulphonyl chloride (MS)

The mechanism of condensation with nucleotide 5'- or 3'-hydroxyls is not known: it appears that DCCI causes the intermediate formation of a trimetaphosphate as the reactive species (Weimann & Khorana 1962):

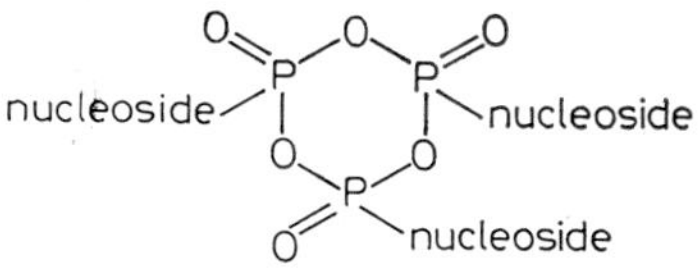

The sulphonyl chlorides produce other, unknown, types of intermediate. Active esters, like *p*-nitrophenyl- or 2,4-dinitrophenyl esters (Borden & Smith 1966) are too inert to serve as activating agents for producing internucleotide links; however, the latter are useful for the preparation of 3′,5′-cyclic mononucleotides in the presence of anhydrous potassium *t*-butoxide.

Recently, Mackey & Gilham (1971) have reported a new scheme which might overcome at least some of the difficulties encountered with conventional condensation methods and protecting groups in the ribopolynucleotide series. They use stepwise elongation in the $5' \rightarrow 3'$ direction with 2′(3′) methoxyethyl nucleoside 5′-diphosphate in conjunction with the enzyme polynucleotide phosphorylase from *Escherichia coli*, for example (*j*):

(*j*) 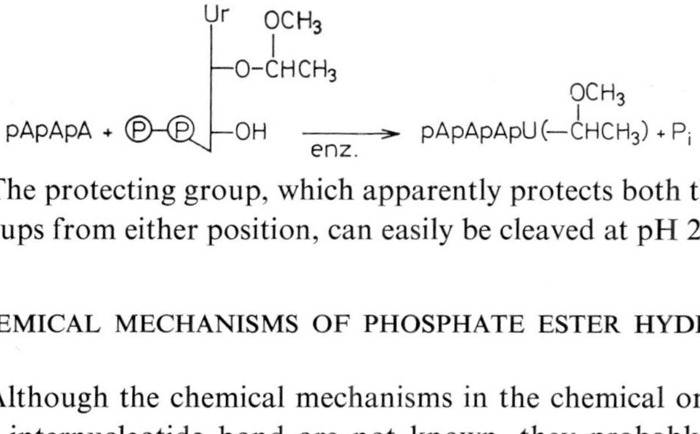

The protecting group, which apparently protects both the 3′- and 2′-hydroxyl groups from either position, can easily be cleaved at pH 2 or 3 (15–90 minutes).

CHEMICAL MECHANISMS OF PHOSPHATE ESTER HYDROLYSIS

Although the chemical mechanisms in the chemical or enzymic synthesis of the internucleotide bond are not known, they probably have some (inverse) relationship to the hydrolysis of phosphate esters. This subject has been thoroughly examined by Haake & Westheimer (1961) and Westheimer (1968) and I wish to conclude with a brief survey.

The currently accepted mechanism is portrayed as follows (*k*):

(*k*) 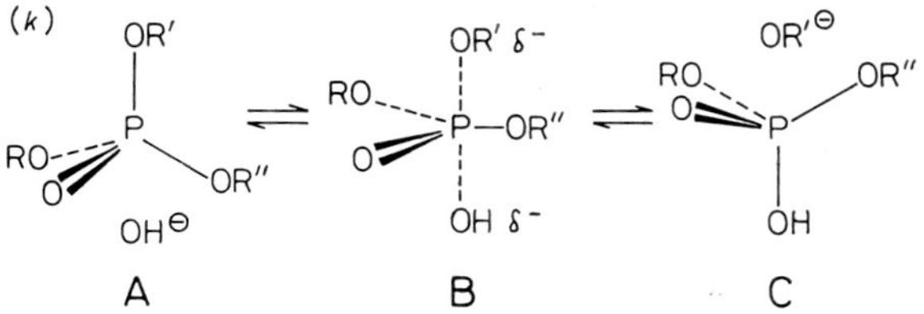

The tetrahedral phosphoric acid triester with sp^3 hybridized bonds is attacked by the nucleophil HO^- through one of the sides of the tetrahedron (A). The transition state (B) is a trigonal bipyramid with sp^2 hybrid bonds in the x,y plane and long, p_z,d bonds to the nucleophil and the leaving group in the z direction. Inversion and rehybridization produces the product of hydrolysis (C) in a manner analogous to an S_N2 reaction at a tetragonal carbon atom. The leaving group, —OR′, will be that group with the most electronegative substituent, which at the same time stabilizes the negative charge and makes use of the extension of the p_z,d phosphorus orbital along the z axis.

Special stability of cAMP

Although 3′,5′-cyclic adenosine monophosphate (cAMP) is a high-energy phosphate compound, its alkaline hydrolysis is extremely slow—much slower than that of 2′,3′-cyclo-nucleotides. The rate ratio of 5-membered:6-membered ring phosphate esters is approximately 10^8. Westheimer (Haake & Westheimer 1961; Westheimer 1968) explains this by easy accommodation of the 90° equatorial to apical angle of the transition state (B) into a 5-membered but not into a 6-membered ring. This is illustrated by mechanisms (*l*) for the 2′,3′-, and (*m*) for 3′,5′-cyclo-nucleotide hydrolysis.

(*l*)

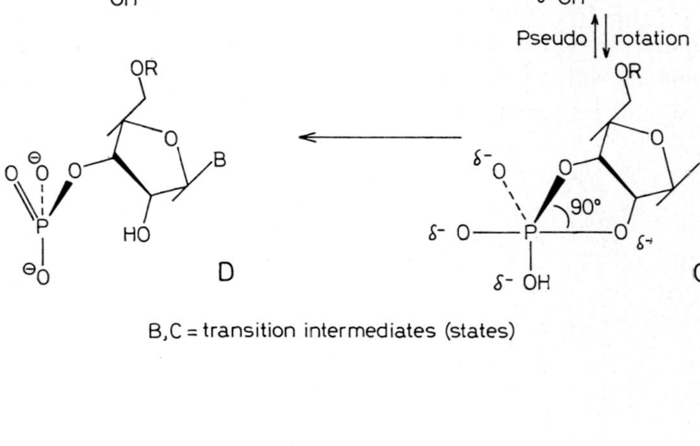

B,C = transition intermediates (states)

(4)

(m)

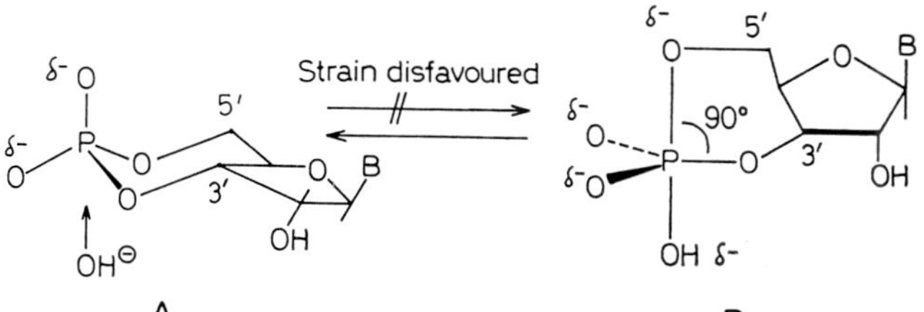

A

B
(transition intermediate)

References

Bodánszky, M. (1955) *Nature (Lond.)* **175**, 685

Bodánszky, M. & du Vigneaud, V. (1959) *Nature (Lond.)* **183**, 1324

Bodánszky, M., Szelke, M., Tömörkény, E. & Weisz, E. (1955) *Chem. Ind. (Lond.)* 1517

Boissonnas, R. A., Guttmann, St., Jaquenoud, P. A. & Waller, J. P. (1955) *Helv. Chim. Acta* **38**, 1491

Borden, R. K. & Smith, M. (1966) *J. Am. Chem. Soc.* **89**, 3241

Brenner, M. (1967) in *Peptides* (Beyerman, H. C., van de Linde, A. & Maassen van den Brink, W., eds.), p. 1, North-Holland, Amsterdam

Camble, R., Garner, R. & Young, G. T. (1968) *Nature (Lond.)* **217**, 247

Farrington, J. A., Kenner, G. W. & Turner, J. M. (1955) *Chem. Ind. (Lond.)* 601

Garner, R. & Young, G. T. (1971) *J. Chem. Soc. Sect. C Org. Chem.* 50

Gawne, G., Kenner, G. W. & Sheppard, R. (1969) *J. Am. Chem. Soc.* **91**, 5669

Gawne, G., Kenner, G. W. & Sheppard, R. (1971) in *The Peptides* 1969 (Scoffone, E., ed.), p. 23, North-Holland, Amsterdam

Goodman, M. & Glaser, C. (1970) in *Peptides: Chemistry and Biochemistry* (Weinstein, B. & Lande, S., eds.), p. 267, Dekker, New York

Goodman, M. & McGahren, W. J. (1965) *J. Am. Chem. Soc.* **87**, 3028

Haake, P. C. & Westheimer, F. H. (1961) *J. Am. Chem. Soc.* **83**, 1102

Hanford, B. O., Jones, J. H., Young, G. T. & Johnson, T. F. W. (1965) *J. Chem. Soc. (Lond.)* 6814

Ingold, C. K. (1953) *Structure and Mechanism in Organic Chemistry*, Cornell University Press, Ithaca, N.Y.

Iselin, B., Rittel, W., Sieber, P. & Schwyzer, R. (1957) *Helv. Chim. Acta* **40**, 373

Jakubke, H.-D. (1965) *Z. Naturforsch.* **20b**, 273

Jakubke, H.-D. & Jeschkeit, H. (1969) *Aminosäuren, Peptide, Proteine*, Akademie-Verlag, Berlin

Jakubke, H.-D. & Voigt, A. (1966) *Chem. Ber.* **99**, 2419

Jakubke, H.-D., Voigt, A. & Burkhardt, S. (1967) *Chem. Ber.* **100**, 2367

Jones, J. H. (1970) in *Amino-acids, Peptides, and Proteins*, Vol. 2, Specialist Periodical Reports, The Chemical Society, London

Kemp, D. S. (1972) in *The Peptides* 1971 (Nesvadba, H., ed.), North-Holland, Amsterdam. In press

Kemp, D. S. & Chien, S. W. (1967) *J. Am. Chem. Soc.* **89**, 2743

Khorana, H. G. (1961) *Some Recent Developments in the Chemistry of Phosphate Esters of Biological Interest*, Wiley, New York

KHORANA, H. G. (1968) *Pure Appl. Chem.* **17**, 349
LAW, H. D. (1970) *The Organic Chemistry of Peptides*, Wiley-Interscience, London
LIBEREK, B. (1963) *Tetrahedron Lett.* 925
LIPMANN, F. (1971) *Science (Wash. D.C.)* **173**, 875
LLOYD, K. & YOUNG, G. T. (1971) *J. Chem. Soc. Sect. C Org. Chem.* 2890
MACKEY, J. K. & GILHAM, P. T. (1971) *Nature (Lond.)* **233**, 551
MERRIFIELD, R. B. (1963) *J. Am. Chem. Soc.* **85**, 2149
MERRIFIELD, R. B., STEWART, J. M. & JERNBERG, N. (1966) *Anal. Chem.* **38**, 1905
NEUBERGER, A. (1948) *Adv. Protein Chem.* **4**, 297
PLESS, J. & BOISSONNAS, R. A. (1963) *Helv. Chim. Acta* **45**, 1609
RUDINGER, J. (1963) *Pure Appl. Chem.* **7**, 335
SCHAFER, D. J., YOUNG, G. T., ELLIOTT, D. F. & WADE, R. (1971) *J. Chem. Soc. Sect. C Org. Chem.* 46
SCHRÖDER, E. & LÜBKE, K. (1965) *The Peptides*, Vol. 1, Academic Press, New York
SCHWARZ, H. & BUMPUS, F. M. (1959) *J. Am. Chem. Soc.* **81**, 890
SCHWYZER, R. (1952) *Helv. Chim. Acta* **35**, 1903
SCHWYZER, R. (1953) *Helv. Chim. Acta* **36**, 414
SCHWYZER, R. (1966) *Naturwissenschaften* **53**, 189
SCHWYZER, R. (1967) in *Peptides* (BEYERMAN, H. C., VAN DE LINDE, A. & MAASSEN VAN DEN BRINK, W., eds.), p. 197, North-Holland, Amsterdam
SCHWYZER, R. & HÜRLIMANN, CH. (1954) *Helv. Chim. Acta* **37**, 155
SCHWYZER, R. & LI, C. H. (1958) *Nature (Lond.)* **182**, 1669
SCHWYZER, R. & SIEBER, P. (1956*a*) *Chimia* **10**, 265
SCHWYZER, R. & SIEBER, P. (1956*b*) *Angew. Chem. Int. Ed. Engl.* **68**, 518
SCHWYZER, R. & SIEBER, P. (1963) *Nature (Lond.)* **199**, 172
SCHWYZER, R. & SIEBER, P. (1966) *Helv. Chim. Acta* **49**, 134
SCHWYZER, R., ISELIN, B. & FEURER, M. (1954) *Chimia* **8**, 264
SCHWYZER, R., ISELIN, B. & FEURER, M. (1955*a*) *Helv. Chim. Acta* **38**, 69
SCHWYZER, R., ISELIN, B., FEURER, M. & KÄGI, H. (1955*b*) *Helv. Chim. Acta* **38**, 80
SCHWYZER, R., FEURER, M. & ISELIN, B. (1955*c*) *Helv. Chim. Acta* **38**, 83
SCHWYZER, R., ISELIN, B., RITTEL, W. & SIEBER, P. (1956) *Helv. Chim. Acta* **39**, 872
SHEEHAN, J. C. & HESS, G. P. (1955) *J. Am. Chem. Soc.* **77**, 1067
STEWART, J. M. & YOUNG, J. D. (1969) *Solid Phase Peptide Synthesis*, Freeman, San Francisco
WESTHEIMER, F. H. (1968) *Acc. Chem. Res.* **1**[3], 70
WEIMANN, G. & KHORANA, H. G. (1962) *J. Am. Chem. Soc.* **84**, 4329
WIELAND, T. & JAENICKE, F. (1956) *Liebigs Ann. Chem.* **599**, 125
WIELAND, T. & RACKY, W. (1968) *Chimia* **22**, 375
WIELAND, T., SCHÄFER, W. & BOKELMAN, E. (1951) *Liebigs Ann. Chem.* **573**, 99
WÜNSCH, E. (1971) *Angew. Chem. Int. Ed. Engl.* **83**, 773
YOUNG, G. T. (1967) in *Peptides* (BEYERMAN, H. C., VAN DE LINDE, A. & MAASSEN VAN DEN BRINK, W., eds.), p. 55, North-Holland, Amsterdam
ZACHAU, H. G. (1960) *Chem. Ber.* **93**, 1822, 1830

Discussion

Levitt: Presumably ribosomes have been designed to synthesize polypeptide chains with maximum efficiency and accuracy. What ribosomal factors would you consider least essential in the biosynthesis of polypeptides?

Schwyzer: If we focus our attention on the formation of the peptide bond, it would seem to be essential to have activated amino acids (or peptides) held in the correct position by some sort of recognition site in order to facilitate the reaction and to give the right sequence. I don't believe that we could do away with any of the components of the transcription machinery without disturbing the process.

Levitt: In the mechanisms you described in your talk, it seemed that the 2′-hydroxy group of the 3′-terminal adenosine of the tRNA was quite important. Could this hydroxy group be left out?

Schwyzer: The reactive element in peptide bond biosynthesis is an active ester of the growing peptide chain with the 3′- and/or 2′-hydroxy group of the 3′-adenyl residue of tRNA. Zachau (1960) investigated the reactivity of a series of amino acid esters structurally related to this element. He found, for example, that valyl-tRNA and valyl-AMP react equally well with hydroxylamine, excluding a special activation by the macromolecular character of the tRNA. According to the general theory of active esters, activation by the neighbouring hydroxy and ether oxygens was observed. We seem to be dealing with typical, albeit not extremely reactive, active esters. Additional activation could occur through neighbouring group effects stemming from the surface of the ribosome, i.e. the peptidyl transferase molecule (Schwyzer 1967).

Baddiley: Neighbouring groups can play quite an important part in the reactivity of the ester linkage. In fact some years ago we made amino acyl esters of ethylene glycol (Shabarova *et al.* 1962) and these are considerably more reactive chemically towards bases than are the ethyl esters. So an adjacent hydroxy group can, in a sense, activate the ester linkage, and so can an adjacent phosphate group. I would imagine that all these centres play some part in increasing reactivity.

Katchalski: In protein biosynthesis a new peptide bond is formed by coupling of the carboxy group of a growing peptide chain attached to the appropriate tRNA, with the amino group of a single amino acid attached via its own carboxy group to the corresponding tRNA. The activated carboxy group of the aminoacyl-tRNA is considerably more reactive chemically than the corresponding linkage in the polypeptidyl-tRNA. From the thermodynamic point of view, it would thus be advantageous for protein biosynthesis to occur via reaction of the aminoacyl-tRNA with the terminal α-NH_2 group of the polypeptidyl-tRNA. Why is it that nature has chosen the energetically 'wrong' mechanism?

Schwyzer: A few years ago, I suggested explanations for this paradox (Schwyzer 1967). In prokaryotes, the amino group of the first amino acid, methionine, is formylated. This enhances the reactivity of the carboxy–ester group and excludes the participation of the (protected) amino group in peptide bond formation (which would result in the wrong polarity of chain sense). In eukaryotes, this amino group is *not* protected. Apparently, these creatures have developed other means of assuring the correct sequence. Provision of a strategically positioned proton on the surface of the enzyme peptidyl transferase could at the same time decrease the nucleophilicity of the methionine and increase the reactivity of the carboxylate group ($-I$ effect of $-NH_3^+$). A basic group near the amino end of the second amino acid would enhance its nucleophilicity and decrease carboxylate reactivity, as I pointed out in my paper.

Katchalski: I posed the question because I think that nature has most likely chosen the above-mentioned mechanism of peptide growth, in spite of its energetically expensive character, in order to ensure that the correct amino acid sequence is obtained. I feel, therefore, that it may be rather risky to use simple thermodynamic arguments in any attempt to explain highly specific biological processes.

Revel: There seems to be a mechanism in bacterial ribosomes that prevents the amino group of the peptide from becoming an acceptor for the carboxy group of the amino acid. We know that during initiation of peptide synthesis, this amino group has to be blocked by formylation. One could imagine that this blocking is required to prevent the type of reaction that Professor Katchalski mentioned. However, the specificity of the ribosomal enzyme peptidyl transferase would probably prevent reaction of the peptidyl amino group with the carboxy group of the amino acid. Indeed, in animal ribosomes formylation of the initiator ('peptidyl') tRNA is absent.

References

SCHWYZER, R. (1967) Biological and chemical synthesis of polypeptides. In *Peptides* (Proc. Eighth European Peptide Symposium, Noordwijk) (BEYERMAN, H. C., VAN DE LINDE, A. & MAASSEN VAN DEN BRINK, W., eds.), p. 197, North-Holland, Amsterdam

SHABAROVA, Z. A., HUGHES, N. A. & BADDILEY, J. (1962) The influence of adjacent phosphate and hydroxyl groups on amino acid esters. *Biochem. J.* **83**, 216–219

ZACHAU, H. G. (1960) Reaktionsfähige Aminosäureester als Modelle der Aminoacyl-ribonucleinsäure. *Chem. Ber.* **93**, 1822; 1830

On the function of DNA-dependent RNA polymerase

R. SCHÄFER, W. ZILLIG and K. ZECHEL

Max-Planck-Institut für Biochemie, München

Abstract The present state of knowledge about transcription is reviewed and our own work on the following discussed in detail: the role of subunits of bacterial RNA polymerase in the transcription process; the mechanism of initiation, elongation and termination in RNA synthesis; the influence of ionic strength at different stages of the transcription process; and the regulatory changes of the host's transcription system after infection of *E. coli* with bacteriophage T4.

Transcription, as the first of the processes of gene expression, leads to the biosynthesis of RNA molecules each of which is complementary to only one of the two strands of a portion of the template DNA. That is, *in vivo* transcription is absolutely asymmetrical (Marmur & Greenspan 1963; Hayashi *et al.* 1963). Every species of RNA has a characteristic length, and a defined beginning and end. Each is the transcription product of a corresponding piece of the DNA, termed the operon by geneticists, which includes defined initiation and termination sites. RNA synthesis is unidirectional, from the 5′ towards the 3′ end, and irreversible (Bremer *et al.* 1961; Maitra & Hurwitz 1965). Repression or derepression as species-specific regulatory phenomena, and cyclic AMP-mediated stimulation of the transcription of catabolite operons as an example of the class-specific regulation of gene expression (Zubay *et al.* 1970), have pleiotropic, that is coordinated, effects on all gene products of one operon via transcription. The element of the operon over which such control mechanisms exert their effects on transcription has been directly identified by genetic analyses and is called the promoter.

In vitro transcription of native double-stranded DNA by highly purified DNA-dependent RNA polymerase (or transcriptase), for example from *E. coli*, shares many features with *in vivo* RNA synthesis. Thus, under proper conditions, it is also strictly asymmetrical, its products have defined lengths, beginnings and ends, and they have even been identified with RNA species synthesized *in vivo* in a number of instances, for example by hybridization competition experiments. The rate of transcription *in vitro* approaches that measured *in vivo*. The regulatory mechanisms mentioned above have also been studied *in vitro* (Zubay *et al.* 1970), where they appear to control the initiation of transcription.

From the nature of the product and the existence of reading units in the DNA which probably correspond to the operons of the geneticists a number of components of the system and of elementary processes in transcription must be postulated. Signals for the recognition of the beginning and end of a reading unit must exist. The polymerase must be able to bind to the DNA, recognize at least the promoter, select the codogenic strand, initiate and promote RNA synthesis and finally recognize a terminator, release its product and become available again for a second round of transcription. Thus, the process may be formally subdivided into *primary binding* to the template, *initiation* as the sum of single steps following thereafter and leading to the formation of the first internucleotide bond, *elongation* or RNA synthesis, including translocation of the enzyme along the template, and finally *termination*, including release.

STRUCTURE OF RNA POLYMERASE

The so-called holoenzyme consists of four different subunits. The monomeric form, which is probably the active one, has the composition $(\beta'\beta\alpha_2)\sigma$ (Burgess 1969; Zillig *et al.* 1970*b*). Subunit σ can be removed by chromatography on phosphocellulose (Burgess *et al.* 1969). The remaining core enzyme is unable to initiate correctly although it still binds to the DNA. Initiation capacity is restored by the addition of σ. Thus σ may be considered an initiation factor (Burgess *et al.* 1969). The enzyme may be dissociated into its subunits and reassembled in active form from a stoichiometric mixture of separated subunits (Heil & Zillig 1970). The part played by the individual components at different points in transcription (partial functions) has been studied using this and related techniques (Zillig *et al.* 1970*a*, *b*; Heil & Zillig 1970). β' is required for the so-called primary binding of enzyme to the template, and β is involved in initiation and elongation. β contains binding sites for both α and σ and for the polymerase-specific inhibitors of transcription, rifampicin and streptolydigin. The difference in properties between the polymerases from *E. coli* infected with phage T4 and normal *E. coli* indicates that α may be involved in promoter recognition specificity.

BINDING TO TEMPLATE

Primary binding of the RNA polymerase to DNA does not require σ; it occurs just as effectively at low and at high temperatures, and it saturates a number of sites along the template, the number depending inversely on ionic strength, being much higher at low ionic strength than the number of promoters (Pettijohn & Kamija 1967). It is not quite clear yet whether primary binding or a specific high affinity fraction of it is a prerequisite for initiation. When core enzyme was primarily bound to an excess of template A, template B added and initiation factor σ provided, together with substrates, then the product of one round of transcription was preferentially synthesized from template A (Zillig *et al.* 1970*b*) (Table 1). This indicates that the core enzyme on excess template

TABLE 1

Dependence of RNA product on the order of addition of different templates to core enzyme before the formation of initiation complex with σ

	% *of total RNA product hybridized to*:	
Template A	*ss T3-DNA*	*ss T4-DNA*
(*a*) T3-DNA	79	21
T4-DNA	2.5	97.5
T3- + T4-DNA	21.5	78.5
	ss T4-DNA	*ss T5-DNA*
(*b*) T4-DNA	66.5	33.5
T5-DNA	27.9	72.0
T4- + T5-DNA	40.2	59.8

Twenty-three μg (*a*) or 10 μg (*b*) of core enzyme were bound to (*a*) 50 μg T3- or T4-DNA or (*b*) 60 μg T4- or T5-DNA (template A) for 5 min at 37 °C in a final volume of 0.5 ml containing 0.008 M-Mg-acetate, 0.05 M-NH_4Cl and 0.03 M-tris acetate, pH 7.9. The same amount of template B was added, keeping the conditions constant. Twenty μg (*a*) or 10 μg (*b*) of σ was added either (*a*) two minutes after, or (*b*) together with template B. One minute later 25 μg heparin was added (to exclude reinitiation) and at either the same time (*b*) or 0.5 min later a substrate salt mix was added. After a further 20-min incubation to allow RNA synthesis the DNA was destroyed by incubation with 20 μg DNase (pancreatic); after heating with 0.01 % SDS, 0.3 ml of the RNA was hybridized to 20 μg each of template A or B (excess) by the method of Bolle *et al.* (1968).

ss: single-stranded.

A is primarily bound to sites from which it initiates either directly or by slippage and not by free redistribution after combination with factor σ. The significance of this result is reduced by the finding that core enzyme is largely consumed by sites on the DNA which do not contribute to transcription after the addition of σ. Complete enzyme, formed by adding σ to core enzyme before the template

TABLE 2

Dependence of σ stimulation of core enzyme on order of addition of core enzyme, σ and template DNA

Present during 5-min preincubation period	*Added at start of incubation period*	*nmol AMP incorporated into RNA*
Core enzyme, DNA	—	0.41
Core enzyme, DNA	σ	0.65
Core enzyme, σ	DNA	2.74
σ, DNA	Core enzyme	1.44
Core enzyme	σ, DNA	1.02

Core enzyme (6 μg) and/or σ (2.5 μg) and/or T5-DNA (50 μg) were incubated in 0.15 ml of 0.05 M-NH_4Cl, 0.01 M-Mg-acetate and 0.03 M-tris acetate, pH 7.9, for 5 min at 37 °C. 0.1 ml of a mixture of substrates (final concentration 1 mM each) and salts (as above) plus the missing components were added. After incubation for 10 min for RNA synthesis, samples were prepared for counting.

is made available, is not consumed by such sites (Table 2). Thus, either the primary binding sites of core and complete enzyme are partially different, or complete enzyme is bound reversibly to those sites which irreversibly consume core enzyme. The conclusion drawn from the two-template experiment on the relation of primary binding to initiation is therefore only valid for a rather small fraction of the primarily bound core enzyme.

INITIATION

The insertion of *initiation steps* between binding and the formation of the first internucleotide bond was made necessary by the indirect demonstration that a highly temperature-dependent event occurs, after binding and before the involvement of substrate, which at low temperature is rate-limiting for transcription (Walter *et al.* 1967). A more direct demonstration of such steps was made independently in our laboratory (Zechel *et al.* 1970; Zillig *et al.* 1970*b*) and by Bautz & Bautz (1970) on the basis of results obtained by Lill *et al.* (1970) and, in another way, by Hinkle & Chamberlin (1970). Whereas free and primarily bound enzyme (that is bound in the absence of σ, or at less than 10 °C, or in excess to 'correct' sites) are instantaneously inactivated by heparin, which as a strong polyanion competes for the template-binding site of the enzyme, or by rifampicin, which blocks initiation by binding to the β-subunit of the polymerase, complete enzyme at high temperature and in the presence of an excess of template forms a tight complex which is very stable towards heparin and fairly stable towards rifampicin. The number of sites on the DNA available for the formation of this complex is low. Bautz & Bautz (1970) attempted to determine this number by counting labelled end-groups of RNA initiated from such sites. The approximate agreement with the expected number of promoters led them to assume that these sites are the promoters. Thus, for example, about 25 'promoters' were counted on phage T4-DNA, 35 on T5-DNA and two to three on T7-DNA. It must be stressed, however, that this assay assumes that only one strand is transcribed from one promoter in enzyme excess in the presence of the inhibitor.

We have used another simple approach for determining the number of promoters on T5-DNA (and T4-DNA). The fraction of T5-DNA transcribed *in vitro* was measured by hybridizing T5-DNA with the transcription product in excess. The sum of the length of all RNA species obtained in this way was divided by the average chain length determined by end-group plus internal labelling. We find that only about 56%, referred to one strand of T5-DNA, is transcribed under our optimum *in vitro* conditions and, from the average chain length of about 4000 nucleotide residues for pppA plus pppG-initiated chains, the number of active promoters per DNA molecule is about 17 (6 for T4-DNA). This would mean that in the conditions used by Bautz & Bautz (1970) about two chains were initiated from one promoter. We find that in excess of enzyme, up to 20 RNA chains are transcribed from each reading unit in conditions excluding consecutive initiation and reinitiation, i.e. when heparin was added to the preformed enzyme-DNA complex 10 minutes before the substrates. This means that up to 20 polymerase molecules per promoter were able to form a heparin-stable initiation complex during preincubation. This number is independent of the time of preincubation from 30 seconds onwards, indicating that the stable complex is formed very quickly.

Recent work by Blattner *et al.* (1972) shows that the genetically defined promoter to the left of the immunity region of phage lambda is several hundred nucleotide residues to the right of the starting point of transcription of the *N* gene messenger. This forces us to assume that before it initiates RNA synthesis the polymerase has to recognize two sites consecutively, first the promoter and then, after it has slipped over the intervening non-transcribed piece of DNA, a site where it actually starts synthesis with either pppA or pppG, and which may be termed the *initiator*. The length of this *preinitiator* piece is quite sufficient for the binding of the above five to nine or even more polymerase molecules which could have slipped in through the promoter, thus allowing a speedy initiation of several chains from one initiator even when no free polymerase is coming in. Our chain counts, together with the results of Blattner *et al.* (1972), strongly support this 'preinitiator' model.

If this model is correct, attempts to isolate the promoter by DNase digestion of the stable complex and isolation of the protected piece may yield the initiator or even segments of the preinitiator region. Until we are able to fix the different states it will be difficult to decide what we are looking at.

A more detailed investigation of the heparin-protected complexes shows that they may be subdivided into at least three classes. We have shown (Zillig *et al.* 1970*b*) that about half to two-thirds of the complexes formed on T5-DNA are very stable, with a half-life of more than 100 minutes in heparin. The other half to one-third on T5-DNA decays with first-order kinetics and a half-life of 60 seconds in the presence of heparin (Fig. 1). The first class of protected com-

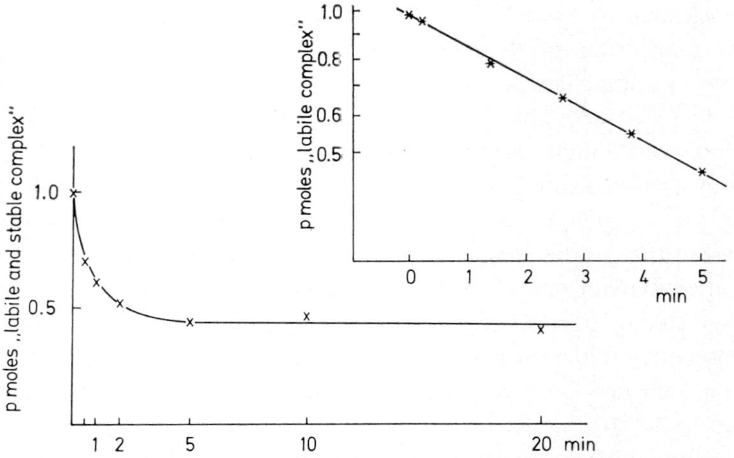

FIG. 1. Kinetics of decay of initiation complexes in the presence of heparin. To a mixture containing 0.006 M-Mg-acetate, 0.05 M-NH_4Cl, 0.03 M-tris acetate, 200 µg/ml of T5-DNA and 20 µg complete enzyme, 100 µg/ml of heparin was added after preincubation for 15 min at 37 °C; at various times thereafter 0.05 ml aliquots were added to a mixture containing 0.03 M-Mg-acetate, 0.13 M-NH_4Cl and 1 mM each of UTP, GTP, CTP and [^{14}C]ATP to allow synthesis for 10 min. These samples were then prepared for counting. Insert: the time-dependence of the decay curve of the heparin-'labile' complex.

plexes may be called the stable, the second the unstable or labile class. The ratio of the two classes for one template is independent of the time of pre-incubation before heparin is added and of the concentration of enzyme, from DNA excess up to a large excess of enzyme (Fig. 2). Thus no interrelationship between precursor and product (two states at one site) could be established. Moreover, the considerable difference in stabilities excluded an assumption of fast equilibria. Since there is usually less of the labile complex than of the stable one, the assumption that the labile complex might be the preinitiator-bound class is also invalid.

It appears that the two classes define two types of high affinity binding sites, for example promoters and preinitiators, between which the enzyme is distri-

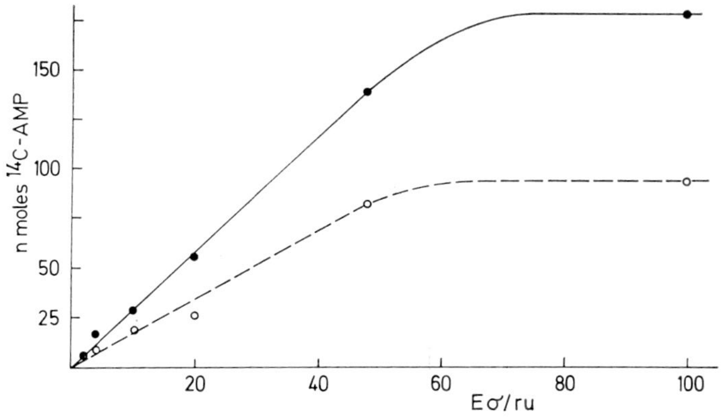

FIG. 2. Dependence of amount of 'labile' and 'stable' initiation complex on amount of enzyme.

Conditions as in Fig. 1, but the activity under heparin added either at the same time as the substrates or 10 min before the substrates was measured with increasing amounts of complete enzyme (mol. wt. 480 000 daltons). Dotted line: difference of both values, or 'labile complex'. Solid line: 10 min preincubation values, or 'stable complex'. Eσ/ru: no. of complete enzyme molecules per reading unit of the DNA.

buted according to their numerical ratio. Besides their striking difference in stability towards heparin, the two classes differ in other respects. Whereas the formation of the stable class is not influenced by the addition of excess core enzyme, the latter competes with complete enzyme for the formation of the labile class (Zillig *et al.* 1970*b*). Since initiation factor σ is absolutely and stoichiometrically required for the formation of both classes of *active* complexes, this leads to a reduction of the labile complex until finally a plateau is reached corresponding to the stable complex. Thus σ is not required for *recognition* of the 'labile sites'. It is, however, necessary for the act of initiation. This, as well as the lack of positive evidence for the existence of different σ factors with different promoter specificities, allows one to speculate that σ might be a large effector rather than an 'adaptor'. The labile class then appears weaker in the sense that σ might be less tightly bound, making possible its substitution or removal by heparin; alternatively this class might require an additional factor, similar to the catabolite gene activator protein (CAP; Zubay *et al.* 1970), for stabilization. Another difference in the two classes of complexes is that the amount of stable complexes is much reduced with increasing ionic strength, whereas the amount of labile complexes is even increased, up to an optimum at around 0.4 (Fig. 3).

The additional distinction of two different types within the stable class of complexes was established by a less direct approach. It has been known since

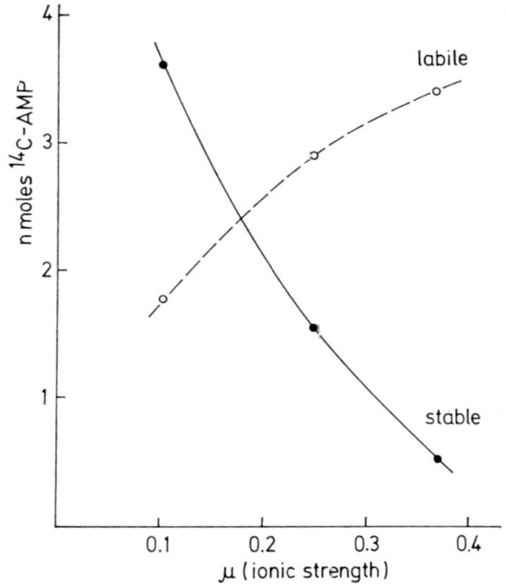

FIG. 3. Dependence of amount of heparin 'labile' and 'stable' initiation complex on ionic strength.

Conditions as in Figs. 1 and 2, except that the 'labile' and 'stable' complexes were formed at increasing ionic strength (μ is calculated from the concentrations of the salts present in the preincubation and synthesis mixtures, i.e. Mg acetate, NH_4Cl and tris acetate). The concentration of complete enzyme is 200 μg/ml and of T5-DNA 100 μg/ml.

Bremer & Konrad's work (1964) that the rate of RNA synthesis at low ionic strength slows down rapidly, finally reaching a plateau which for T5-DNA occurs between 10 and 15 minutes after the start of incubation at 37 °C. The rate of growth of the average single chain in these conditions is 65 nucleotides per second, not significantly different from that at high ionic strength; moreover, every polymerase molecule, whether template is in excess or not, initiates only once. From the rate of chain growth and the average chain length, the synthesis phase should take only about two minutes and we wondered why the plateau was reached so slowly. When we measured the kinetics of chain initiation by labelling 5′ terminal γ-phosphates with pppA as well as with pppG, it became clear that the time course of initiation was almost the same as that of RNA synthesis (Fig. 4). This appeared to be in striking contradiction to the extremely fast formation of the heparin-stable initiation complex. However, the time course of initiation was just as slow when heparin was added to the preformed initiation complex slightly before or together with the substrate. Thus it appeared that the first stable state, which is formed too fast to be measured, is the precursor of a second one, the transition from one to the other state being rate-limiting at low ionic strength. The possibility that this transition

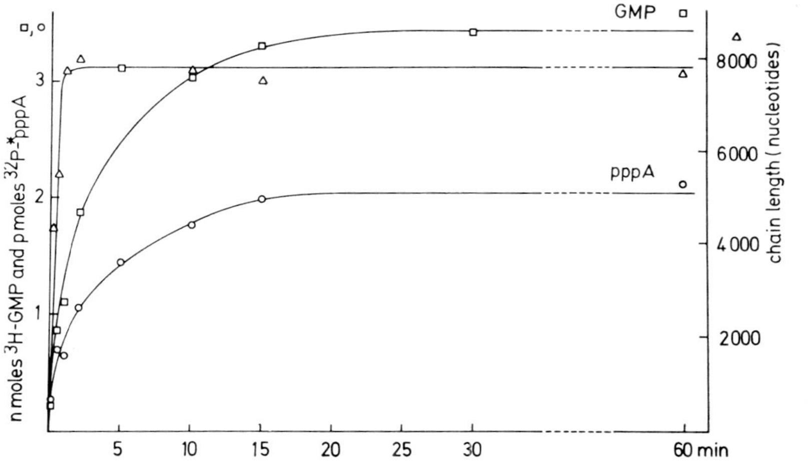

FIG. 4. Kinetics of initiation and RNA synthesis on T4-DNA at low ionic strength. Incubation mixture: 0.006 M-Mg-acetate, 0.05 M-NH_4Cl, 0.03 M-tris acetate, pH 7.9, 1 mM each of UTP, CTP, GTP (^3H-labelled) and ATP (γ-^{32}P-labelled), and 90 µg of T4-DNA/ml and 5 µg of complete enzyme/ml. Aliquots of 0.2 ml were prepared for counting at various times after incubation began at 37 °C. —□—: RNA synthesis (^3H incorporation). —○—: initiation (^{32}P incorporation). —△—: average chain length (nucleotides).

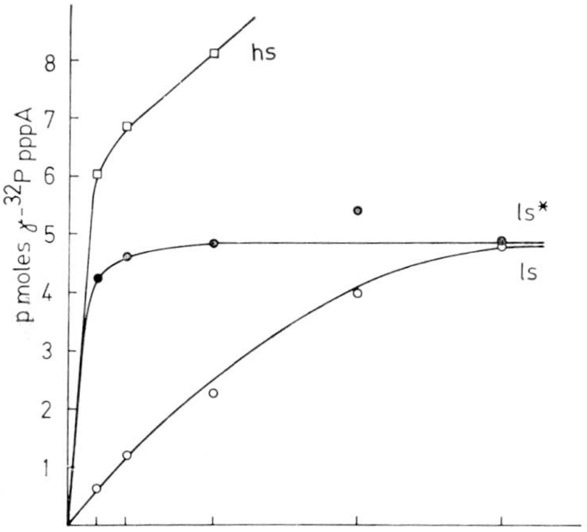

FIG. 5. Comparison of kinetics of initiation on T5-DNA at high ionic strength (□) and low ionic strength with (●) and without (○) preincubation in absence of substrate. Conditions as in Fig. 4, except that 125 µg of T5-DNA and 25 µg of complete enzyme/ml were used. One of the low salt mixtures was preincubated for 10 min at 37 °C before the addition of substrates; the high salt mixture contained 0.07 M-Mg-acetate and 0.13 M-NH_4Cl.

corresponds to the binding of the initiating pppA or pppG, first demonstrated by Anthony *et al.* (1966), was excluded by the striking acceleration of initiation observed after preincubation in the *absence* of substrate (Fig. 5). Thus, the so-called stable class of initiation complexes must consist of two consecutive states: a so-called delayed starter and, as its transition product, a so-called immediate starter. The half-life for the first-order transition of the delayed into the immediate starter state is around 3.5 minutes at low ionic strength and 37 °C (Fig. 6) and decreases quickly with increasing ionic strength.

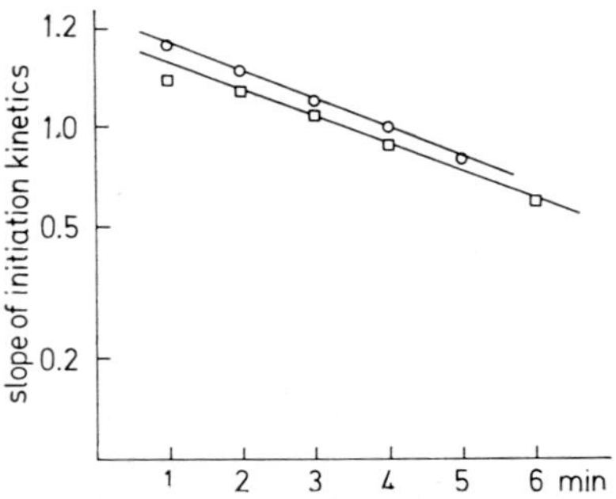

FIG. 6. Dependence of the logarithm of the slope of initiation kinetics on T5-DNA on time after incubation begins at 37 °C in the presence of heparin. —O—: pppA initiation; —□—: pppG initiation.

Conditions as in Fig. 4, except that 125 μg T5-DNA and 100 μg of complete enzyme/ml were used. DNA and the enzyme were preincubated for 2 min at 37°C; afterwards 100 μg of heparin was added for 1 min, followed by substrates (either ATP or GTP were γ-^{32}P-labelled).

We have already demonstrated that the formation of the 'heparin-stable initiation complex' shows a temperature-dependence similar to a DNA melting curve with a T_m of 17.5 °C for the *E. coli* enzyme on T4-DNA at low ionic strength (Zillig *et al.* 1970*b*). Since a sharp temperature profile is observed only after long preincubation with heparin (10 min or more) (Zechel 1971), this must represent the temperature-dependence of the transition of the delayed into the immediate starter state, which should be a highly cooperative conformational change. The participation of enzyme and DNA in the process remains to be clarified, as well as the question of whether these are two states at one site or whether a site change, for example from preinitiator to initiator, is involved. Nor is it known yet whether the labile class which is no longer observed after prolonged incubation with heparin undergoes a similar state transition before initiation.

The final step preceding the formation of the first internucleotide bond is the binding of pppA or pppG to the 'immediate starter' (Anthony *et al.* 1966), which will not be discussed in detail here.

Thus, initiation consists of at least three and maybe four consecutive events:

(1) The fast formation of a heparin-sensitive starter complex (the decay of the 'labile complex' in the presence of heparin occurs with first-order kinetics);

(2) Its rapid transition into a protected state called the delayed starter;

(3) A slow, highly cooperative transition into a second protected state, called the immediate starter;

(4) The binding of purine ribonucleoside triphosphates to this to form the complex ready for synthesis.

σ is required for the second step. Recognition of the promoter might occur at the first or second step, recognition of the initiator at the third.

RNA SYNTHESIS PHASE

As shown by Goldthwait *et al.* (1970), RNA polymerase possesses two different binding sites for ribonucleoside triphosphates: one, active in the absence of Mg^{2+} ions, binds purine ribonucleoside triphosphates only, and the other, requiring Mg^{2+} ions, has a ten times higher affinity for purine ribonucleoside triphosphates than the first, and also binds pyrimidine ribonucleoside triphosphates. The first site appears to be engaged in initiation and is inactivated by rifampicin; the second is considered to be the polymerization site.

Polymerization begins with the formation of the first internucleotide bond. Factor σ is not released until some ten nucleotide residues have been assembled (Krakow & von der Helm 1970). Krakow concluded from these results that since σ was replaced by the product, it had been linked to the product site of the enzyme. If the complex engaged in synthesis is called the active one, one might exaggerate by saying that σ plays the same part in the initiation complex as RNA plays in the active complex. The addition of each nucleotide residue to the 3'-hydroxyl end of the growing chain is accompanied by the translocation of the enzyme along the template by one nucleotide pair. The exchange of pyrophosphate residues between triphosphates and free pyrophosphate which has been observed during synthesis is not caused by a reversal of RNA synthesis (pyrophosphorolysis) (Maitra *et al.* 1967). It involves only the 3' terminal residue. Krakow has shown that it occurs in only one of two states during elongation, possibly related to two elongation sites on the enzyme corresponding to the tRNA-binding sites of the ribosome (Krakow & Fronk 1969). The nature of the active complex and, specifically, the state of the DNA under it have not yet been elucidated. Strand separation has been assumed to be necessary for initiation and it appeared to be established for the active complex by the work of Hayashi (1965). Since this may, however, be a special case for a supercoiled ring template (φX174-RFI) and since other models, especially that of Florentiev & Ivanov (1970), are not in contradiction with the facts, this question remains unanswered.

TERMINATION

Two mechanisms for termination have been pinpointed so far. One of these, discovered by Roberts (1969), requires the participation of a specific termination factor, ρ, and is observed only at sufficiently low ionic strength (below 0.1). Since under these conditions RNA synthesis stops after one round per enzyme molecule in the absence of this factor, the reduction of the chain length and overall synthesis by the action of ρ can only be explained by assuming that ρ termination leads to a dead-end state. This raises the question of whether additional factors are required to keep the system going.

The second mechanism does not require additional factors. At high ionic strength it leads to the liberation of RNA and enzyme, thus allowing continued reinitiation and, therefore, synthesis. On T5-DNA the average chain length is at least twice that of ρ-terminated RNA. Millette *et al.* (1970) showed that the 2.2×10^6 mol. wt. product obtained under these conditions from T7-DNA possesses a defined 3′ terminus, clearly indicating specific and not statistical termination. This mechanism is partly blocked at low ionic strength. It has been shown by Millette *et al.* (1970) for T7-DNA and in our laboratory for T7- and T5-DNA that the products obtained at low and at high ionic strength are identical. From filtration experiments on nitrocellulose filters and from sucrose gradient centrifugation with double labels it became clear that the transcription product formed at low ionic strength on excess DNA is released as well as the RNA formed at high ionic strength. The enzyme, however, remains fixed to the template and can be removed only by increasing the ionic strength. Only in excess of enzyme does the RNA begin to remain bound to the template.

These results, together with the findings about the influence of ionic strength on initiation (see p. 49), explain the long-known effects of ionic strength on transcription: when the polymerase recognizes the 'salt terminator' at low ionic strength it can release its transcription product but it stays tightly bound to the terminator. Thus synthesis stops after one round. Additional enzyme molecules coming up behind never reach the terminator and are therefore unable to release their products. When the ionic strength is increased, synthesis resumes, with immediate reinitiation by the released enzyme molecules (Fig. 7). If the ionic strength is sufficiently high from the beginning, continued synthesis of an identical product can take place.

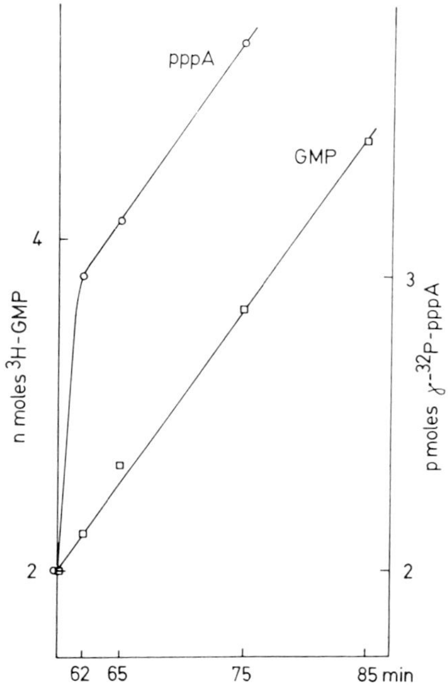

FIG. 7. Comparison of reinitiation and resumption of RNA synthesis after increase of ionic strength, in the plateau of RNA synthesis obtained at low ionic strength.

Conditions as in Fig. 5. After 60 min at 'low salt', the ionic strength was increased to 0.07 M-Mg-acetate and 0.13 M-NH_4Cl. —□—: RNA synthesis ([^{3}H]GMP incorporation). —○—: initiation ([γ-^{32}P]AMP incorporation).

Measurements of the kinetics of pyrophosphate accumulation in the absence and presence of pancreatic RNase led Krakow (1966) to postulate that the levelling off in the kinetics of RNA synthesis observed at low ionic strength was due to product-inhibition. Since, using Krakow's assay, we also see a clear although higher plateau in the presence of RNase, and since pyrophosphate accumulation in these conditions begins again when the salt concentration is increased, this conclusion is unwarranted. It remains true, however, that the product, as a polyanion, and like external RNA added to the system, consumes non-initiated and also—at high ionic strength—released enzyme in competition with the template. This might be one reason (besides oxidation, surface denaturation and thermal inactivation of σ) why even at high ionic strength the synthesis rate slowly decreases with time of incubation.

It appeared interesting to follow the fate of factor σ. During RNA synthesis at low ionic strength, σ is found in the free state on sucrose gradients shortly after initiation. After release of the RNA, however, σ is very tightly bound to the unreleased enzyme, maybe again at the product site.

It can be shown, however, by the use of catalytic amounts of σ, that the presence of σ in the dead-end complex of enzyme with terminator is not required for low-salt stability.

A special case is that of transcription of T4-DNA. We showed that the amount of product transcribed at low ionic strength is only a fraction of that transcribed at high ionic strength (Qasba & Zillig 1969). This result has been confirmed. Moreover, in contrast to the results on T5- and T7-DNA, the average chain length at low ionic strength (7500 nucleotide residues) is only 60% of that at high ionic strength (12 500 nucleotide residues). The product is not released even in a large excess of DNA. Reinitiation does not immediately follow the resumption of synthesis after an increase in salt concentration. Thus here at low ionic strength synthesis is blocked at a new type of specific site during propagation and it resumes, at least in part, without reinitiation, when the block is removed. The blocked state is distinguished from the active complex by being accessible to heparin, which inhibits resumption of synthesis after an increase in ionic strength. Since factor ρ exerts its activity on T4-DNA at low ionic strength, the ρ termination site must be situated, on average, promoter-proximal to this low-salt blocking site.

KAPPA, A NEW FACTOR BLOCKING RNA SYNTHESIS

When factor ρ was being isolated by the method of Roberts (1969), a late fraction was eluted from the phosphocellulose column (0.5 M-phosphate) which also greatly reduced RNA synthesis on T4-, T5-, and T7-DNA even at high ionic strength. By DEAE-chromatography (elution in a gradient at about 0.2 M-phosphate), sucrose gradient centrifugation and electrophoresis on cellulose acetate, a nearly pure RNase-, DNase- and protease-free single inhibitor protein was isolated. In sodium dodecyl sulphate (SDS)-polyacrylamide gels the molecular weight of the peptide chain was about 17 000 (Fig. 8). By analytical equilibrium sedimentation a molecular weight of 34 000 (3.3S) was found, indicating that the native protein in solution is a dimer.

This protein, termed kappa, inhibits RNA synthesis at both low and high ionic strength down to a plateau of about 40% of that observed in the absence of the factor at low ionic strength (Figs. 9, 10). This means that, in the presence of saturating amounts of κ, only one round of synthesis occurs even at high ionic strength. The average chain length of the product from T5-DNA is 1500 nucleotides against 4000 nucleotides for normal T5-RNA. Double-labelling experiments showed that the product formed in the presence of κ, unlike the product formed at low salt in its absence, remains fixed to the enzyme on the

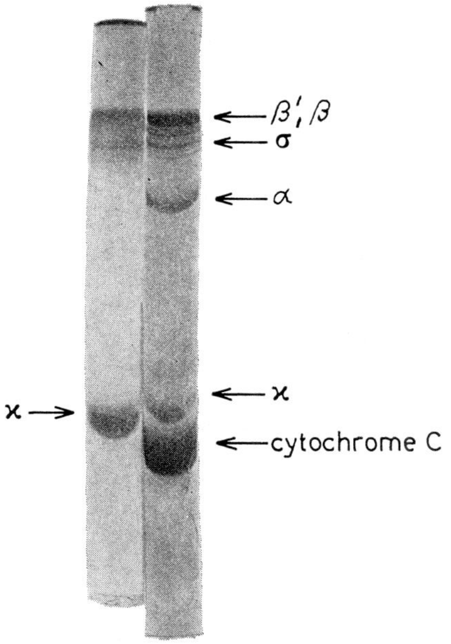

FIG. 8. Polyacrylamide gel disc electrophoresis of κ and markers (RNA polymerase and cytochrome *c*) in 20% gels, by the method of Laemmli (1970).

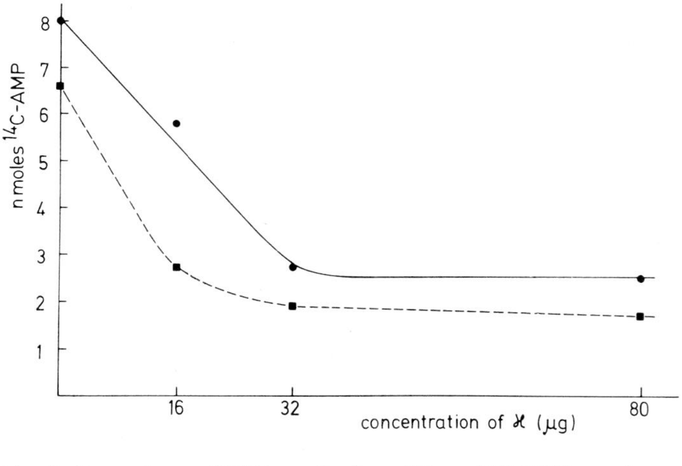

FIG. 9. Dependence of RNA synthesis on T5- and T7-DNA, on the amount of κ present. Fifty μg of T7-DNA (—●—) and 25 μg of T5-DNA (--■--) are incubated with 1 μg and 5 μg, respectively, of complete enzyme for 30 min in 'medium salt' conditions with increasing amounts of κ.

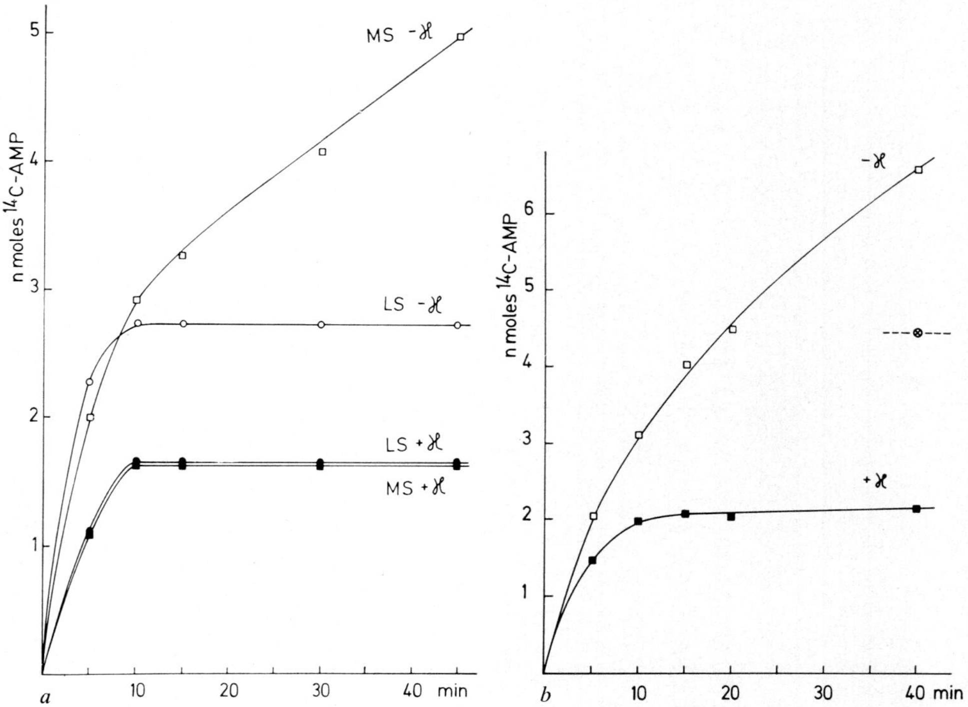

FIG. 10. Kinetics of RNA synthesis on T5-DNA in the absence of and with saturating amounts of κ at low, medium and high ionic strength.

(*a*) Low and medium salt (LS, MS); (*b*) high salt.

Conditions as given in previous legends. 0.2 ml aliquots containing 25 μg T5-DNA, 5 μg of complete enzyme and 30 μg of κ were removed from the incubation mixture at various times.

——⊕——: the low salt plateau of *a* without κ.

template. Sucrose gradient centrifugation (a single homogeneous peak with reduced sedimentation velocity) and hybridization competition with normal T5-transcription products (Figs. 11, 12) indicate that κ, like ρ, blocks RNA synthesis at defined sites along the template.

The amount of κ required for maximum inhibition is proportional to the amount of template and not that of enzyme. It is, however, much higher than would be expected for a few inhibitor-binding sites per reading unit, at least at low ionic strength. At low and medium ionic strength κ binds DNA to nitrocellulose filters quantitatively; at high ionic strength it binds with much reduced efficiency. It co-sediments with DNA in sucrose gradients. From the amount bound in saturation on T5-DNA, as well as from the amount required for maximum inhibition of RNA synthesis on T5- and T7-DNA, we calculated that one molecule of κ (in dimeric form) was bound for each 75 to 100 nucleotide

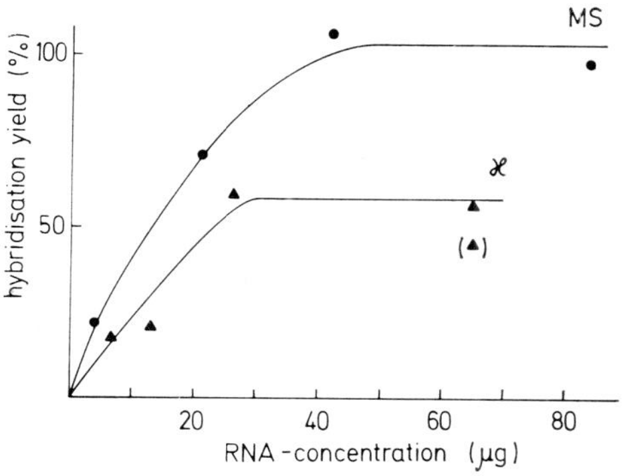

FIG. 11. Dependence of the hybridization yield of RNA, synthesized from T5-DNA in the absence (—●—) and presence (—▲—) of κ at medium ionic strength (0.03 mol Mg-acetate, 0.13 mol NH_4Cl), on RNA concentration. Filters carried 5 μg of T5-DNA. The technique was as described by Spiegelman & Gillespie (1965). The plateau obtained with RNA synthesized in the absence of κ is taken as 100%.

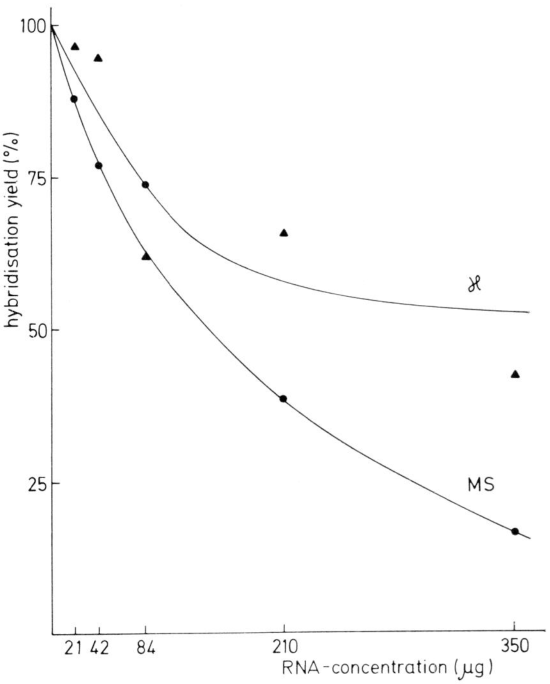

FIG. 12. Hybridization competition of non-labelled RNA, synthesized from T5-DNA in the presence (▲) and absence (●) of κ, with ^{14}C-labelled 'medium salt' RNA.
Conditions as in Fig. 11.

pairs of the DNA. This is very tight packing and appears inconsistent with the assumption of specific inhibitor-binding sites and with both a measured 60% reduction in the average chain length and the very sharp levelling off at 40% RNA synthesis in κ saturation. To reconcile these seemingly inconsistent facts one might assume that at low ionic strength κ is bound to DNA at many sites, maybe even in the tightest possible packing, including a small number of specific sites. Only the latter contribute to the inhibition observed when the kinetics of synthesis levels off, at low as well as at high salt concentrations. The unspecifically bound κ molecules might interfere with the rate of chain growth, as indicated by the reduction in initial synthesis rate, which is not paralleled by a reduction in γ-^{32}P-labelled starts (5′ terminals). However, these molecules do not interfere with the amount finally synthesized.

From the rather high yield of κ (120 mg from 2 kg of *E. coli*, wet weight) we assume that the amount present in the cell is quite large. The significance of our observations for *in vivo* synthesis of RNA remains unclear. κ might be a factor regulating the rate of transcription by interference with chain elongation, but it is also possible that its effect on transcription *in vitro* is unrelated to its real function in the cell. We believe, however, that one can assume that it is involved in DNA function or replication.

REGULATION OF TRANSCRIPTION

It is outside the scope of this paper to discuss in detail the modes of regulation of transcription and the participation of additional factors therein. A short summary from which repression is omitted must suffice.

The classical hypothesis that specific σ factors determine the recognition of corresponding promoter classes (Burgess *et al.* 1969) has been neither disproved nor substantiated. On the other hand, evidence is accumulating for the thesis that σ is an effector and that other elements of the enzyme are involved in promoter recognition. This evidence is based especially on the finding that, in several systems where a time control of transcription specificity is known, the core enzyme is structurally modified, resulting in a changed enzyme specificity (Walter *et al.* 1968; Goff & Weber 1970; Seifert *et al.* 1971; Losick *et al.* 1970). At least in one case, namely that of CAP (see p. 47), it is clear that additional factors also participate in the regulation of transcription of classes of operons (Zubay *et al.* 1970).

All these regulatory mechanisms concern initiation. A possible mechanism for the regulation of chain elongation, and especially for the coupling of transcription with translation, remains to be substantiated.

Finally, transcription may be regulated via termination, by ρ, possibly by factors controlling the effect of ions on 'salt termination', and perhaps by κ. If this turns out to be an all-or-none effect and not a rate regulation, the evolution of a meaningful sequential order of genes would have to be postulated.

CONCLUSION

The existence of a number of small RNA polymerases (T7, Chamberlin *et al.* 1970; T3, Dunn *et al.* 1971; and *Neurospora*-mitochondrial, Küntzel & Schäfer 1971) which are capable of correct initiation, elongation and termination in rigid transcription systems containing one or a few reading units poses the question: why do the normal bacterial, blue-green algal, and even, as far as one can see, eukaryotic RNA polymerases show such a complex but uniform quaternary structure? The reason cannot be the complexity of function, which is the same in all cases. One is tempted to speculate, in the light of growing evidence for the structural flexibility of RNA polymerase (modification, etc.), that the real reason is a requirement for functional flexibility, since a normal polymerase has to be able to transcribe or not transcribe, in regulated fashion, a large number of operons ($\sim 10^4$ in *E. coli*) or operon classes. Although factual knowledge is growing, we are still far from a real understanding of the transcription process.

References

ANTHONY, D. A., ZESZOTEK, E. & GOLDTHWAIT, D. A. (1966) *Proc. Natl. Acad. Sci. U.S.A.* **56**, 1026

BAUTZ, E. K. F. & BAUTZ, F. A. (1970) *Nature (Lond.)* **226**, 1219

BLATTNER, F. R., DAHLBERG, J. E., BOETTIGER, J. K., FIANDT, M. & SZYBALSKI, W. (1972) *Nat. New Biol.* **237**, 232

BOLLE, A., EPSTEIN, R. H., SALSER, W. & GEIDUSCHEK, E. P. (1968) *J. Mol. Biol.* **31**, 325

BREMER, H. & KONRAD, M. W. (1964) *Proc. Natl. Acad. Sci. U.S.A.* **51**, 801

BREMER, H., KONRAD, M. W., GAINES, K. & STENT, G. S. (1961) *J. Mol. Biol.* **13**, 540

BURGESS, R. R. (1969) *J. Biol. Chem.* **244**, 6160, 6168

BURGESS, R. R., TRAVERS, A. A., DUNN, J. J. & BAUTZ, E. K. F. (1969) *Nature (Lond.)* **221**, 43

CHAMBERLIN, M., MCGRATH, J. & WASKELL, L. (1970) *Nature (Lond.)* **228**, 227

DUNN, J. J., BAUTZ, F. A. & BAUTZ, E. K. F. (1971) *Nat. New Biol.* **230**, 94

FLORENTIEV, V. L. & IVANOV, V. I. (1970) *Nature (Lond.)* **228**, 519

GOFF, C. G. & WEBER, K. (1970) *Cold Spring Harbor Symp. Quant. Biol.* **35**, 101

GOLDTHWAIT, D. A., ANTHONY, D. A. & WU, C. W. (1970) in *RNA Polymerase and Transcription* (SILVESTRI, L., ed.), p. 10 (1st Lepetit Colloquium), North-Holland, Amsterdam

HAYASHI, M. (1965) *Proc. Natl. Acad. Sci. U.S.A.* **54**, 1736
HAYASHI, M., HAYASHI, M. N. & SPIEGELMAN, S. (1963) *Proc. Natl Acad. Sci. U.S.A.* **50**, 664
HEIL, A. & ZILLIG, W. (1970) FEBS (*Fed. Eur. Biochem. Soc.) Lett.* **11**, 165
HINKLE, D. C. & CHAMBERLIN, M. (1970) *Cold Spring Harbor Symp. Quant. Biol.* **35**, 65
KRAKOW, J. S. (1966) *J. Biol. Chem.* **241**, 1830
KRAKOW, J. S. & FRONK, E. (1969) *J. Biol. Chem.* **244**, 5988
KRAKOW, J. S. & HELM, K. VON DER (1970) *Cold Spring Harbor Symp. Quant. Biol.* **35**, 73
KÜNTZEL, H. & SCHÄFER, K. P. (1971) *Nat. New Biol.*, **231**, 265
LAEMMLI, U. K. (1970) *Nature (Lond.)* **227**, 680
LILL, H., LILL, U., SIPPEL, A. & HARTMANN, G. (1970) in *RNA Polymerase and Transcription* (SILVESTRI, L., ed.), p. 55 (1st Lepetit Colloquium), North-Holland, Amsterdam
LOSICK, R., SONENSHEIN, A. L., SHORENSTEIN, R. G. & HUSSEY, C. (1970) *Cold Spring Harbor Symp. Quant. Biol.* **35**, 443
MAITRA, U. & HURWITZ, J. (1965) *Proc. Natl. Acad. Sci. U.S.A.* **54**, 815
MAITRA, U., NAKATA, Y. & HURWITZ, J. (1967) *J. Biol. Chem.* **242**, 4908
MARMUR, J. & GREENSPAN, C. M. (1963) *Science (Wash. D.C.)* **142**, 387
MILLETTE, R. L., TROTTER, C. D., HERRLICH, P. & SCHWEIGER, M. (1970) *Cold Spring Harbor Symp. Quant. Biol.* **35**, 135
PETTIJOHN, D. & KAMIJA, T. (1967) *J. Mol. Biol.* **29**, 275
QASBA, P. K. & ZILLIG, W. (1969) *Eur. J. Biochem.* **7**, 315
ROBERTS, J. W. (1969) *Nature (Lond.)* **224**, 1168
SEIFERT, W., RABUSSAY, D. & ZILLIG, W. (1971) *FEBS (Fed. Eur. Biochem. Soc.) Lett.* **16**, 175
SPIEGELMAN, S. & GILLESPIE, D. (1965) *J. Mol. Biol.* **12**, 829
WALTER, G., ZILLIG, W., PALM, P. & FUCHS, E. (1967) *Eur. J. Biochem.* **3**, 194
WALTER, G., SEIFERT, W. & ZILLIG, W. (1968) *Biochem. Biophys. Res. Commun.* **30**, 240
ZECHEL, K. (1971) Doctoral thesis, University of Munich
ZECHEL, K., RABUSSAY, D. & ZILLIG, W. (1970) *Hoppe Seyler's Z. Physiol. Chem.* **351**, 136
ZILLIG, W., FUCHS, E., PALM, P., RABUSSAY, D. & ZECHEL, K. (1970*a*) in *RNA Polymerase and Transcription* (SILVESTRI, L., ed.), p. 151 (1st Lepetit Colloquium), North-Holland, Amsterdam
ZILLIG, W., ZECHEL, K., RABUSSAY, D., SCHACHNER, M., SETHI, V. S. & SEIFERT, W. (1970*b*) *Cold Spring Harbor Symp. Quant. Biol.* **35**, 47
ZUBAY, G., SCHWARTZ, D. & BECKWITH, J. (1970) *Cold Spring Harbor Symp. Quant. Biol.* **35**, 433

Discussion

Levitt: Is magnesium more effective than sodium at low concentration? Is there a divalent cation requirement?

Zillig: Magnesium is definitely required for transcription, but small (catalytic) amounts are sufficient and, beyond this, optimum conditions can be achieved by the addition of monovalent ions. By low I mean about 0.1 ($\mu = \frac{1}{2} \Sigma C_iZ_i^2$) and by high 0.37.

Levitt: Did you say that non-specific binding is stronger at low ionic strength?

Zillig: Yes. A larger number of polymerase molecules can be bound to DNA at low ionic strength, but more transcription occurs at high ionic strength because there is no termination block.

Lowey: Is anything known about the three-dimensional structure of the polymerase molecule by electron microscopy, for example?

Zillig: Several years ago we published pictures of hexagons with a hole in the middle which we thought represented the enzyme, but these turned out to be pictures of a minor contaminant.

Lowey: Have there been any hydrodynamic studies?

Zillig: My collaborator Rabussay has looked at the enzyme with small angle diffraction X-ray and obtained data on mass, size and form (Pilz *et al.* 1972).

Katchalski: Has the enzyme been crystallized?

Zillig: Several claims have been made. At the moment we have something that looks crystalline but we cannot grow the crystals. So far only core enzyme which can be obtained in very pure state has been subjected to crystallization.

Lifson: If the polymerase can act on different operons, which part of what determines which operon will be chosen?

Zillig: That is a very difficult question. We would have to go deeply into regulation. The old σ hypothesis postulated the existence of a number of classes of reading units corresponding to classes of promoters to which classes of homologous σ factors belong. No evidence at the moment substantiates this thesis. By the way, this would be a very uneconomic way of regulating transcription. A number of external factors such as CAP are involved in operon recognition. The basic structure of the polymerase can be modified chemically, for example by attaching adenosine and phosphate residues to the α subunits, and the specificity of the enzyme changes as a consequence. There is no clear answer yet.

Katchalski: Is it a general phenomenon to have enzymes queuing for their turn to participate in DNA transcription?

Zillig: It could be of advantage that such storage sites for polymerase allow initiation of transcription of several RNA molecules per reading unit whether free enzyme is available or not.

Gilbert: Oscar Miller's photographs of the ribosomal genes in action suggest that there is a line of polymerases on the DNA between the regions of active synthesis.

Katchalski: Is this true only for transcription or would it be true in other cases for other polymerases?

Gilbert: It has been an attractive hypothesis that DNA replication might be done by a bunch of polymerases, but I think the evidence is now swinging the other way. One can now do experiments with the correct DNA polymerase, *pol III*, to see if it works rapidly enough, putting in 1000 bases per second, so that a single enzyme at the growing point will suffice. Protein synthesis does use multiple ribosomes on a single piece of messenger, each following the other at short intervals.

Katchalski: But this does not seem to be true for repressors.

Gilbert: For the *lac* repressor there seems to be only one protein sitting on the DNA blocking that particular operon. For the λ repressor, it seems to be possible for two more molecules (tetramers) to polymerize along the DNA and block an appreciably larger region.

Revel: Does the polymerase 'waiting line' model perhaps contradict the kinetics of RNA synthesis as published for the tryptophan operon (Morse *et al.* 1969)? Two minutes seem to elapse between each RNA chain initiation. If a train of polymerases were waiting, you shouldn't get such a time lag.

Zillig: That depends on the time needed for the state transitions in initiation. If one of them becomes rate limiting for transcription it would determine the distance between polymerase molecules on the DNA.

Gilbert: I don't think the two-minute lag between initiations is true. That entire argument was originally based on the claim that there were waves appearing in the rate of mRNA synthesis, corresponding to a several-minute delay between initiations. My impression is that there is no longer any evidence for that hypothesis. Baker & Yanofsky (1970) showed that polymerases transcribing the tryptophan operon initiate very rapidly, once every 11 seconds on the average.

Zillig: The experiments I described were done with heparin. The kinetics of initiation with heparin show the same slow initiation as those without the inhibitor. Thus the rate-limiting step must follow the formation of the heparin-stable preinitiation complex. We know that rifampicin does not wait for the complex between enzyme and DNA to decay. Rifampicin must bind to the preinitiation complex itself, just inhibiting the forward reaction of that complex. Rifampicin-sensitive enzyme competes with rifampicin-resistant enzyme in the presence of the inhibitor, apparently for 'promoter' sites.

Katchalski: Does unwinding of DNA occur during the various processes discussed? Is the unwinding induced by the enzyme, are other factors involved, or is it a spontaneous random process?

Zillig: Models have been presented which would not require this unwinding. Facts so far neither prove nor disprove any of the possibilities you mention.

EVOLUTION OF BIOLOGICAL MACROMOLECULES

[As Professor S. Spiegelman was unable to attend the symposium, his evolution experiments with phage Qβ were summarized by Professor M. Eigen (see

Schmitt 1970 for a review of Professor Spiegelman's work). Professor Eigen then discussed his own recent theory of evolution, as published in *Naturwissenschaften* (Eigen 1971). A modified translation of the summary of this paper follows.—*Eds.*]

Eigen: The focal point lies in the question '*Can biology be explained by the present concepts of physics*?'

The answer, so far as it can be given briefly, would have to be: in those biological processes and phenomena extensively investigated so far, there are no indications that physics, as presently understood, should not be in a position to do so, even if there are limits for detailed description, as with the macroscopic phenomena of the inanimate world. But those limits are imposed only by the complexity of the phenomena and not by the principles. The same applies to the fact that certain laws of physics familiar to us may express themselves in a special, characteristic form in life phenomena. Above all one has to mention the—physically derivable—value concept for the theory of information production which controls the optimization process of evolution, or the preferred temporal direction of this process which has its origin in the stability criteria of the thermodynamic theory of irreversible processes, thus making evolution a basically 'inevitable' event.

Darwin's theory appears as a derivable optimization principle bound to certain physical requirements, but not as an irreducible phenomenon confined to the biosphere alone. According to the stability criterion of Glansdorff & Prigogine (1971) it is connected with the thermodynamic theory of stationary states. Concepts like selection strain and selection value can be physically objectified and quantitatively formulated if one accepts defined dynamic conditions such as constant 'fluxes' or 'forces'. The value concept expressed here provides the basis for an information theory which includes a description of information production. 'Information' is here a 'molecular' property based on the dynamic theory of matter, and 'valued' by its ability to conserve or reproduce itself. In particular it is able to differentiate energetically degenerate states and is characterized by a primary self-organization of matter according to the purely functional aspect of self-reproduction under variable external conditions.

Stochastic theory provides a foundation for the 'mean value' behaviour of the phenomenological equations in the case of great numbers of molecules. But it extends well beyond the statements of phenomenological theory. As evolutionary developments throughout have their origin in single events, which are 'amplified' by the growth process and so mapped macroscopically, they show—in comparison with closed systems existing in equilibrium—a considerably increased uncertainty, not only of the individual structures but also of the 'historical' course of the events. Of course, this theory removes also the prejudice widespread amongst biologists, particularly in regard to this 'uncertainty' of the single event, that Darwin's principle, formulated as 'survival of the fittest', contains only the trivial tautology: 'survival of the survivor'. Such an interpretation would only then be justified if 'fittest' were solely determined by the fact of 'survival' as the result of random events. Such a case is singular and hardly refers to natural conditions. Not only would it require a completely exact, i.e. perfect, reproduction, but amongst other things also a complete degeneracy of all selection values—in other words the absence of any differentiation of the dynamic properties of the competing species. Only then, i.e. in case of such a complete degeneracy, could a random fluctuation in the concentration of an individual species increase as a consequence of the autocatalytic mechanism of reproduction, so that selection of one species is possible. But in this case one could not have predicted by any means which of the species were to be selected. All reproduction processes actually occurring are affected by a certain rate of miscopying because of the finiteness of the energies of interaction. Above that they generally differentiate with regard to their selection values. But then selection is always characterized by an optimization procedure, where the concept 'fittest' is allied with a value optimum under certain constraints given in the form of inequalities (which sometimes prevent a true maximum being reached). Fluctuations play an important role because those mutations with a selective advantage appear first of all in the form of a single copy. The modification of the deterministic behaviour due to fluctuations is the more important the smaller the 'selective advantage' favouring survival is.

It has been demonstrated that the *formation of 'viable' structures* (probably forerunners of unicellular organisms) is connected with special conditions of the selection mechanism, which cannot be fulfilled by the nucleic acids or the proteins alone. The selection mechanism, which is based on the reproduction process, itself has to be of non-linear nature. Not only nucleic acids but also proteins can reproduce themselves on a 'quasi-linear' reaction mechanism. However, at the same time there arise systems which contain too little information, as in the case of nucleic acids, or too much, as in the case of proteins. 'Too little' signifies that the sequences competing with each other are not in a position to collect enough *reproducible* information for the coding of selection-favourable functions. On the other hand, too much information means that the probability for a self-promoting mutation becomes too small or that the system is unable to free itself from the network of 'parasitic' couplings. But it is also possible that a system containing nucleic acids as well as proteins could utilize for a stable selection the functional advantages of both types of subtsances. The advantageous characteristics are:

(*a*) The inherent capability of self-instruction of the nucleic acids with whose help it is possible to reproduce not only each fully-developed information state but also any further change.

(*b*) The enormous functional capacity of the proteins (specific recognition, catalysis and regulation) which is indispensable for the coupling and correlation of single reactions in the synthesis of organized functional units.

The hierarchy of reaction cycles resulting from such coupling between nucleic acids and proteins and exhibiting non-linear growth kinetics already displays the essential characteristics of a 'living' system and is 'open' for further evolution up to the living cell. The origin of such a self-reproducing 'hypercycle' depends on the development of a specific code system with unique assignments.

Various assumptions can be tested by evolution experiments as first carried out by S. Spiegelman with Qβ phages. The theory developed in the present paper provides the basis for carrying out reproducible measurements and their quantitative evaluation.

The results can be summarized thus:

(1) Detailed analysis of the reproduction mechanism of nucleic acids and proteins does not offer any support for the hypothesis that life phenomena involve any forces or interactions not otherwise known in physics. The selection behaviour which is characteristic for the evolution of living systems appears already at this stage as an inherent property of certain matter under special reaction conditions.

(2) Every system obtained by mutation and selection is indeterminate with regard to its individual structure; nevertheless the resulting process of evolution is determinate, i.e. regulated by physical laws. The occurrence of a mutation with selective advantage corresponds to an instability and as such can be explained with the aid of Glansdorff and Prigogine's principle for stationary irreversible thermodynamic processes. The optimization process of evolution is therefore inevitable, but with regard to the selection of the individual route, undetermined.

(3) Finally, the origin of life depends on a series of characteristics which collectively can be physically clearly defined. The requirements for the formation of these characteristics have probably been fulfilled step by step so that the 'origin of life', like the evolution of the species, cannot be represented as a single act of creation.

Gilbert: Is there any non-linearity in the Qβ model?

Eigen: The replication kinetics of the plus and minus strands of phage RNA can be described by a system of pairs of linear differential equations, each pair relating to a different mutant. Of course this system becomes non-linear by the constraints superimposed, e.g. constant population density. With the help of our theory we were able to evaluate Dr Spiegelman's experiments and give the true rate constant for the plus and minus strands. Non-linear replication kinetics does not occur if the enzyme, i.e. the replicase, is provided in the test tube.

Flory: Your theory has the flavour of the theory of branching processes.

Eigen: Yes. Many such models were played on a computer. The non-linear equations usually involve oscillating solutions.

Flory: The unpredictability of the survival of the fittest follows the statistics of branching processes, a well-developed subject. The problem finds analogy in the observation that certain surnames vanish as one generation succeeds another. The theory of branching processes predicts that eventually only one name would survive for the entire human race—in a steady population, of course.

Eigen: This is a question of the degeneracy of the rates of the replication processes for the different names. Only when there is a degeneracy would a few names (or one) become redundant.

Flory: According to the theory, you can't predict those names.

Eigen: For non-degenerate processes the rates and precision of replication and the lifetimes of the different species determine the selective advantage. What was surprising in our experiments was how much individuality is already in RNA molecules. It is not simply a single strand with a uniform replication rate and lifetime. This structural diversity has played a very important role in the early adaptation of enzymes. Such a diversity is not found with DNA structures.

Lifson: The supercycle, or hypercycle, is essentially a structure which should compete, according to phenomenological considerations, with other different supercycles. Apparently the principle of selection leading to a unique optimum will still be correct for this supercycle. This is in accord with the fact that all living organisms have some basic properties in common, such as the chirality of amino acids and of sugars, but then there comes a point where diversity is preferred to uniformity. In principle why does evolution prefer diversity at a certain point when uniformity was preferred at an earlier point?

Eigen: First, there is a very simple principle behind any 'self-' or 'complementary' recognition mechanism. This principle is based on the preferred hydrogen bonding between the bases A and U, or G and C, present in the nucleic acids (RNA). Evolution based on this type of replication mechanism only is very limited, due to the relatively frequent errors. The next step therefore would have to involve the evolution of a code whereby a new molecule with higher recognition power could be brought in (e.g. the enzymes). There are many possible ways to originate a code, as different codes might have been present originally. The hypercycle I described is a tool for unique selection of a universal code (and chirality) due to its non-linear selection behaviour. It is only one (perhaps the most simple) among several possibilities. Once a grown hypercycle has become dominant (due to its non-linear growth kinetics), its further evolution requires 'compartmentalization' as well as 'individualization'. Those individualized systems then will again follow a linear reproduction kinetics, which more easily tolerates the coexistence of different descendants. Moreover these complex systems have many ways of evolving interactions which extend the possibility of the coexistence of diversity.

Katchalski: In Spiegelman's experiments, what are the molecular mechanisms involved?

Eigen: We will try to find out. The answer will come when we know the sequence.

Szwarc: When material is transferred from one test tube to the next in Spiegelman's experiments (see Schmitt 1970, p. 939), one gets back to the initial concentration of the reagent. Does this mean that in every initial concentration of the medium the species which has survived gets even more growth?

Eigen: Yes. The distribution of mutants in each subsequent step results from the distribution produced in the preceding tube. This may be a new starting condition at each step, otherwise you would always repeat the first step. Of course you could do the whole experiment in one tube. We thought of building a machine where we have only one reaction vessel and a circulating solution in which the proper concentrations are controlled. But it turns out that this procedure requires much more substance than the serial transfer (even if you reconstitute the enzyme which is not stable for longer than about 10–30 minutes). So really the best way is to transfer to fresh solutions. One can easily automate these transfer experiments and one can adapt the time intervals of transfer exactly to the growth. This was not yet done in Spiegelman's experiments; as a consequence his curve drifts away so that the number of transfers to reach a final product has no absolute meaning. If you want to compare different experiments they have to be done under fixed and reproducible conditions, to which theoretical analysis can be applied.

Levitt: Didn't Dr Spiegelman also try changing the concentration of the nucleotides by adding more and more cytidine to the reaction mixture?

Eigen: Yes, he has done different experiments with different selection constraints. For instance, he fixed the enzyme to a membrane which resulted in mutants of larger molecular weights (i.e. of higher stickiness). He could also induce higher resistivity to inhibitors (such as ethidium bromide). However starvation experiments, e.g. a reduction in the concentration of CTP, did not simply result in a reduced production of the polymers. Apparently a more complex regulation mechanism at the enzyme level is involved.

References

BAKER, R. & YANOFSKY, C. (1970) Transcription initiation frequency for the tryptophan operon of *Escherichia coli. Cold Spring Harbor Symp. Quant. Biol.* **35**, 467

EIGEN, M. (1971) Selforganization of matter and the evolution of biological macromolecules. *Naturwissenschaften* **58**, 465–523

GLANSDORFF, P. & PRIGOGINE, I. (1971) *Thermodynamic Theory of Structure, Stability and Fluctuations*, Wiley-Interscience, New York

MORSE, D. E., MOSTELLER, R. D. & YANOFSKY, C. (1969) *Cold Spring Harbor Symp. Quant. Biol.* **34**, 725–740

PILZ, I., KRATKY, O. & RABUSSAY, D. (1972) *Eur. J. Biochem.* **28**, 205–220

SCHMITT, F. O. (ed.) (1970) *The Neurosciences, 2nd Study Program*, Rockefeller University Press, New York

Protein synthesis machinery and the regulation of messenger RNA translation

MICHEL REVEL, YORAM GRONER, YAAKOV POLLACK, RUTH SCHEPS and HANNA BERISSI

Department of Biochemistry, Weizmann Institute of Science, Rehovoth

Abstract Studies of the very complex ribosome machinery show how cellular protein synthesis is regulated. Initiation signals in mRNA are recognized by ribosome-associated protein initiation factors, which by promoting ribosome binding to mRNA determine whether a particular cistron of mRNA is to be translated and at which frequency, thereby providing a control of gene expression at the translational level.

The machinery responsible for protein synthesis in living cells appears to be a complicated metabolic pathway (see Fig. 1 and, for review, Lengyel & Soll 1969) involving the different cellular RNA species (messenger RNA templates, over 40 transfer RNA molecules, three ribosomal RNA's) and almost 100 protein molecules (20 aminoacyl tRNA synthetases, the 56 proteins of the ribosome and a dozen protein factors involved in the different functions of the ribosome). This complexity contrasts with the relative simplicity of the reaction carried out, namely the formation of peptide bonds between amino acids. In fact polypeptide chains with defined amino acid sequences can also be formed by a much simpler, purely enzymic pathway. Thus, the small polypeptides of the antibiotic gramicidin S are synthesized by two enzymes (Kleinkauf & Gevers 1969) which provide the necessary information for the ordering and sequential polymerization of amino acids. As in ribosomal protein synthesis, amino acids are first activated by reaction with ATP to form aminoacyl adenylates, which are then bound to the enzyme instead of being transferred to tRNA. The amount of information stored in such an enzymic system is, however, limited and chains of no more than 15 amino acids are synthesized. A bacterial cell such as *Escherichia coli* possesses the information to assemble over 1000 different large proteins. This information, transferred genetically by the DNA, is retrieved according to the needs of the cell by a controlled transcription of the genes into messenger RNA templates. These bind to the non-specialized ribosomal machinery which decodes (translates) the nucleotide sequence into the correct amino acid sequence.

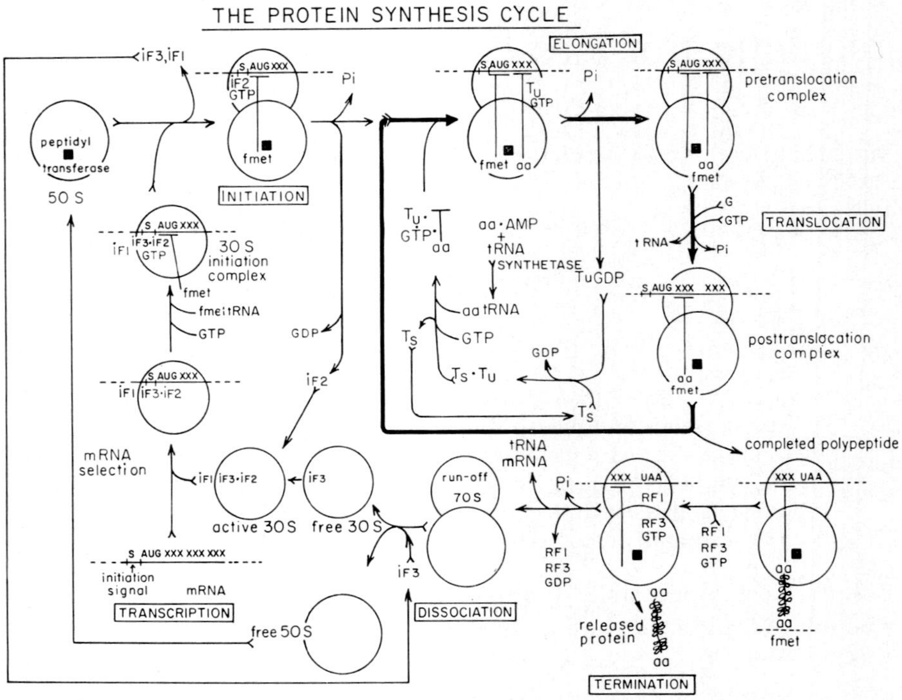

FIG. 1. A schematic representation of the steps in mRNA translation. Circles: ribosomes; ⊤: tRNA.

UNIFORM NOMENCLATURE FOR BACTERIAL TRANSLATION FACTORS:

	Abbreviation	*Other nomenclatures replaced*
Initiation factors		
Initiation factor 1	IF1	F1, A, FI
Initiation factor 2	IF2	F2, C, FIII
Initiation factor 3	IF3	F3, B, FII
Elongation factors		
Elongation factor Tu	EF Tu	Tu, Fl_u, S3
Elongation factor Ts	EF Ts	Ts, Fl_s, S1
Elongation factor G	EF G	G, FII, S2
Release factors		
Release factor 1	RF1	R1
Release factor 2	RF2	R2
Release factor 3	RF3	S, α

(For references see Lengyel & Soll 1969; for the 21 proteins of the 30S ribosome and the 34 proteins of the 50S see Kaltschmidt & Wittmann 1970).

Decoding is not, in principle, different from the transcription reaction, since it results from base pairing between three nucleotides forming a codon in mRNA and three complementary nucleotides in the anticodon of the tRNA.

Once the correspondence between amino acids and their specific tRNA's has been established by the aminoacyl-tRNA synthetases, one may wonder why the complicated ribosomal machinery is needed—could not a simple enzyme translate mRNA? The answer is that not all the information needed for protein formation is expressed by codon–anticodon pairing. We wish to illustrate here, through a few examples taken from *E. coli*, that the ribosome and its associated protein factors participate actively in the decoding process. Punctuation signals exist in mRNA (built-in in either the nucleotide sequences or in the secondary and tertiary structure of the RNA) which, when recognized by protein factors, instruct the ribosome to start and stop the translation. Most important for the cell biologist, ribosomes do not translate all mRNA equally, and regulation of protein synthesis is established by factors which control the selection on mRNA of the cistron to be translated.

ROLE OF RIBOSOME IN DECODING mRNA

The ribosome appears to be essential for the correct recognition of the anticodon. Studies by equilibrium dialysis (Uhlenbeck *et al.* 1970) without ribosomes show that oligonucleotides can bind in many different places on the tRNA molecule, and not only through the three bases of the anticodon. On the ribosome surface, however, the trinucleotide codon interacts very specifically only with the anticodon (Dube *et al.* 1969) and this allows an aminoacyl tRNA to be assigned to each of the 61 sense codons.

It is the 30S ribosomal subunit which carries the decoding site and the proteins of this subunit directly influence the codon-anticodon interaction. This was originally demonstrated by Davies *et al.* (1964), who found that streptomycin causes certain code words to be misread. Mutants resistant to streptomycin were shown by Strigini & Gorini (1970) to have a 30S subunit with altered decoding properties. Ribosomes from *strA* mutants of *E. coli* strongly restrict the binding of suppressor tRNA's to nonsense or mis-sense codons. The mutation affects one 30S protein, P10 (Ozaki *et al.* 1969). Another mutation, *ram* (which affects P4a), has an effect opposite to that of *strA* (Rosset & Gorini 1969). Yet another protein, P7, was shown by Nomura *et al.* (1969) to function in decreased misreading of codons. These different proteins appear therefore to cooperate in controlling decoding: if the control is too loose, mistakes in translation occur; if it is too stringent, certain classes of tRNA can never enter the 30S subunit.

Although non-enzymic binding of aminoacyl tRNA to 30S can be obtained *in vitro*, during protein synthesis the tRNA's do not bind by themselves but as a complex with the elongation factor Tu and GTP (see Fig. 1 and Lengyel &

Soll 1969). The need for the elongation factor in the coding process is not yet understood. One function may be to increase the rate of binding to the *in vivo* rate of 15 amino acids per second at 37 °C (Lacroute & Stent 1968). Another possible function may be to prevent binding of the wrong aminoacyl tRNA: tRNA interacts through its aminoacyl-ACC end with the 50S ribosome (Gilbert 1963) independently of the decoding which occurs on the 30S. Since decoding must proceed by trial and error, the presence of Tu may prevent binding to the 50S until the anticodon-codon matching is achieved. Only then would the GTP be hydrolysed and Tu-GDP released, allowing the formation of the peptide bond (Lengyel & Soll 1969). Ribosomal proteins and elongation factors would therefore both actively participate in the decoding process.

RECOGNITION OF PUNCTUATION SIGNALS BY RIBOSOME-ASSOCIATED PROTEIN FACTORS

The ribosome does not start translation of mRNA at the 5′ end but at specific 'start' signals (ribosomal binding sites) present at the beginning of each cistron (Steitz 1969; Revel *et al.* 1969). Termination signals appear to be separated from the next start signal by intercistronic regions (Jeppesen *et al.* 1970) (Fig. 2). Recognition of these punctuation signals is a function of ribosome-associated factors.

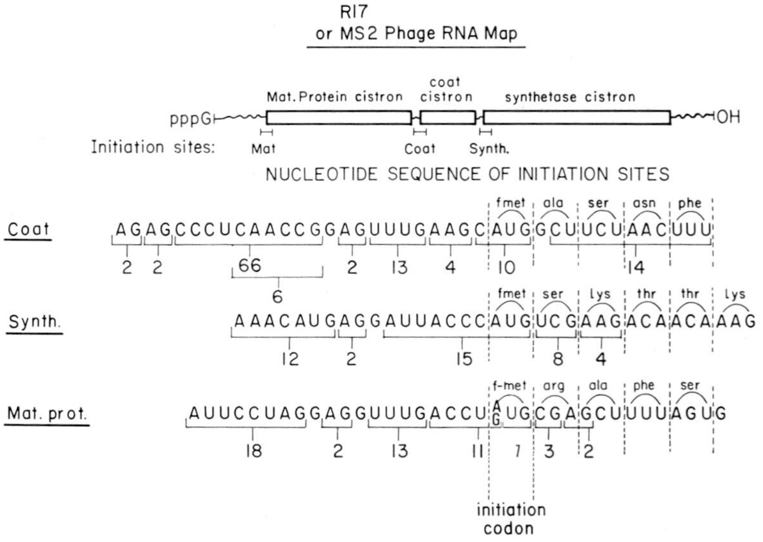

a

FIG. 2. The initiation sites on MS2 RNA. (*a*) Data from Steitz (1969), Jeppesen *et al.* (1970) and Berissi *et al.* (1971).

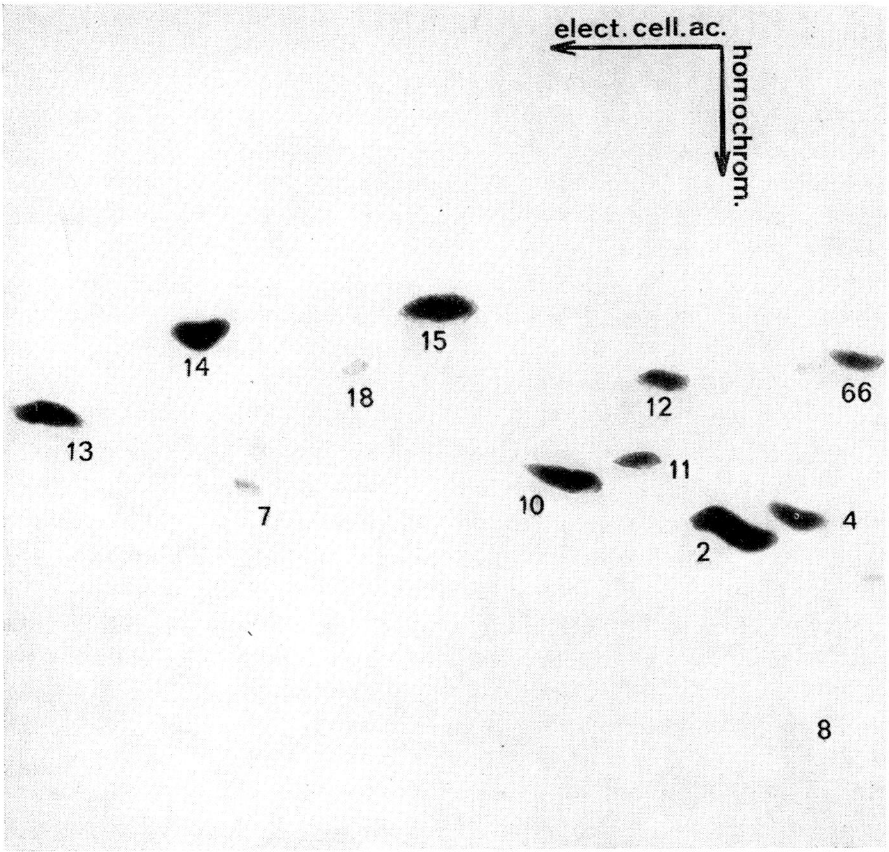

b

(*b*) Autoradiograph of the T1 oligonucleotides from ribosome-protected RNase A fragments of MS2 (Revel *et al.* 1972).

TERMINATION SIGNAL

Termination of the polypeptide chain occurs when one of the three nonsense codons, UAG, UAA or UGA, appears in the mRNA. These codons are not recognized by tRNA but by protein factors: RF1 (for UAG, UAA) or RF2 (for UAA, UGA) (Caskey *et al.* 1969). Direct binding of these triplets to the protein factors was demonstrated by Capecchi & Klein (1969). The binding of RF1 or RF2 to the ribosome in response to the nonsense codons (which also requires the presence of RF3 and GTP) triggers the transfer of the peptide chain from tRNA to a water molecule and release from the ribosome (Fig. 1). The termination signal is therefore a clear example of information in mRNA which is not expressed by codon-anticodon pairing.

INITIATION SIGNAL

On termination, the ribosome is released from mRNA and dissociated into subunits by interaction with an initiation factor (Davis 1971). The start of a new polypeptide chain occurs by the formation of a complex between (1) the free native subunits produced by dissociation, (2) the initiator tRNA species (which appears universally to be the formylatable (*N*-formyl) Met-tRNA species) and (3) a methionine codon (AUG) present at the beginning of a cistron of mRNA (see Fig. 2). Selection of these macromolecular elements and interaction between them is entirely determined by the protein initiation factors IF2(C), IF1(A) and IF3(B) (Wahba *et al.* 1969; Revel *et al.* 1969). First, these factors associate specifically with free 30S subunits, but not with 70S (Parenti-Rosina *et al.* 1969; Sabol & Ochoa 1971). Interaction of IF3 and IF1 on the 30S ribosomal surface competes with the binding of streptomycin to 30S (Lelong *et al.* 1972). Secondly, factor IF2 interacts with IF3 and this complex binds GTP and fMet-tRNA strongly, to the exclusion of all other aminoacyl tRNA's (Groner & Revel, 1971; Rudland *et al.* 1971). This property ensures that fMet-tRNA will be the only tRNA used to form an initiation complex. Finally recognition of the initiation signal on mRNA is also a function of initiation factors. All three factors appear to promote 30S ribosome binding to mRNA (Revel *et al.* 1969; Sabol *et al.* 1971). The first complex forms without fMet-tRNA, which is, however, absolutely required for the formation of the 70S-mRNA complex (see Fig. 1).

The specificity of mRNA recognition by ribosomes was investigated in template competition experiments (Revel *et al.* 1970*a*). With the initiation factors, ribosomes recognize and bind natural mRNA's, such as phage MS2 RNA or T4 mRNA, preferentially to synthetic polynucleotides. Factor IF3 is required for this specific recognition, whereas with IF2 ribosome binding appears rather unspecific. Use of oligonucleotide competitors of defined base sequence showed that what is recognized in native mRNA is more than the codon AUG, which strongly suggests the existence of some initiation signal which would be found near an initiator AUG codon, but not near the other internal AUG sequences in the mRNA.

SELECTION OF INITIATION SITES ON mRNA

The role of initiation factor IF3 in the recognition of the initiation signal was further investigated by studying the selection of ribosomal binding sites on phage MS2 RNA (Berissi *et al.* 1971). MS2 RNA, whose highly ordered structure was disrupted by mild formaldehyde treatment to allow independent

TABLE 1

Fractionation of IF3 activity on DEAE-Sephadex

NH_4Cl concentration of eluate (M)	*Nomenclature*	*[^{14}C]valine incorporation with: MS2 RNA (c.p.m.)*	*T4 mRNA (c.p.m.)*	*T4 mRNA / MS2 RNA*	*Specific activity [^{14}C]valine incorporation with T4 mRNA (pmoles/µg protein)*
0.05	F3·B1	879	1695	1.9	32
0.10	F3·B2	1120	1500	1.3	86
0.15	F3·B3	439	1657	3.8	70*
0.20	F3·B4	25	780	31.0	21†
—	Crude factor	—	—	—	0.35

* Further purified on hydroxyapatite.
† Further purified on phosphocellulose.
Protein synthesis was carried out as described previously (Revel *et al.* 1970*b*).

The proteins washed off *E. coli* MRE 600 ribosomes by 2 M-NH_4Cl were chromatographed on DEAE-cellulose. The fractions corresponding to IF3 activity (Groner *et al.* 1972*a*) were rechromatographed on DEAE-Sephadex A50 with an 0.02–0.3 M-NH_4Cl gradient in buffer 20 mM-potassium phosphate, pH 7.2, 0.2 mM-$MgCl_2$, 7 mM-β-mercaptoethanol and 5% glycerol. Each fraction was assayed for its activity in stimulating translation of MS2 RNA and T4 mRNA, in the presence of purified IF1 and IF2. The four subfractions of IF3 obtained were further purified as indicated (Groner *et al.* 1972*a*). Background incorporation with IF1 + IF2 alone (100 c.p.m.) was subtracted.

TABLE 2

Translation of MS2 RNA

Net stimulation with factor:	*[^{35}S]formylmethionine labelled peptides: I (Coat) (c.p.m.)*	*II (Synthetase) (c.p.m.)*	*III (c.p.m.)*	*Ratio Coat / Synthetase*
F3·B1	2700	4400	550	0.6
F3·B2	1500	500	250	3.0
F3·B3	1100	1100	350	1.0

Protein synthesis reaction mixture of 0.2 ml (Revel *et al.* 1971) contained 370 µg ribosomes, 5 µg IF1, 12 µg IF2, 17 µg IF3-B1, 8.5 µg IF3-B2, or 7.5 µg IF3-B3 (see Table 1), 15 µl supernatant, 50 µg MS2 RNA treated with formaldehyde (Lodish 1970*a*) and 7.5×10^5 c.p.m. [^{35}S]fMet-tRNA (4000 c.p.m./pmole). After 20 minutes at 37 °C, proteins were acid-precipitated and submitted to trypsin and chymotrypsin fingerprinting (Lodish 1970*b*). Radioactivity in each peak was computed after electrophoresis at pH 1.9 on thin-layer DEAE cellulose plates. Incorporation with IF1 and IF2 alone (I = 1100; II = 3500 and III = 200) was subtracted. Peptides were identified by further pronase digestion: peptide I yielded only fMet and fMet-Ala while peptide II yielded fMet, fMet-Ser and small amounts of deformylated products.

access to each of the three cistrons (Lodish & Robertson 1969), was used to study ribosome binding with IF2 in the presence or absence of IF3. The heavily labelled [^{32}P]MS2 RNA fragments, protected by the ribosome against RNase degradation, were analysed according to the method of Steitz (1969) in R17 RNA (see Fig. 2). Without IF3, ribosome binding occurred to the coat, synthetase and maturation cistrons in a ratio of 1:2:1. Some unidentified binding sites were also observed. Addition to the binding reaction of a highly purified IF3 preparation (F3.B2 in Table 1) selectively stimulated ribosome attachment to the initiation site for coat (Fig. 3*a*), resulting in a marked shift in ribosome distribution (coat: synthetase: maturation cistron 1:0.8:0.4). The same IF3.B2 preparation selectively stimulated translation of coat protein from the formaldehyde-treated MS2 RNA (Table 2), while in a crude extract of *E. coli* this template mainly produces synthetase (Lodish 1970*a*). These results strongly

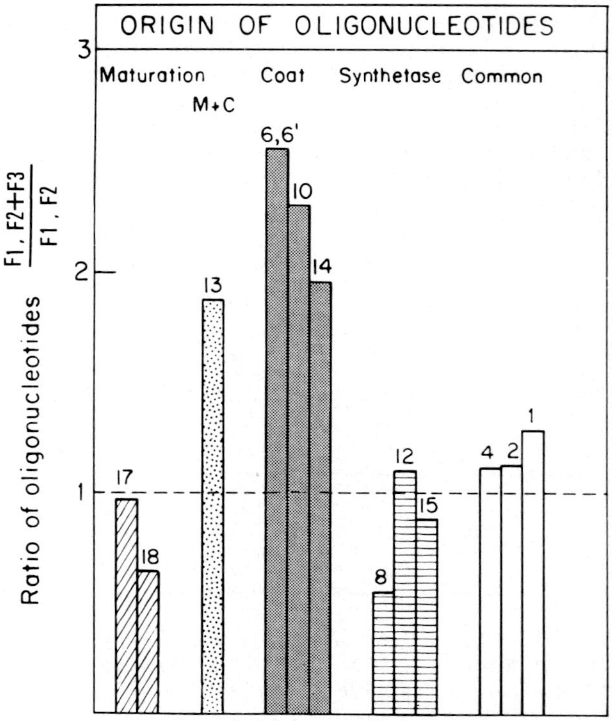

FIG. 3. Diagram representing the ratio of the relative amount of specific RNase T1 oligonucleotides from the three initiation sites of MS2 RNA (as numbered in Fig. 2) obtained when ribosomes were bound in the presence of initiation factor IF3.B2 (Table 1), in addition to IF1 and IF2, to the amount observed in the absence of IF3. Experimental details in Berissi *et al.* (1971). IF3.B2 stimulates selectively ribosome binding to the initiation site for coat protein cistron.

support the model in which IF3 promotes the recognition of a particular nucleotide sequence in the initiation site for coat protein. A direct interaction of IF3 with oligonucleotide 66 (see Fig. 2) of the initiation site was recently reported (Revel *et al.* 1972). Since IF3 readily binds to the 30S (Sabol & Ochoa 1971) and IF2 is required for mRNA-ribosome binding (Greenshpan & Revel 1969), it is however likely that mRNA interacts with more than one site of the ribosome-factor complex. Furthermore, analysis of the MS2 RNA protected fragments shows that most 30S subunits are not bound at the regular initiation sites (M. Revel & H. Berissi, unpublished results). This may suggest that the ribosome first binds the RNA and then slides until it reaches the correct site where IF3 recognizes the initiation signal, which allows the 30S ribosome to bind fMet-tRNA and the 50S subunit.

How are the other cistrons of MS2 RNA translated? Table 2 shows that other IF3 fractions (such as IF3.B1) can be isolated from ribosomes which are much more active for the translation of the synthetase cistron. A functional heterogeneity of IF3 activity is seen (Revel *et al.* 1970*b*; Lee-Huang & Ochoa 1971) which also extends to other mRNA's. Thus, fraction IF3.B4 in Table 1 has little activity for MS2 RNA but actively translates T4 mRNA. The recognition by ribosomes of different mRNA is therefore determined by the IF3 fraction added.

Physiological changes in the activity of the mRNA discriminating factor can be observed. After infection with T4 ribosomes are much less active in the translation of MS2 RNA (Hsu & Weiss 1969), *E. coli* mRNA (Dube & Rudland 1970) and early T4 mRNA (Pollack *et al.* 1970) but very active in the translation of late T4 mRNA. The change can be shown to result from a change in IF3 activity (Pollack *et al.* 1970; Lee-Huang & Ochoa 1971).

MOLECULAR BASIS FOR THE HETEROGENEITY OF INITIATION FACTOR IF3

In all experiments IF3 activity is defined by complementation with IF2 and IF1 for the initiation of translation of natural mRNA, but the molecular basis for the different IF3 activities is not yet fully understood. A purified IF3 appears as a basic protein, which in sodium dodecyl sulphate-polyacrylamide gel behaves as a polypeptide of mol. wt. 22 000. This protein has ribosome dissociation activity (Sabol *et al.* 1971). It is possible that several species of this protein exist in *E. coli*. An alternative possibility is that IF3 activity actually arises from more than one polypeptide chain. In line with this idea, we have recently (Groner *et al.* 1972*a*) isolated new protein factors, designated interference factors, which combine with the 22 000 mol. wt. protein and change its activity

TABLE 3

Factor *i* activity

Conditions	*T4 mRNA*	*MS2 RNA*	*Formaldehyde treated MS2 RNA*
	[^{35}S]formylmethionine incorporation, c.p.m.		
Control	36 700	11 300	13 800
+ factor *i*, 5 μg	38 500	4 200	15 200

80 μg T4 mRNA, 16 μg MS2 RNA or formaldehyde-treated (Lodish 1970*a*) MS2 RNA were incubated for 30 min at 37 °C in 0.05 ml reaction mixture (Revel *et al.* 1971) with 3 μg IF1, 15 μg IF2 and 0.5 μg IF3, 100 μg ribosomes, 5 μl high speed supernatant and 5 μg of [^{35}S]-formylmethionyl-tRNA (12 500 c.p.m./μg) to measure amino terminal labelling of the protein product. A background of 2000 c.p.m. incorporated in the absence of template was subtracted.

toward various mRNA's. One such protein, factor *i* (Table 3), has been purified to homogeneity and shown to have a molecular weight of 74 000. It binds readily to IF3 and native IF3-*i* complexes are found in *E. coli* ribosome-associated proteins. This protein inhibits ribosome binding to the coat protein cistron of MS2 RNA but stimulates ribosome binding to the synthetase initiation site (Fig. 4). Factor *i* also affects the translation of T4 mRNA, stimulating initiation of certain cistrons but inhibiting that of others. The role of factor *i* in *E. coli* is not yet clear. Recent results show that other similar interfering factors can be isolated which differ from factor *i* in both physical and biological properties. One such factor inhibits translation of all MS2 RNA cistrons but initiates actively on T4 mRNA. It is tempting to assume that these interference factors act by binding with IF3 to the ribosome and increasing its affinity for some mRNA's while decreasing it for others. The heterogeneity of IF3 activity may serve in *E. coli* to control the translation of different classes of mRNA. Evidence for translation control in the repression of certain operons of *E. coli* has recently been reported (Lavallé & de Hauwer 1970).

Factor *i* appears to have an important function in the regulation of RNA phage development. Not only does it stimulate the formation of the MS2 phage synthetase protein but it is itself the largest of the three host proteins (Kamen 1971) which combine with the phage synthetase to form the complete replicating enzyme (Groner *et al.* 1972*b*). This translation control factor therefore also has a function in the replication of the phage RNA.

Variation in the messenger specific factors may result from several mechanisms. In *E. coli* maintained in the stationary phase we observed a large decrease in the amount of inactive factor IF3 (Scheps & Revel 1972). Since in these conditions the amount of factor *i* does not vary, the ratio of IF3-*i* to free IF3 increases. Consequently the ribosomes translate much less MS2 RNA than T4

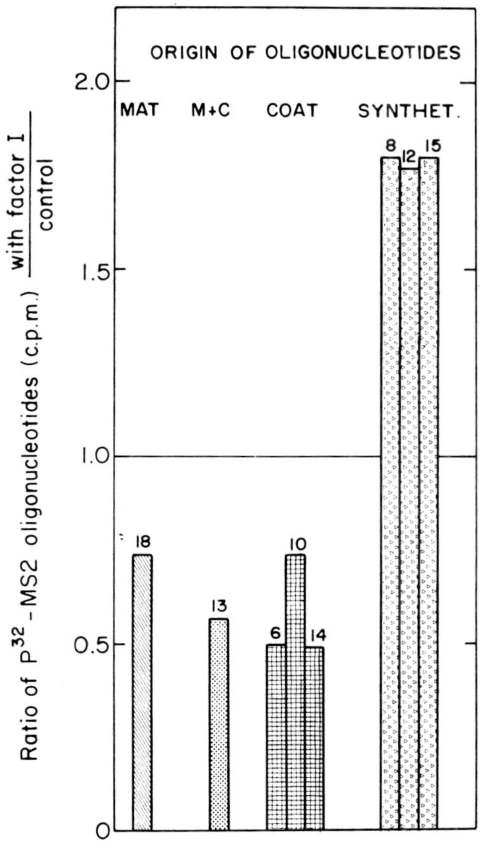

FIG. 4. Diagram representing the ratio of the relative amount of RNase T1 oligonucleotides from the three initiation sites of MS2 RNA (Fig. 2) obtained when ribosomes were bound with factor *i* in addition to IF1, IF2 and IF3, to the amount observed in the absence of factor *i*. Factor *i* stimulates ribosome binding to the synthetase cistron.

mRNA. In T4-infected *E. coli*, on the other hand, there are no apparent changes in the amount of IF3 but we observe a chemical modification of factor *i* (H. Inouye, R. Scheps, Y. Pollack & M. Revel, unpublished). The protein purified from T4-infected cells migrates more rapidly to the anode at pH 8.5 on immunoelectrophoresis against anti-*i*. The modification may be related to the change in template specificity in the infected factors (Pollack *et al.* 1970).

MECHANISMS OF TRANSLATION CONTROL

Initiation factors appear, therefore, to play an important role in the mRNA selection process, but control of translation can be exerted by several mechanisms. The secondary and tertiary structure of the mRNA plays an important role, as shown by Lodish (1970*a*) with phage f2 RNA. In native f2 RNA, only the coat protein cistron is fully accessible to ribosomes. Unfolding the f2 RNA by formaldehyde or fragmentation of the RNA renders all these cistron initiation sites available. As mentioned earlier, the structure of the 30S ribosome itself may be the determining factor: Lodish (1970*b*) showed that ribosomes from *B. stearothermophilus* recognize only the maturation protein cistron of the three f2 cistrons. Even when *E. coli* IF3 is used, *B. stearothermophilus* 30S cannot bind to the coat protein cistron (R. Scheps & M. Revel, unpublished results). In a single bacterial species there is, however, no evidence yet for a functional heterogeneity of the ribosomal particles themselves, apart from the heterogeneity imposed by the initiation factors and other regulatory factors (Kurland *et al.* 1969) which continuously bind and leave the ribosome during protein synthesis. Finally, translation could be controlled by the availability of particular minor tRNA species and it is possible that degeneracy of the code is necessary for such regulatory processes (Anderson & Gilbert 1969).

Control of gene expression is therefore exerted not only at the level of transcription by the amount of template RNA produced, but also at the level of translation. Even if the chemistry of polypeptide synthesis is relatively simple, it appears that the regulation of this synthesis, working through mechanisms coded by the genome to meet the needs of the cell, requires the very complicated ribosomal machinery which we know today.

ACKNOWLEDGEMENTS

The excellent assistance of Mrs P. Federman is gratefully acknowledged. This work was supported in part by a grant from the Service d'Aide à la Recherche et au Développement Français.

References

Anderson, W. F. & Gilbert, J. M. (1969) Translation control of in vitro hemoglobin synthesis. *Cold Spring Harbor Symp. Quant. Biol.* **34**, 585–588

Berissi, H., Groner, Y. & Revel, M. (1971) Effect of a purified initiation factor F3(B) on the selection of ribosomal binding sites on phage MS2 RNA. *Nat. New Biol.* **234**, 44–47

Capecchi, M. & Klein, H. (1969) Characterization of three proteins involved in polypeptide chain termination. *Cold Spring Harbor Symp. Quant. Biol.* **34**, 469–477

CASKEY, T., SCOLNICK, C., TOMPKINS, R., GOLSTEIN, J & MILMAN, G. (1969) Peptide chain termination, codon, protein factor and ribosomal requirements. *Cold Spring Harbor Symp. Quant. Biol.* **34**, 479–491

DAVIES, J., GILBERT, W. & GORINI, L. (1964) Streptomycin, suppression and the code. *Proc. Natl Acad. Sci. U.S.A.* **51**, 883–890

DAVIS, B. (1971) Role of subunits in the ribosome cycle. *Nat. New Biol.* **231**, 153–157

DUBE, S. & RUDLAND, P. S. (1970) Control of translation by T4 phage: altered binding of disfavoured messenger. *Nature (Lond.)* **226**, 820–823

DUBE, S., RUDLAND, P., CLARK, B. F. C. & MARCKER, K. (1969) A structure requirement for codon-anticodon interaction on the ribosome. *Cold Spring Harbor Symp. Quant. Biol.* **34**, 161–166

GILBERT, W. (1963) Protein synthesis in *E. coli. Cold Spring Harbor Symp. Quant. Biol.* **28**, 287–297

GREENSHPAN, H. & REVEL, M. (1969) Initiator protein dependent binding of mRNA to ribosomes. *Nature (Lond.)* **224**, 331–335

GRONER, Y. & REVEL, M. (1971) A novel form of initiation factors from *E. coli* which binds formylmethionyl tRNA and GTP: 'F2–F3 complex'. *Eur. J. Biochem.* **22**, 144–152

GRONER, Y., POLLACK, T., BERISSI, H. & REVEL, M. (1972*a*) Characterization of cistron specific factors for the initiation of messenger RNA translation in *E. coli. FEBS (Fed. Eur. Biochem. Soc.) Lett.* **21**, 223–228

GRONER, Y., SCHEPS, R., KAMEN, R., KOLAKOFSKY, D. & REVEL, M. (1972*b*) Host subunit of Qβ replicase is translation control factor *i. Nat. New Biol.* **239**, 19–20

HSU, W. T. & WEISS, S. (1969) Selective translation of T4 template RNA by ribosomes from T4 infected *E. coli. Proc. Natl Acad. Sci. U.S.A.* **64**, 345–351

JEPPESEN, P., STEITZ, J., GESTELAND, R. & SPAHR, P. (1970) Gene order in the bacteriophage R17 RNA: 5′-A protein-coat protein-synthetase-3′. *Nature (Lond.)* **226**, 230–237

KALTSCHMIDT, E. & WITTMANN, H. G. (1970) Ribosomal proteins, XII. Number of proteins in small and large ribosomal subunits of *E. coli* as determined by two-dimensional gel electrophoresis. *Proc. Natl Acad. Sci. U.S.A.* **67**, 1276–1282

KAMEN, R. (1971) Characterization of the subunits of Qβ replicase. *Nature (Lond.)* **228**, 527–533

KLEINKAUF, H. & GEVERS, W. (1969) Non-ribosomal polypeptide synthesis: The biosynthesis of a cyclic peptide antibiotic, gramicidin S. *Cold Spring Harbor Symp. Quant. Biol.* **34**, 805–813

KURLAND, C., VOYNOW, P., HARDY, S., RANDALL, L. & LUTTER, L. (1969) Physical and functional heterogeneity of *E. coli* ribosomes. *Cold Spring Harbor Symp. Quant. Biol.* **34**, 17–24

LACROUTE, F. & STENT, G. (1968) Peptide chain growth of β-galactosidase in *E. coli. J. Mol. Biol.* **35**, 165–170

LAVALLÉ, R. & DE HAUWER, G. (1970) Repression by tryptophan at the level of transcription and translation in *E. coli. Cold Spring Harbor Symp. Quant. Biol.* **35**, 491–494

LEE-HUANG, S. & OCHOA, S. (1971) Messenger discriminating species of initiation factor F3. *Nat. New Biol.* **234**, 236–239

LELONG, J. C., COUSIN, M., GROS, F., MISKIN, R., VOGEL, Z., GRONER, Y. & REVEL, M. (1972) Protection of *E. coli* ribosomes against streptomycin by purified initiation factors. *Eur. J. Biochem.* **27**, 174–180

LENGYEL, P. & SOLL, D. (1969) Mechanism of protein biosynthesis. *Bact. Rev.* **33**, 264–301

LODISH, H. F. (1970*a*) Secondary structure of bacteriophage f2 RNA and the initiation of in vitro protein synthesis. *J. Mol. Biol.* **50**, 689–702

LODISH, H. F. (1970*b*) Specificity in bacterial protein synthesis: role of initiation factors and ribosome subunits. *Nature (Lond.)* **226**, 705–708

LODISH, H. F. & ROBERTSON, H. D. (1969) Regulation of in vitro translocation of bacteriophage f2 RNA. *Cold Spring Harbor Symp. Quant. Biol.* **34**, 655–673

NOMURA, M., MIZUSHIMA, S., OZAKI, M., TRAUB, P. & LOWRY, C. V. (1969) Structure and function of ribosomes and their macromolecular components. *Cold Spring Harbor Symp. Quant. Biol.* **34**, 49–51

OZAKI, M., MIZUSHIMA, S. & NOMURA, M. (1969) Identification and functional characterization of the protein controlled by the streptomycin-resistant locus in *E. coli. Nature (Lond.)* **222**, 333–339

PARENTI-ROSINA, R., EISENSTADT, A. & EISENSTADT, J. (1969) Isolation of protein initiation factors from 30 S ribosomal subunits. *Nature (Lond.)* **221**, 363–365

POLLACK, Y., GRONER, Y., AVIV, H. & REVEL, M. (1970) Role of initiation factor B(F3) in the preferential translation of T4 late mRNA in T4 infected *E. coli. FEBS (Fed. Eur. Biochem. Soc.) Lett.* **9**, 218–221

REVEL, M., HERZBERG, M. & GREENSHPAN, H. (1969) Initiator protein dependent binding of messenger RNA to the ribosome. *Cold Spring Harbor Symp. Quant. Biol.* **34**, 261–275

REVEL, M., GREENSHPAN, H. & HERZBERG, M. (1970*a*) Specificity in binding of *E. coli* ribosomes to natural mRNA. *Eur. J. Biochem.* **16**, 117–122

REVEL, M., AVIV, H., GRONER, Y. & POLLACK, Y. (1970*b*) Fractionation of translation initiation factor B(F3) into cistron specific species. *FEBS (Fed. Eur. Biochem. Soc.) Lett.* **9**, 213–217

REVEL, M., GREENSHPAN, H. & HERZBERG, M. (1971) *E. coli* initiation factors in the binding of ribosomes to messenger RNA. *Methods Enzymol.* **20**, 261–280

REVEL, M., GRONER, Y., POLLACK, Y., BERISSI, H. & HERZBERG, M. (1972) Initiation factors of protein synthesis and the control messenger RNA translation. In *Functional Units in Protein Biosynthesis* (FEBS Symposium), Academic Press, New York

ROSSET, R. & GORINI, L. (1969) A ribosomal ambiguity mutation. *J. Mol. Biol.* **39**, 95–112

RUDLAND, P., WHYBROW, W. & CLARK, B. F. C. (1971) Recognition of bacterial initiator tRNA by an initiation factor. *Nat. New Biol.* **231**, 76–78

SABOL, S. & OCHOA, S. (1971) Ribosomal binding of labelled initiation factor F3. *Nat. New Biol.* **234**, 233–235

SABOL, S., SILLERO, M., IWASAKI, K. & OCHOA, S. (1971) Purification and properties of initiation factor F3. *Nat. New Biol.* **228**, 1269–1275

SCHEPS, R. & REVEL, M. (1972) Deficiency in initiation factors of protein synthesis in stationary phase *E. coli. Eur. J. Biochem.* **29**, 319-342

STEITZ, J. A. (1969) Polypeptide chain initiation: nucleotide sequences of the three ribosomal binding sites on bacteriophage R17 RNA. *Nature (Lond.)* **224**, 957–964

STRIGINI, P. & GORINI, L. (1970) Ribosomal mutations affecting efficiency of amber suppression. *J. Mol. Biol.* **47**, 517–530

UHLENBECK, O., BALLER, J. & DOTTY, P. (1970) Complementary oligonucleotide binding to the anticodon loop of fMet-tRNA. *Nature (Lond.)* **225**, 508–511

WAHBA, A., IWASAKI, K., MILLER, M., SABOL, S., SILLERO, M. & VASQUEZ, C. (1969) Initiation of protein synthesis in *E. coli. Cold Spring Harbor Symp. Quant. Biol.* **34**, 291–299

Discussion

Tissières: In order to understand fully the mechanism of protein synthesis, we need detailed information on the structure of ribosomes. Indeed, Dr Revel has told us of a number of specific interactions taking place on the ribosome during protein synthesis, and we must assume that there are a number of specific sites, for binding transfer and messenger RNA's, initiation, elongation and termination factors. In addition, there must be sites on the 50S and 30S ribosomes able to recognize each other when the two subparticles get together. These sites must involve some specific ribosomal proteins, and perhaps in some cases and to some extent ribosomal RNA as well. Also a restricted number of sites may have enzymic activity, as for instance the peptidyl transferase on the 50S ribosome. Thus, the role or function of several, or perhaps most, ribosomal proteins, can well be to build up specific sites needed during the cycle of protein synthesis and to confer the proper shape on the machinery. The solving of the structure of ribosomes at a sufficient resolution with physical means will probably not be done for quite some time. However, various approaches will probably yield essential information in the near future. These are mainly the studies of interaction of specific ribosomal proteins with ribosomal RNA's and RNA fragments on the one hand and the use of bifunctional reagents on the other, which should give precious information on the topography of ribosomes, and also tell us where on the ribosomes the factors bind.

Katchalski: If most ribosomal proteins play a role in protein biosynthesis, and if in addition they have to interact with each other or with the ribosomal RNA, each protein must be enormously complicated because it has to contain different recognition and functional sites.

Changeux: But what is the state of aggregation of this system? When the system is functioning, are some of its components going on and out?

Revel: Each ribosomal protein needs only limited information, as most of them work in complexes with other proteins. Certainly the initiation factors *by themselves* do very little, unless they bind to the appropriate site on the ribosomal surface. We know that they bind to very specific sites on the ribosome, because they compete with antibiotics like streptomycin (cf. Lelong *et al.* 1972). Their active site may therefore involve other proteins of the ribosome. For this reason, not every protein needs to have all the information as they can share it with other proteins of the ribosome. This may be true also for specific functions of the ribosomes themselves: some functions might be carried out by more than one protein. The work of G. Stoffler and H. G. Wittmann (1971, unpublished findings) shows that binding of antibiotics is not a function of one protein on the ribosome. At least two or three different proteins participate in the binding of one antibiotic. So active sites must extend on different proteins.

Luzzati: I didn't understand your point, Dr Katchalski.

Katchalski: It was a simple one. If one separates all the various proteins in the various subunits of the ribosome, as Dr Nomura and Dr Tissières did, one gets about 55 proteins. Somehow these have to combine in the presence of ribosomal RNA to make the ribosome. So they must recognize ribosomal RNA, they must recognize one another in order to have a definite structure, and they have to have recognition sites. If in addition some of them have catalytic activity or any other special activity involved in protein biosynthesis, they must have more functions than the usual active protein.

Luzzati: Isn't that question equivalent to a more general problem which has been discussed many times—why are proteins so big?

Gilbert: There is also the possibility that there may not be 55 different enzymic reactions. Clusters of proteins may finally form single active sites. In the final dissection of the functions of ribosomes, one may see ten proteins all going together to construct a proper tRNA binding site, and we will be able to associate groups of proteins with different enzymic reactions.

On the other side of that question, I was going to add another facet. Single proteins may possibly have the same function in several totally different biosynthetic systems. Dr Revel told us that one of the components of phage Qβ synthetase was factor *i* which is involved in the initiation of protein synthesis. In Qβ synthetase, in order of size, there is the *i*-factor, which is the largest, the phage-directed component next, and two lighter components which are also needed for activity. Klaus Weber and his co-workers have shown that the third component is the Tu factor from protein synthesis, and the smallest one is the Ts factor from protein synthesis. However these both seem to be necessary in the Qβ synthetase for the enzymic function of making a copy of phage RNA or making a poly(G) copy on a poly(C) template (Blumenthal *et al.* 1972).

Silman: Lipmann's system for the biosynthesis of gramicidin S has maybe 12 sites. How big is the macromolecular entity and can it bind several amino acids simultaneously?

Eigen: The enzyme for gramicidin recognizes only five different amino acids. It keeps the pentapeptide bound to the enzyme until it makes another one and then cyclizes both to the decapeptide.

Luzzati: Dr Revel, you mentioned 55 proteins and ten factors. How are these elements identified and characterized, by biochemical and genetic techniques?

Revel: There is much genetic information on the loci of some proteins of the ribosome. In *E. coli*, a cluster of genes for at least 18 ribosomal proteins is known. For example, the genes and corresponding proteins for streptomycin resistance (protein P10 or S12) and for spectinomycin resistance (P4 or S5) were identified (for review cf. Nomura 1970): in both cases it has been shown that the amino acid sequence of the protein is modified in the antibiotic-resistant mutant (Wittmann 1972). There are other examples, such as the ribosomal ambiguity mutation (*ram*; protein P4a or S4) (Zimmermann *et al.* 1971). Certain proteins differ from one *E. coli* strain to another. Among the factors, only mutants in elongation factor G have been identified (Tocchini-Valentini *et al.* 1969).

Changeux: This system has not evolved very much from bacteria to mammals. How much is known about the sequence of the proteins involved? Have they the same sequence in different groups: in bacterial cells and in mammalian cells?

Revel: The ribosomal proteins of 70S bacterial ribosomes, of 70S ribosomes from chloroplasts, or of 80S ribosomes from plants or from animals, are very different: there is no immunological cross-reaction between these proteins (Wittmann 1972). The factors are also different, with different molecular weights, although the functions which are carried out for protein synthesis seem to be quite similar. There are some similarities between ribosomal proteins of related bacterial species and between 80S ribosomes of several higher plants (Wittmann 1972). Ribosome-associated factors from different bacterial species also show limited immunological cross-reaction (R. Scheps & M. Revel, unpublished findings). In conclusion, the functions seem to be conserved by evolution but the primary structure of the ribosomal proteins has not been.

References

BLUMENTHAL, T., LANDERS, T. & WEBER, K. (1972) *Proc. Natl Acad. Sci. U.S.A.* **69**, 1313-1317

LELONG, J. C., COUSIN, M., GROS, F., MISKIN, R., VOGEL, Z., GRONER, Y. & REVEL, M. (1972) *Eur. J. Biochem.* **27**, 174–180

NOMURA, M. (1970) *Bacteriol. Rev.* **34**, 228–277

TOCCHINI-VALENTINI, G., FELICETTI, L. & RINALDI, G. (1969) *Cold Spring Harbor Symp. Quant. Biol.* **34**, 463–468

WITTMANN, H. G. (1972) in *Functional Units in Protein Biosynthesis* (FEBS Symposium), Academic Press, New York

ZIMMERMANN, R. A., GARVIN, R. T. & GORINI, L. (1971) *Proc. Natl Acad. Sci. U.S.A.* **68**, 2263–2266

Bacterial cell wall biosynthesis

JAMES BADDILEY

Microbiological Chemistry Research Laboratory, School of Chemistry, University of Newcastle upon Tyne

Abstract The walls of Gram-positive bacteria consist of peptidoglycan and either teichoic acids or acidic polysaccharides; occasionally the three polymers occur together. The biosynthesis of these components from nucleotide precursors occurs in the cytoplasmic membrane. Residues are transferred from appropriate nucleotides to isoprenoid phosphate intermediates and then to the growing polymers. It is proposed that in all cases the lipid is undecaprenol, and it has been shown that the lipid phosphate molecules are shared through a common pool during simultaneous synthesis of several different wall polymers. It is suggested that the rate of synthesis of walls of constant composition is controlled by regulation of the amount of undecaprenol phosphate, whereas the relative amounts of different wall polymers could be regulated partly by control of the availability of appropriate nucleotide precursors. Although satisfactory evidence for the regular arrangement of polymers in layers of discreet regions of walls is lacking, it has been shown that there is a considerable degree of order in the mutual attachment of polymers to each other. Thus one teichoic acid chain is attached to only one glycan chain in those walls examined so far, indicating the existence of an accurately controlled process of synthesis.

In the Gram-positive bacteria the cytoplasm and chromosomal material is surrounded by a membrane composed of glycerides, glycolipids, phospholipids and membrane teichoic acid together with protein. Surrounding the membrane is a thicker more rigid structure, the wall, and in some cases this may itself be surrounded by a capsule. In the Gram-negative bacteria the situation is more complex, the main morphological difference being the presence of two membranes, one outside the wall and the other on the inside corresponding to the cytoplasmic membrane in Gram-positive bacteria. As less is known about structure and biosynthesis of wall and membrane components in Gram-negative organisms, attention will be confined here mainly to the Gram-positive bacteria.

At least 50% of the weight of the wall is composed of a complex polymer usually called mucopeptide or peptidoglycan; the rest of the wall is either teichoic acid or acidic polysaccharide or both of these. Peptidoglycans are two-dimensional or possibly three-dimensional polymers. In one direction extend the glycan chains which consist of a regular alternating sequence of *N*-acetyl-glucosamine and its 3-*O*-lactyl ether, *N*-acetylmuramic acid. The glycosidic linkages between these have the β-configuration and involve in both cases the hydroxyl groups at the 4-position. The length of the glycan chains varies from one organism to another and also may depend upon growth conditions; values in the general order of ten repeating disaccharide units have been reported. *O*-Acetyl groups may occur on some of the muramic acid residues and lactone structures have been found occasionally, but the sugars have a remarkable uniformity of composition. These glycans are cross-linked, to a varying extent in different organisms, by peptide chains. The structure of the peptidoglycan of *Staphylococcus aureus* is given in Fig. 1; the peptide chains extend from the carboxyl groups of muramic acid residues, and cross-linkage occurs through five glycine residues attached to the ε-amino groups of lysine in neighbouring peptide chains. The peptide chains in different peptidoglycans vary considerably in their composition and in the amount of cross-linkage.

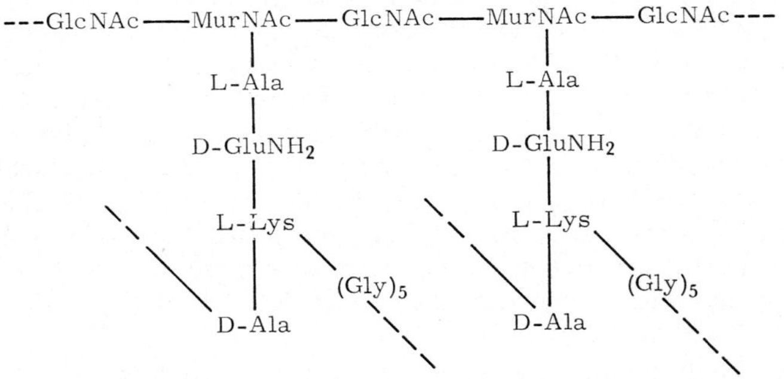

FIG. 1. Structure of peptidoglycan from *Staphylococcus aureus*. GlcNAc = *N*-acetylglu-cosamine; MurNAc = *N*-acetylmuramic acid.

The acidic polysaccharides in walls have not been studied extensively, but those examined are linear polymers containing simple sugars and uronic acids; they are sometimes called teichuronic acids. The teichoic acids on the other hand have been studied in considerable detail, both in their structure and in their biosynthesis. They are polymers containing glycerol phosphate or ribitol phosphate residues and are found in membranes, walls and capsules of bacteria. The simplest members are linear polymers in which adjacent glycerol or ribitol residues are joined through phosphodiester linkages at positions 1 and 3 or 1

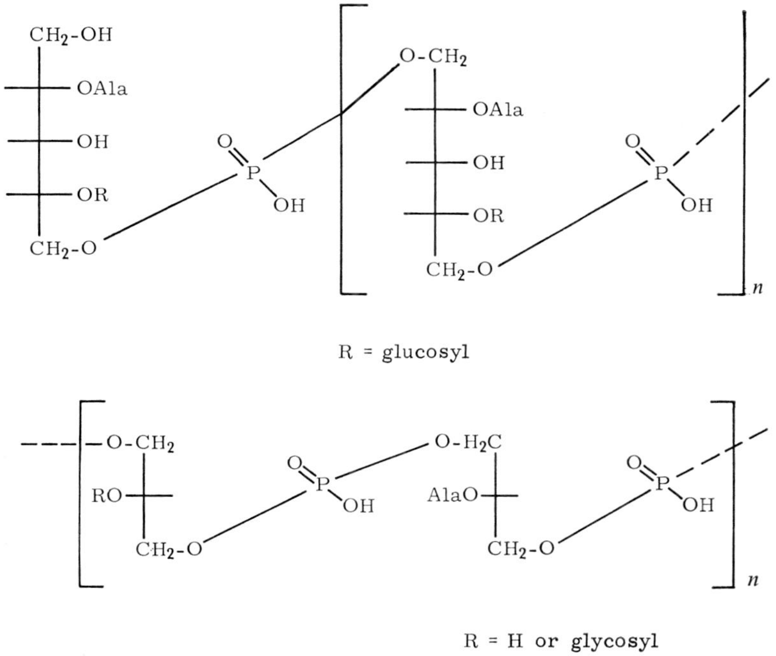

FIG. 2. Glycerol and ribitol teichoic acids.

and 5 respectively. Many teichoic acids possess glycosyl and D-alanyl residues on hydroxyl groups of the polyol residues. Examples of glycerol and ribitol teichoic acids are given in Fig. 2. Variations on these structures are numerous, and include examples (Fig. 3) in which glycosyl residues and sugar 1-phosphate residues occur in the main chain of the polymer. Although not by definition teichoic acids, sugar 1-phosphate polymers are found in cell walls of bacteria, and the biosynthesis of these has been studied in some detail.

In Gram-negative bacteria peptidoglycan occurs in the wall, and although teichoic acids are absent the outer membrane contains lipopolysaccharide. This consists of complex polysaccharide chains, probably cross-linked through phosphodiester groups, attached to a glycolipid called lipid A (Fig. 4).

It is believed that peptidoglycan is responsible for the strength and shape of the wall, whereas teichoic acids and acidic polysaccharides are concerned with cation balance and the maintenance of high and controlled concentrations of bivalent cations in the region of the membrane; it is possible that lipopolysaccharides have a similar role to that of the teichoic acids. Polysaccharides and teichoic acids are covalently attached to peptidoglycan in the wall through a terminal phosphodiester linkage involving the hydroxyl at position 6 on a

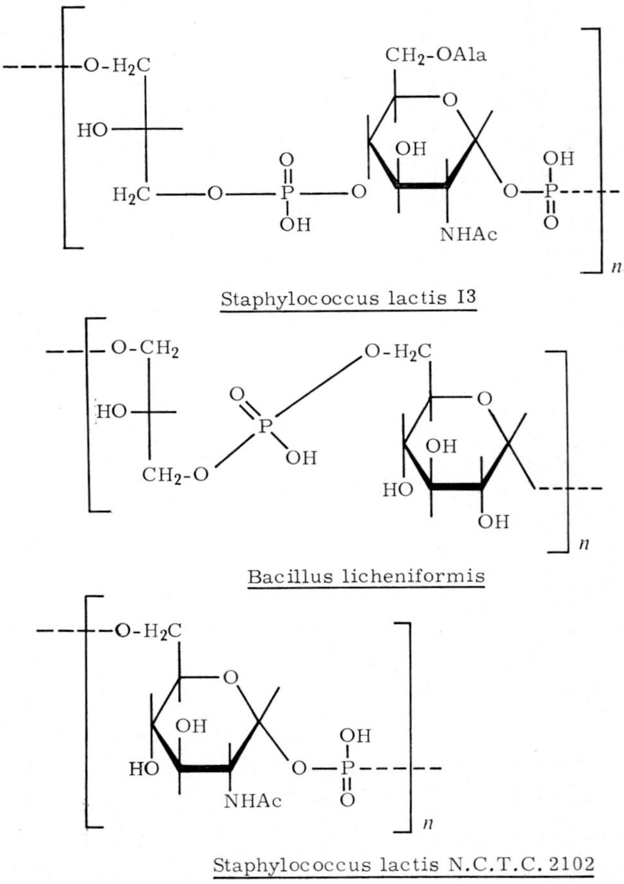

FIG. 3. Polymers containing sugar and sugar 1-phosphate residues in the polymer chain.

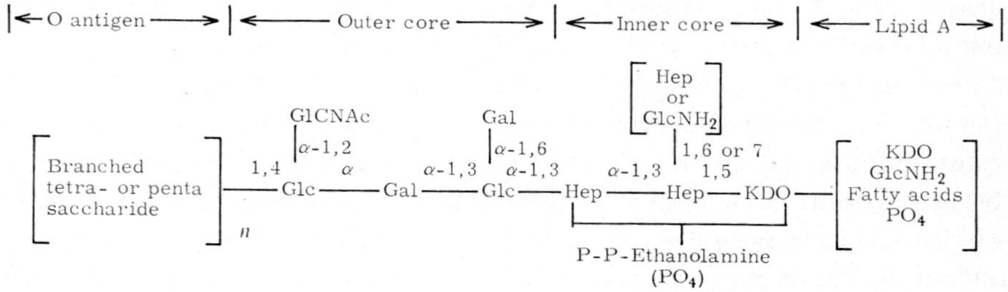

FIG. 4. General structure of a lipopolysaccharide.

muramic acid residue of the peptidoglycan (Ghuysen *et al.* 1965; Knox & Hall 1965; Button *et al.* 1966).

This discussion will be confined to the biosynthesis of those wall polymers that have been examined in greatest detail, namely peptidoglycans, teichoic acids and sugar 1-phosphate polymers in Gram-positive bacteria.

For an understanding of bacterial cell wall biosynthesis we need to know, for each constituent polymer, the nature of the precursors, the mechanism of attachment of monomer units to polymer chains, and the direction of growth of the chains. Moreover, it is necessary to know the chemical nature of the linkages between polymers and the mechanism of formation of these linkages, the site of synthesis and the process of translocation from inside to outside the cell, and finally the mechanisms that control the rate of synthesis and the composition of the wall, i.e. the relative amounts of each constituent polymer. Although not all these questions have been answered, substantial progress has been made towards an understanding of some of them.

The biosynthesis of peptidoglycan differs substantially from protein synthesis, and in fact it can continue in whole cells in conditions where, either because of a nutritional deficiency or through the inhibitory action of chloramphenicol (Mandelstam & Rogers 1958; Hancock & Park 1958), protein synthesis does not occur. When suitable staphylococci are incubated with penicillin, the antibiotic interferes with the synthesis of peptidoglycan and several nucleotide precursors accumulate (Park 1952). One of these nucleotides (Fig. 5) possesses an *N*-acetylmuramic acid residue to which is attached a pentapeptide containing an amino acid sequence clearly related to that of the peptidoglycan itself; the other nucleotides are similar, but possess incomplete peptide sequences. Polymeric material resembling peptidoglycan is synthesized in cell-free systems

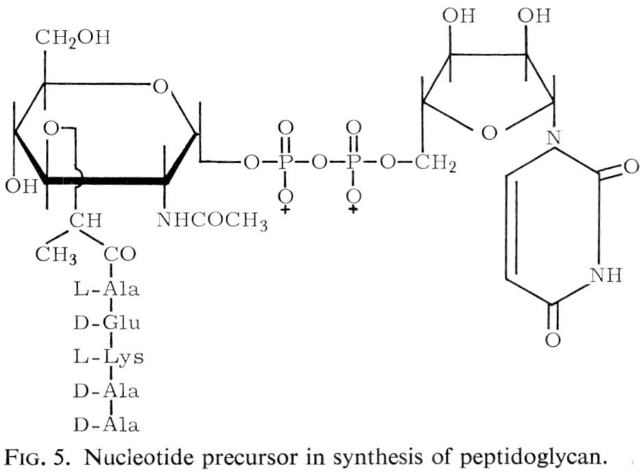

FIG. 5. Nucleotide precursor in synthesis of peptidoglycan.

containing the muramyl nucleotide, UDP-*N*-acetylglucosamine, ATP, Mg^{2+} and glycine (Chatterjee & Park 1964; Meadow *et al.* 1964); the insoluble enzyme preparation required for this synthesis is, in fact, fragmented cytoplasmic membrane, and for the incorporation of the glycine residues transfer-RNA is required (Matsuhashi *et al.* 1965).

The cell-free synthesis of peptidoglycan or closely related material is consistent with the overall reaction outlined in Fig. 6. However, the nucleotides

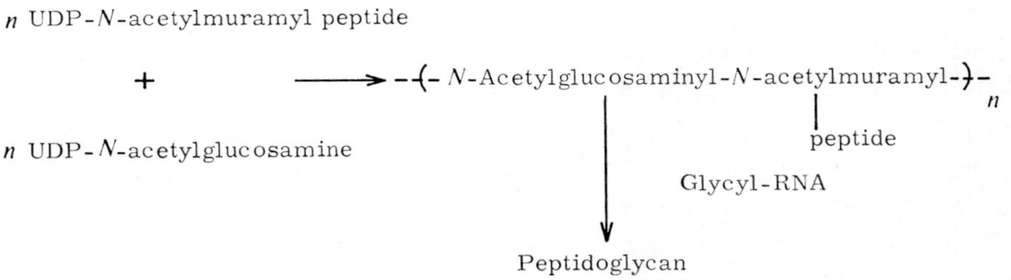

FIG. 6. Overall synthesis of peptidoglycan.

produced in the reaction are not those expected from a simple polymerization, and by using different procedures for making the enzyme system (Struve & Neuhaus 1965) it can be concluded that an acceptor of the *N*-acetylmuramyl peptide residue must be included before attachment of the new unit to the growing polymer chain occurs. In fact, it is now known that synthesis proceeds via lipid intermediates (Anderson *et al.* 1965; Matsuhashi *et al.* 1965) according to the scheme given in Fig. 7. The monophosphate (Fig. 8) of the C_{55} isoprenoid alcohol, undecaprenol, acts as an acceptor of the *N*-acetylmuramyl peptide residue from the nucleotide, giving a lipid pyrophosphate containing the

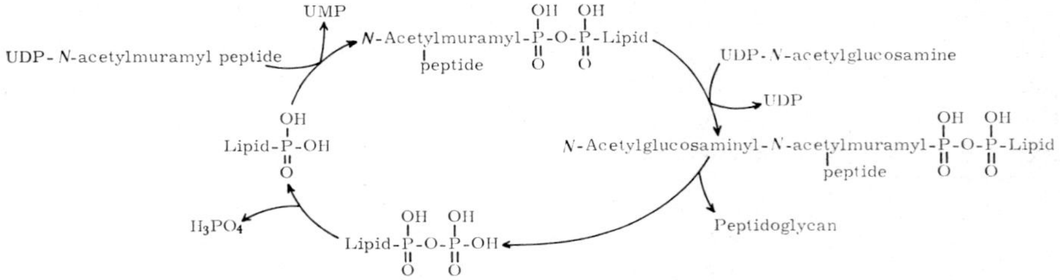

FIG. 7. Biosynthetic pathway to peptidoglycan through lipid intermediates.

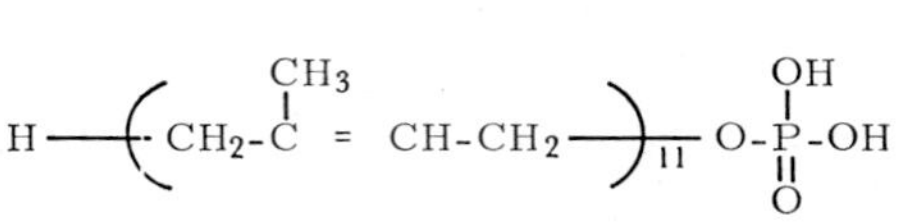

FIG. 8. Undecaprenol phosphate.

N-acetylglucosaminyl-*N*-acetylmuramyl peptide residue. It is likely that up to this point peptidoglycan synthesis is closely similar in all bacteria, but different routes may occur subsequently, according to the composition of the peptide chains in the wall. In staphylococci a pentaglycine chain is built on the ε-amino group of the lysine residue of the lipid intermediate, and tRNA is involved in this step; with other organisms different bridging peptides are attached.

In staphylococci, the structure of the peptide on the uridine diphosphate derivative is L-Ala-D-Glu-L-Lys-D-Ala-D-Ala. The L-Ala, D-Glu and L-Lys residues are attached in that order to UDP-*N*-acetylmuramic acid by soluble enzymes that require bivalent cations and ATP (Ito & Strominger 1962*a*, *b*, 1964). The peptide D-alanyl-D-alanine is synthesized separately from D-alanine and is then attached to the carboxyl end (lysine) of the UDP derivative; this reaction also requires bivalent cations and ATP. It is interesting that nucleic acids do not participate in the synthesis of the pentapeptide chain, specificity for producing the correct sequence residing entirely in the individual enzymes.

In the biosynthetic cycle in Fig. 7 the step after the completion of the pentaglycine bridge is the transfer of the disaccharide peptide from lipid pyrophosphate to peptidoglycan. In fact, the nature of the acceptor has not been established, but it is assumed to be the growing end of a glycan chain in peptidoglycan already present in the particulate enzyme preparation. The product at this stage does not possess peptide cross-linkages, and it differs from the peptidoglycan of the wall in that each peptide chain terminates at its carboxyl end in D-alanyl-D-alanine. In the wall itself only one D-alanine occurs in each chain, and this is linked to the terminal amino group of a pentaglycine grouping on a neighbouring chain. Cross-linkage is achieved through the action of a transpeptidase that catalyses the exchange of the terminal D-alanine for the amino group of the glycine on a neighbouring peptide chain; in the cell-free system free D-alanine is the other reaction product (Izaki *et al.* 1966). The transpeptidase is inhibited by penicillin, and it is probable that this is the main site of action of the antibiotic in its growth-inhibitory effect on living cells.

The biosynthesis of teichoic acids shows a marked similarity to that of the peptidoglycans. The nucleotide precursors are CDP-glycerol and CDP-ribitol (Armstrong *et al.* 1958), and preparations of disrupted cytoplasmic membrane catalyse the synthesis of poly(glycerol phosphate) or poly(ribitol phosphate) from the respective nucleotides (Burger & Glaser 1964; Glaser 1964). When the polyol phosphate polymer has glycosyl substituents on hydroxyl groups of glycerol or ribitol these can be introduced either by substitution on polymer chains or simultaneously during synthesis of the chain from CDP-glycerol or CDP-ribitol (Glaser & Burger 1964; Nathenson *et al.* 1966); thus it has not been possible to decide whether in the living cell the glycosyl substituents are introduced after polymer chain synthesis or as a sequential step occurring during the chain-lengthening process. These synthetic processes, and in fact all those known to be concerned in the synthesis of teichoic acids and peptidoglycans, require bivalent cations frequently in relatively high concentration (10–50 mM). Glucose residues are introduced into teichoic acids from UDP-glucose, and *N*-acetylglucosamine residues from UDP-*N*-acetylglucosamine.

For those teichoic acids in which the sugar residues form a part of the polymer chain, synthesis occurs with a particulate enzyme preparation only when both appropriate nucleotide precursors are present. Thus the polymer from *Bacillus licheniformis* (Fig. 3) requires for its synthesis the presence of both CDP-glycerol and UDP-glucose (Burger & Glaser 1966); the CDP-glycerol provides a glycerol phosphate residue and UDP-glucose provides the glucosyl residue (Hancock & Baddiley 1972). Similarly the teichoic acid from the wall of *Staphylococcus lactis* I3 (Fig. 3) requires for its synthesis both CDP-glycerol and UDP-*N*-acetylglucosamine; the *N*-acetylglucosamine 1-phosphate residue is supplied by UDP-*N*-acetylglucosamine and the D-glycerol 1-phosphate residue by CDP-glycerol (Baddiley *et al.* 1968). Apparently sugar 1-phosphate residues in other teichoic acids and related polymers are transferred as intact units from nucleoside diphosphate sugars, as has been shown for the poly(*N*-acetylglucosamine 1-phosphate) (Fig. 3) from *Staph. lactis* N.C.T.C. 2102 (Brooks & Baddiley 1969*a*).

There is still an inadequate understanding of the details of the mechanism of introduction of the D-alanine ester residues into teichoic acids, especially those in the walls of bacteria. Although a highly specific soluble enzyme exists in bacteria which is capable of bringing about a reaction between D-alanine and ATP to form a reactive complex capable of acylating bases, this enzyme has not yet been shown to be capable of transferring D-alanine to teichoic acids or their precursors (Baddiley & Neuhaus 1960). On the other hand, a D-alanine ligase has been demonstrated recently to bind the amino acid in an esterified form to a membrane preparation in which the acceptor is believed to be the membrane teichoic acid (Reuch & Neuhaus 1971). As little is known about the mechanism of synthesis of membrane teichoic acids it has not yet been possible to decide at what stage the alanine ester residues are introduced, although it seems likely that this occurs after synthesis of the polymer chain.

Lipid intermediates participate in the synthesis of teichoic acids in a manner closely similar to that for the synthesis of peptidoglycans. A polyisoprenol pyrophosphate glycerol is an intermediate in the synthesis of poly(glycerol phosphate) by a particulate enzyme preparation from *B. licheniformis* (Anderson *et al.* 1972), and a polyisoprenol pyrophosphate *N*-acetylglucosamine participates similarly in the synthesis of poly(*N*-acetylglucosamine 1-phosphate) in a staphylococcus (Brooks & Baddiley 1969*b*). The participation of such intermediates in the synthesis of teichoic acids where the sugar residues are a part of the main polymer chain are especially interesting, because their study makes it possible to decide what is the biosynthetic repeating unit, i.e. in what order the components are assembled. Thus in the synthesis of the wall teichoic acid in *Staph. lactis* I3 (Fig. 3) the first intermediate is the lipid pyrophosphate *N*-acetylglucosamine; a glycerol phosphate residue from CDP-glycerol is then attached to the 4-hydroxyl of the *N*-acetylglucosamine residue, giving a lipid intermediate with a fully formed repeating unit. The unit is then transferred to the growing polymer and lipid pyrophosphate is returned to complete the cycle (Fig. 9) (Douglas & Baddiley 1968; Hussey & Baddiley 1972).

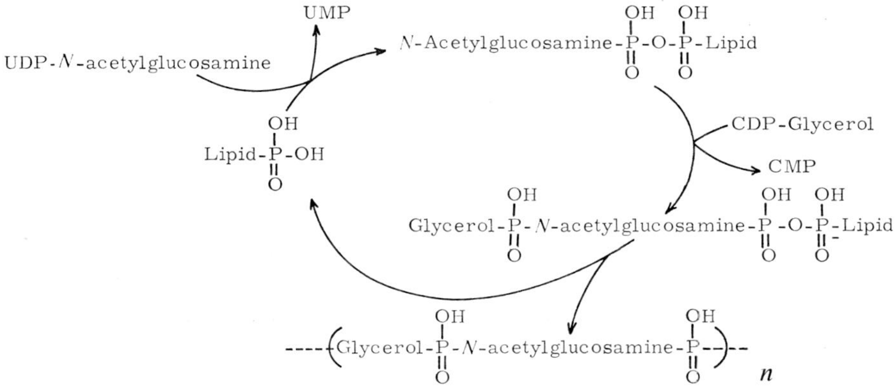

FIG. 9. Lipid intermediates in the synthesis of the wall teichoic acid from *Staphylococcus lactis* I3.

Pulse-labelling experiments show that the first-formed intermediate in the synthesis of the teichoic acid from *B. licheniformis* (Fig. 3) is the polyisoprenol monophosphate glucose (Fig. 10) which then accepts a glycerol phosphate from CDP-glycerol to give the second intermediate with a complete repeating unit. The unit is transferred to growing polymer and polyisoprenol monophosphate is returned to the cycle (Hancock & Baddiley 1972). In both UDP-glucose and the teichoic acid itself the glucose residues have the α-configuration, whereas in the lipid intermediates they have the β-configuration, indicating a double inversion during synthesis. This closely resembles the introduction of glucosyl residues from UDP-glucose into lipopolysaccharide, where a lipid monophosphate glucose containing a β-glucosyl residue participates (Nikaido & Nikaido 1971). It is also interesting that in this teichoic acid synthesis the overall transfer of repeating unit to polymer is a transglycosylation and not a transphosphoryla-

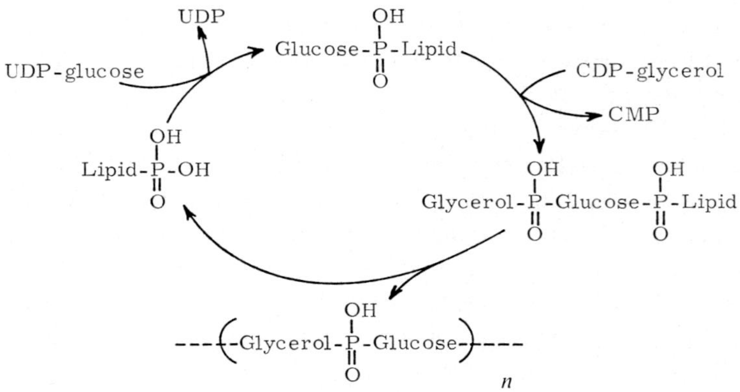

FIG. 10. Transglycosylation through lipid intermediates in the synthesis of a wall teichoic acid in *Bacillus licheniformis*.

tion as it is with other teichoic acid syntheses. A comparable example of the participation of lipid monophosphates is in the biosynthesis of a mannan in *Micrococcus lysodeikticus* where undecaprenol monophosphate mannose is an intermediate (Scher *et al.* 1968).

Although the identity of the lipid required for the synthesis of peptidoglycan has in several cases been established as undecaprenol by direct chemical and physical structural examination, this has not been done for the intermediates in the synthesis of teichoic acids. Nevertheless, its identity with undecaprenol can be deduced from indirect experiments. The reasoning is based on the conclusion that if the two synthetic routes require the same lipid phosphate, and if there is a common pool of this, then with a cell-free system capable of synthesizing both polymers it should be possible to demonstrate interdependence of the two systems. With membrane preparations from *B. licheniformis* it can be shown that the rate of synthesis of peptidoglycan is markedly decreased when nucleotide precursors for teichoic acid are present. Similarly the rate of synthesis of teichoic acid is greatly decreased when nucleotide precursors for peptidoglycan are present. The optimum conditions for the two syntheses differ slightly, and the inhibitory effects are greatest when conditions are adjusted to the optimum for the competing synthesis.

The interdependence of the two systems is confirmed by studying the effect of bacitracin. This antibiotic interferes with the synthesis of peptidoglycan by inhibiting the dephosphorylation of undecaprenol pyrophosphate to the monophosphate, thereby blocking the completion of the lipid phosphate cycle (Siewert & Strominger 1967). As the lipid pyrophosphate is not an intermediate in the teichoic acid cycle bacitracin does not inhibit this in cell-free systems.

When both polymers are synthesized simultaneously the antibiotic would prevent the return of lipid monophosphate to the common pool, and so should deprive the teichoic acid cycle of lipid phosphate and consequently cause inhibition. This is in fact readily demonstrated, and it follows that not only do both syntheses proceed through undecaprenol phosphate, but that there is sharing of these molecules in a common pool (Anderson *et al.* 1972). The scheme for this study is given in Fig. 11.

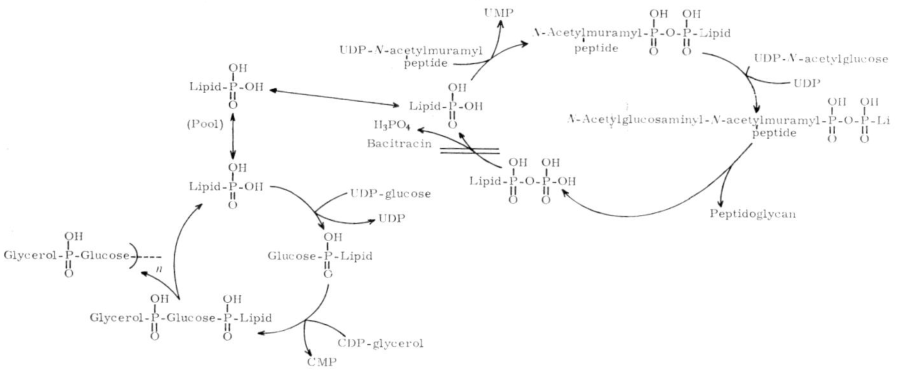

FIG. 11. Simultaneous synthesis of teichoic acid and peptidoglycan in cell-free system from *Bacillus licheniformis*.

In *B. licheniformis* a second teichoic acid occurs in the wall. This is a poly-(glycerol phosphate) and it can be shown that its synthesis, which also proceeds through lipid phosphate intermediates, is similarly affected by the simultaneous synthesis of peptidoglycan and the first teichoic acid. Consequently, all syntheses share undecaprenol phosphate. A similar relationship has been demonstrated for the wall polymers of *Staph. lactis* I3 (Watkinson *et al.* 1971). This close interrelationship between the biosynthesis of wall polymers in bacteria has not yet been extensively studied, but it is possible that all will show a common feature of sharing of undecaprenol phosphate. This is supported by the observations that this lipid phosphate participates in the synthesis of lipopolysaccharides in Gram-negative bacteria (Wright *et al.* 1967) and of the mannan in *M. lysodeikticus* (Scher *et al.* 1968). It is also involved in the synthesis of capsular polysaccharides (Troy *et al.* 1971).

There are two established methods for chain extension of natural polymers, i.e. by the successive addition of units at one or the other end of a chain. The two possibilities are illustrated by glycogen, where chain growth occurs by the addition of glucose units to the non-reducing end of a glycan chain, and proteins where growth occurs at the carboxyl end of peptide chains; in protein synthesis the transfer is of the growing oligopeptide from its compound with RNA to an aminoacyl-RNA. Although the direction of growth of the glycan chains of peptidoglycan is not known, it has been established that oligosaccharide chains of lipopolysaccharides grow from the 'reducing' end by a mechanism analogous to that for peptide synthesis. In this case the undecaprenol pyrophosphate oligosaccharide transfers the oligosaccharide chains to the 'non-reducing' end of an undecaprenol pyrophosphate trisaccharide in order to extend the chain by repeating units containing three sugar residues (Robbins *et al.* 1967). On the other hand, pulse-labelling experiments on the growth of chains in teichoic acids indicate that transfer of residues from the lipid intermediates occurs at the end of the chain remote from that attached to the peptidoglycan (Kennedy & Shaw 1968; Hussey *et al.* 1969; Brooks & Baddiley 1969*a*) (Fig. 12).

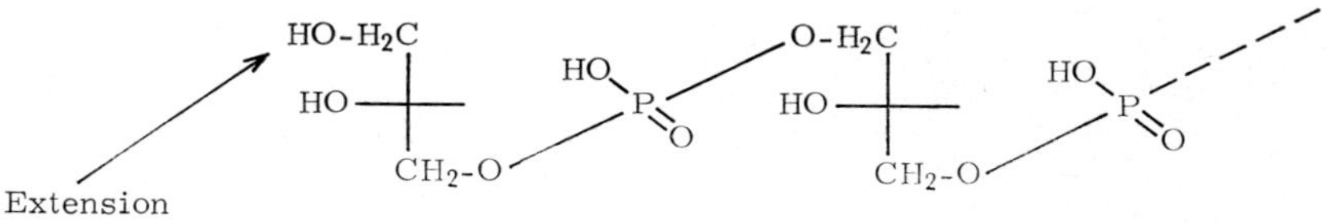

FIG. 12. Growth of a poly(glycerol phosphate).

The synthesis of a cell wall of the correct composition, consisting of complex polymers with generally uniform homogeneous structure, at a rate appropriate to the growth rate and shape of the cell requires rigorous control of the several individual synthetic systems. Although there is little factual information about these control mechanisms, there are indications from experimental observations as to how they might operate.

The conclusion that several and possibly all wall polymers synthesized in the membrane share undecaprenol phosphate during synthesis suggests that control of the availability of the lipid phosphate might provide a good mechanism for the general control of the rate of wall synthesis. In the cell-free system the rate is undoubtedly controlled by the amount of lipid phosphate available for sharing (Anderson *et al.* 1972), and in the living system control of the availability of this common intermediate could be achieved by a balance between two enzymes, undecaprenol phosphokinase and undecaprenol phosphate phosphatase (Higashi *et al.* 1970); thus rate of wall synthesis would be controlled through a single common factor without the general composition of the resulting wall material being altered.

The structural regularity of glycan chains in peptidoglycan, and of the repeating structure of wall teichoic acids and polysaccharides, is ensured by units being built up on undecaprenol phosphate, whereupon transfer to the growing polymer chain occurs only after the completion of the repeating structure on the lipid intermediate. In the experiments demonstrating sharing of undecaprenol phosphate in teichoic acid and peptidoglycan synthesis in *B. licheniformis* (p. 97), return of the lipid phosphate to the common pool occurred less readily than expected. This conclusion was supported by the increase in the inhibitory effects of competitive nucleotides seen after preincubation. A possible explanation of these effects is that lipid phosphate might remain associated with the multi-enzyme complex for the synthesis of one polymer until that polymer chain has been completed, and only then would it be released for return to the common pool. If this explanation is correct, it would ensure that incomplete chains do not become incorporated into the wall structure. An ordered system of synthesis of polymer chains is supported by studies on peptidoglycan-teichoic acid complexes. The polypeptide component of the peptidoglycan in walls of *Staph. lactis* I3 can be completely removed by use of an enzyme from a flavobacter. The soluble products include all of the glycan chains, and about 40% of these have teichoic acid attached to them; in this fraction each glycan chain is attached to only one teichoic acid chain. All of the teichoic acid is attached to glycan and the average sizes of the two polymer chains are respectively 24 and 9 repeating units (unpublished work with Drs J. Heckels and A. R. Archibald). Thus there is a considerable degree of order in the teichoic acid-peptidoglycan complex, suggesting a corresponding degree of order in its synthesis. It is proposed (Anderson *et al.* 1972) that the multi-enzyme systems responsible for the synthesis of the two or more wall polymers are situated close together in the membrane, and that they share a small pool of undecaprenol phosphate, with possibly only one molecule being shared between each group of enzymes.

The relative amounts of different polymers in a wall can vary in relation to changes in the composition of the medium. Bacilli grown in a chemostat where growth is limited by phosphate availability may substitute wall teichoic acid by a phosphate-free acidic polysaccharide; similarly under conditions of magnesium limitation of growth they may produce larger amounts of wall teichoic acid, presumably in order to bind the limited amount of magnesium more effectively (Tempest *et al.* 1968). It is not known how such gross changes in wall composition are controlled, but it may be significant that the pyrophosphorylase responsible for the synthesis of CDP-glycerol from glycerol phosphate and CTP is powerfully and probably allosterically inhibited by UDP-*N*-acetylmuramyl pentapeptide, a precursor of peptidoglycan synthesis (R. J. Anderson, H. Hussey & J. Baddiley, unpublished).

References

ANDERSON, J. S., MATSUHASHI, M., HASKIN, M. A. & STROMINGER, J. L. (1965) Lipid phosphoacetylmuramyl pentapeptide and lipid phosphodisaccharide pentapeptide: presumed membrane transport intermediates in cell wall synthesis. *Proc. Natl Acad. Sci. U.S.A.* **53**, 881–889

ANDERSON, R. J., HUSSEY, H. & BADDILEY, J. (1972) The mechanism of wall synthesis in bacteria. The organization of enzymes and isoprenoid phosphates in the membrane. *Biochem. J.* **127**, 11–25

ARMSTRONG, J. J., BADDILEY, J., BUCHANAN, J. G., CARSS, B. & GREENBERG, G. R. (1958) Isolation and structure of ribitol phosphate derivatives (teichoic acids) from bacterial cell walls. *J. Chem. Soc.* 4344–4345

BADDILEY, J. & NEUHAUS, F. C. (1960) The enzymic activation of D-alanine. *Biochem. J.* **75**, 579–587

BADDILEY, J., BLUMSOM, N. L. & DOUGLAS, L. J. (1968) The biosynthesis of the wall teichoic acid in *Staphylococcus lactis* I3. *Biochem. J.* **110**, 565–571

BROOKS, D. & BADDILEY, J. (1969*a*) The mechanism of biosynthesis and direction of chain extension of a poly(*N*-acetylglucosamine 1-phosphate) from the walls of *Staphylococcus lactis* N.C.T.C. 2102. *Biochem. J.* **113**, 635–642

BROOKS, D. & BADDILEY, J. (1969*b*). A lipid intermediate in the synthesis of poly(*N*-acetylglucosamine 1-phosphate) from the walls of *Staphylococcus lactis* N.C.T.C. 2102. *Biochem. J.* **115**, 307–314

BURGER, M. M. & GLASER, L. (1964) The synthesis of teichoic acids. I. Polyglycerol phosphate. *J. Biol. Chem.* **239**, 3168–3177

BURGER, M. M. & GLASER, L. (1966) The synthesis of teichoic acids. V. Polyglucosylglycerol phosphate and polygalactosylglycerol phosphate. *J. Biol. Chem.* **241**, 494–506

BUTTON, D., ARCHIBALD, A. R. & BADDILEY, J. (1966) The linkage between teichoic acid and glycosaminopeptide in the walls of a strain of *Staphylococcus lactis*. *Biochem. J.* **99**, 11–14C

CHATTERJEE, A. N. & PARK, J. T. (1964) Biosynthesis of cell wall mucopeptide by a particulate fraction from *Staphylococcus aureus*. *Proc. Natl Acad. Sci. U.S.A.* **51**, 9–16

DOUGLAS, L. J. & BADDILEY, J. (1968) A lipid intermediate in the biosynthesis of a teichoic acid. *FEBS (Fed. Eur. Biochem. Soc.) Lett.* **1**, 114–116

GHUYSEN, J. M., TIPPER, D. J. & STROMINGER, J. L. (1965) Structure of the cell wall of *Staphylococcus aureus*, strain Copenhagen IV. The teichoic acid-glycopeptide complex. *Biochemistry* **4**, 474–485

GLASER, L. (1964) The synthesis of teichoic acids. II. Polyribitol phosphate. *J. Biol. Chem.* **239**, 3178–3186

GLASER, L. & BURGER, M. M. (1964) The synthesis of teichoic acids. III. Glucosylation of polyglycerophosphate. *J. Biol. Chem.* **239**, 3187–3191

HANCOCK, I. C. & BADDILEY, J. (1972) Biosynthesis of the wall teichoic acid in *Bacillus licheniformis*. *Biochem. J.* **127**, 27–37

HANCOCK, R. & PARK, J. T. (1958) Cell-wall synthesis by *Staphylococcus aureus* in the presence of chloramphenicol. *Nature (Lond.)* **181**, 1050–1052

HIGASHI, Y., STROMINGER, J. L. & SWEELEY, C. C. (1970) Biosynthesis of the peptidoglycan of bacterial cell walls. XXI. Isolation of C_{55}-isoprenoid alcohol and of lipid intermediates in peptidoglycan synthesis from *Staphylococcus aureus*. *J. Biol. Chem.* **245**, 3697–3702

HUSSEY, H. & BADDILEY, J. (1972) Lipid intermediates in the biosynthesis of the wall teichoic acid in *Staphylococcus lactis* I3. *Biochem. J.* **127**, 39–50

HUSSEY, H., BROOKS, D. & BADDILEY, J. (1969) Direction of chain extension during the biosynthesis of teichoic acids in bacterial cell walls. *Nature (Lond.)* **221**, 665–666

Ito, E. & Strominger, J. L. (1962*a*) Enzymic synthesis of the peptide in bacterial uridine nucleotides. I. Enzymic addition of L-alanine, D-glutamic acid and L-lysine. *J. Biol. Chem.* **237**, 2689–2695

Ito, E. & Strominger, J. L. (1962*b*) Enzymic synthesis of the peptide in bacterial uridine nucleotides. II. Enzymic synthesis and addition of D-alanyl-D-alanine. *J. Biol. Chem.* **237**, 2696–2703

Ito, E. & Strominger, J. L. (1964) Enzymic synthesis of the peptide in bacterial uridine nucleotides. III. Purification and properties of L-lysine-adding enzyme. *J. Biol. Chem.* **239**, 210–214

Izaki, K., Matsuhashi, M. & Strominger, J. L. (1966) Glycopeptide transpeptidase and D-alanine carboxypeptidase; penicillin-sensitive enzymic reactions. *Proc. Natl Acad. Sci. U.S.A.* **55**, 656–663

Kennedy, L. D. & Shaw, D. R. D. (1968) Direction of polyglycerol phosphate growth in *Bacillus subtilis. Biochem. Biophys. Res. Commun.* **32**, 861–865

Knox, K. W. & Hall, E. A. (1965) The linkage between the polysaccharide and mucopeptide components of the cell wall of *Lactobacillus casei. Biochem. J.* **96**, 302–309

Mandelstam, J. & Rogers, H. J. (1958) Chloramphenicol-resistant incorporation of amino acids into staphylococci and cell-wall synthesis. *Nature (Lond.)* **181**, 956–957

Matsuhashi, M., Dietrich, C. P. & Strominger, J. L. (1965) Incorporation of glycine into the cell wall glycopeptide in *Staphylococcus aureus*: sRNA and lipid intermediates. *Proc. Natl Acad. Sci. U.S.A.* **54**, 587–594

Meadow, P. M., Anderson, J. S. & Strominger, J. L. (1964) Enzymic polymerization of uridine diphosphate (UDP)-acetylmuramyl-L-Ala-D-Glu-L-Lys-D-Ala and UDP-acetylglucosamine by a particular enzyme from *Staphylococcus aureus* and its inhibition by antibiotics. *Biochem. Biophys. Res. Commun.* **14**, 382–387

Nathenson, S. G., Ishimoto, N., Anderson, J. S. & Strominger, J. L. (1966) Enzymatic synthesis and immunochemistry of α and β-*N*-acetylglucosaminyl ribitol linkages in teichoic acids from several strains of *Staphylococcus aureus. J. Biol. Chem.* **241**, 651–658

Nikaido, K. & Nikaido, H. (1971) Glucosylation of lipopolysaccharide in *Salmonella*: biosynthesis of O-antigen factor 12_2. II. Structure of the lipid intermediate. *J. Biol. Chem.* **246**, 3912–3919

Park, J. T. (1952) Uridine-5′-pyrophosphate derivatives. III. Amino acid-containing derivatives. *J. Biol. Chem.* **194**, 897–904

Reuch, V. M. & Neuhaus, F. C. (1971) D-Alanine: membrane acceptor ligase from *Lactobacillus casei. J. Biol. Chem.* **246**, 6136–6143

Robbins, P. W., Bray, D., Dankert, M. & Wright, A. (1967) Direction of chain growth in polysaccharide synthesis. *Science (Wash. D.C.)* **158**, 1536–1542

Scher, M., Lennarz, W. J. & Sweeley, C. C. (1968) The biosynthesis of mannosyl-1-phosphorylpolyisoprenol in *Micrococcus lysodeikticus* and its role in mannan synthesis. *Proc. Natl Acad. Sci. U.S.A.* **59**, 1313–1320

Siewert, G. & Strominger, J. L. (1967) Bacitracin: an inhibitor of the dephosphorylation of lipid pyrophosphate, an intermediate in biosynthesis of the peptidoglycan of bacterial cell walls. *Proc. Natl Acad. Sci. U.S.A.* **57**, 767–773

Struve, W. G. & Neuhaus, F. C. (1965) Evidence for an initial acceptor of UDP-NAc-muramyl pentapeptide in the synthesis of bacterial mucopeptide. *Biochem. Biophys. Res. Commun.* **18**, 6–12

Tempest, D. W., Dicks, J. W. & Ellwood, D. C. (1968) Influence of growth conditions on the concentration of potassium in *Bacillus subtilis* var. *niger* and its possible relationship to cellular ribonucleic acid, teichoic acid and teichuronic acid. *Biochem. J.* **106**, 237–243

Troy, F. A., Frerman, F. E. & Heath, E. C. (1971) The biosynthesis of capsular polysaccharide in *Aerobacter aerogenes. J. Biol. Chem.* **246**, 118–133

Watkinson, R. J., Hussey, H. & Baddiley, J. (1971) Shared lipid phosphate carrier in the biosynthesis of teichoic acid and peptidoglycan. *Nat. New Biol.* **229**, 57–59

WRIGHT, A., DANKERT, M., FENNESSEY, P. & ROBBINS, P. W. (1967) Characterization of a polyisoprenoid compound functional in O-antigen synthesis. *Proc. Natl Acad. Sci. U.S.A.* **57**, 1798–1803

Discussion

Changeux: Peptidoglycan is part of the cell wall outside the membrane and the precursors are inside the cell membrane. How do the enzymes work?

Baddiley: As the final product is outside the wall whereas the nucleotide precursors are clearly internal to the membrane structure, it has been widely assumed that the lipid intermediates are concerned in the transportation of monomer or oligomer units through the membrane structure to the outside, but I don't much favour that view. The structure of the lipid intermediates is such that one would not expect them to flip over easily from one side of the membrane to the other, and I think there is growing evidence that flip-flop of that kind doesn't occur readily—at least, not in the time required for this sort of biosynthesis (Kornberg & McConnell 1971). One must then postulate either that there is no lipid bilayer in the region of synthesis or, as we have suggested (Anderson *et al.* 1972), that the polymers are extruded through the lipid bilayer, but this is based on relatively little evidence. In brief, we don't know the answers!

Changeux: Can you invert the membrane? Is it possible to have vesicles with the membranes inside out?

Baddiley: The experiments I have described here are with membrane fragments (particles), and I don't think that one could call them vesicles or liposomes, so we cannot easily do the sort of study you have in mind.

Chapman: Professor Baddiley and I have discussed this question, and we think it might be possible to obtain information about the mechanisms by using model systems, e.g. the black film technique.

I would like to ask Professor Baddiley about the Strominger protein—the highly hydrophobic protein. One problem in separating proteins from membranes is related to the particular extraction method used. In this case the protein was soluble in organic solvents.

Baddiley: The enzymes concerned with biosynthesis, or with many other activities associated with membranes, are clearly bound in or onto the membrane structure, and it doesn't surprise me that some of these turn out to be lipophilic in nature, as this one of Strominger's did.

Chapman: This protein is soluble in benzene. It would be very interesting to do experiments with black film techniques or other model membrane techniques, to see how the lipid bilayer is modified when this particular protein is included in the structure.

Katchalski: Cytochrome *c* when complexed with phospholipids can dissolve in heptane.

Chapman: Cytochrome *c* is soluble in organic solvent when it is interacted with a lipid. This Strominger protein has the isoprenoid lipid attached to it.

Baddiley: As I understand it, the protein structure is somewhat inverted; the R groups are on the outside and the hydrophilic groups are on the inside.

Luzzati: You said that there is growing evidence that lipid molecules do not flip over from one side of the membrane to the other. In fact I am aware of only the one experiment you mentioned (Kornberg & McConnell 1971) on liposomes.

Baddiley: The other experiments are those Professor Chapman and I are doing.

Chapman: The flip-flop experiments have been carried out with sonicated lipid using spin label methods and the flip-flop rate is very slow (less than $2 \times 10^{-5}\ s^{-1}$). Seeman (1972) has also reported experiments carried out by Hsia and Spero with erythrocyte ghosts.

Tanford: The amino acid compositions of quite a few membrane-associated proteins have been determined. But C_{55}-isoprenoid alcohol phosphokinase is the only one known to have a really remarkable excess of hydrophobic residues. That is perhaps not an unexpected general feature of membrane proteins but this is the only one in which the expectation is actually met, and the suspicion arises that perhaps this large number of hydrophobic residues is required to bind the substrate, which in this case is itself hydrophobic, rather than to bind the enzyme to the membrane.

I have recently compared the amino acid compositions of all serum lipoprotein apoproteins with those of normal proteins, and there is no significant bias. The data for the low density lipoprotein are discussed by Margolis & Langdon (1966).

Eigen: How far could those findings be biased by the high evolutionary level of separation techniques for polar solvents? Have enough cases been studied for satisfactory comparisons to be made? Certainly, separation techniques for non-polar solvents are not at the same level as those for polar solvents.

Tanford: Quite a few techniques are available now. A dozen or more such proteins have been obtained in pure form in denatured states. Their amino acid compositions have been determined, and there is not a very significant difference in relative amounts of ionic and hydrophobic amino acids for most of them. Even the so-called proteolipid protein in myelin, which is soluble in chloroform-methanol and is difficult to get out, is not dramatically different in composition from other proteins.

Baddiley: When the other enzymes concerned in these biosynthetic cycles are isolated, solubilized and purified, one may expect some of them to be soluble in lipid solvents.

We have fairly recently succeeded in solubilizing the enzymic system that synthesizes the wall teichoic acid in *Bacillus licheniformis*, dislodging it from the membrane by freeze-thaw techniques. In preliminary experiments it still appears to require the undecaprenol phosphate.

Katchalski: Does the acceptor in your biosynthetic system have to be of a relatively high molecular weight?

Baddiley: The preliminary investigation suggests that it is not a very big molecule.

Flory: You mentioned that the cross-linked mucopeptide is rigid. Can you say more about that?

Baddiley: Since the shape of an organism is at least partly determined by its wall structure, and as most bacteria can survive considerable internal osmotic pressure without great changes in shape, one assumes that the peptidoglycan is fairly strong.

Flory: Strength and rigidity are not quite the same thing. Does the peptidoglycan partake of any of the characteristics that I would associate with rubber elasticity, as might be anticipated from the structure you have described?

Baddiley: The physical properties have not been much studied.

Flory: Is it out of the question to measure such properties? Is the thickness within the black film range?

Baddiley: The peptidoglycan layer is about 10 nm.

Flory: Is there evidence in support of the regular lattice structure you described, or is that just conjecture?

Baddiley: That slightly idealized picture (Fig. 1, p. 88) occurs in some staphylococci as far as one can tell, in that in these organisms the peptidoglycan does appear to be fully cross-linked. With most bacteria, the cross-linkage is far from complete and so there are plenty of open ends.

Flory: Not only that, but unless the linkage is highly directed over a long range it will not give rise to the topological squares you have drawn—especially in three dimensions. Even a thickness of 10 nm is sufficient for the chains to be interconnected through multifunctional junctions at random, without adhering to the regular lattice array depicted by the structures you postulate.

The number of chains traversed along a shortest closed circuit returning to the same multifunctional junction may exceed four, the number stipulated by the structure you show. It should be variable, being any positive even number.

Have these questions of topology been studied?

Baddiley: I don't think so.

Lifson: Is the network of polysaccharide and polypeptide, which is essentially two-dimensional, on the outside of the membrane, or does it form several layers?

Baddiley: There must be several layers, but I don't know how many. Presumably glycan chains are cross-linked in three dimensions, and I think there is growing evidence that secondary modification of these structures occurs within the wall itself. Cross-linking from one polymer chain to another in a three-dimensional sense probably occurs, but no serious study of this has yet been made. X-ray analysis of walls of bacteria has so far been surprisingly unrewarding. At least from preliminary X-ray studies, nothing resembling real regularity of structure in the final wall is seen.

Kellenberger: Several laboratories are trying to investigate the possible regularity of the array of polymer chains and cross-links by electron microscopy. A mercury-containing stain has been worked out, by Formanek & Formanek (1970) and others, which specifically stains parts of the cross-links and therefore should reveal periodicities. With the techniques for observing single heavy atoms (conventional transmission or dark field microscopy), this should make micrographs of murein layers accessible to analysis with optical diffraction.

Murein layers seem rather flexible. The surface tension of the aqueous suspension medium, acting during specimen preparation for the electron microscope, is enough to flatten the murein sacculus completely without breaking it. It is certainly possible to calculate approximately these forces and the flexibility coefficient of the layer. This is not yet done. But by crude comparison we can already say that it is more flexible than any known 'plastic' or any other commonly used material. As for the tensile strength, we can derive nothing from these observations.

As regards the regularity, one has obviously to account for the rounded ends where the periodicity has to be severely modified.

Flory: The kind of irregularity I had in mind would be on a much finer scale than any irregularity required to give curvature. On the matter of deformability being very high, amongst non-biological networks there is virtually no lower limit for the modulus; the compliance may be very high indeed. It would be interesting somehow to gain an idea of the range of deformability either from surface tension forces or, if possible, by mechanical forces, although the latter method may be impracticable. Beyond that, one would like to get a notion of the resistance to deformation, or the elastic modulus. Observations by electron microscopy raise concerns about what happens when the sample is prepared for examination. If the chains are initially well separated from one another and not agglomerated, then when the liquid environment is removed to prepare the sample for electron microscopy, aggregation may occur and structures may develop which are not characteristic of the original material. I think a great deal could be learned just from mechanical measurements. Optical measurements could be useful as well if layers of sufficient thickness could be assembled for examination.

Baddiley: X-ray crystallographers have been trying to prepare layers of peptidoglycan, but so far this has not been very rewarding.

Levitt: Does the inclusion of glycine into peptidoglycan indicate a need for flexibility? Glycine has greater rotational freedom than other amino acids.

Baddiley: There are five glycine residues cross-linking peptidoglycan in *Staphylococcus aureus.*

Levitt: One would expect that network to be much more flexible in one direction than the other. Polysaccharide chains are considerably less flexible than polypeptides with high glycine content.

Baddiley: That theory is fine, except that not all bacterial walls contain glycine; this amino acid is confined mainly to the micrococci and staphylococci.

Flory: I don't think you can draw that conclusion about it being stiffer in one direction than in another. The carbohydrate chains may be 'stiff' in a certain sense, but they can be quite elastic. The relevant parameter is the extensibility. It cancels from the equations of the rubber-elasticity theory, and does not affect the modulus. To differ in different directions would require structural anisotropy and your model does not imply that.

Lifson: To understand the structure one needs to ask what is the relation of structure to function. The structure has been assembled so that it can function as a cell wall, namely be permeable and closed at the same time. Maybe this is why it has to be flexible, and why it has to respond properly to the entrance of the molecule and still prevent the flow of material which has to be preserved.

Baddiley: Yes, I agree. In that connection, one of the functions of teichoic acid is cation binding to the phosphate groups; in fact this is probably the most important function (Heptinstall *et al.* 1970; Hughes *et al.* 1971). The structure of the teichoic acids may fit in ideally with that purpose, with their regular repeating phosphate centres and relatively flexible molecules.

Lifson: Does active transport take place through the membrane?

Baddiley: Yes.

Lifson: Then evidently it requires very specific structural properties.

Engel: Formanek & Formanek (1970) attached a large mercury reagent to lysine residues in bacterial walls. They found a regular network of dark spots in electron micrographs of the ghosts. As I recall, the pictures were interpreted as being superposition of two walls (back and front of the collapsed ghost), the distance between the layer lines being half that expected from the chemical structure. A staggered arrangement of the reagent molecules was assumed.

Baddiley: Was that work carried out on the whole wall, or just on the murein of the wall?

Engel: I think this was on the murein.

Luzzati: I remember several electron micrographs of cell walls which display highly regular hexagonal patterns. In those cases, is the chemical composition the same as you described, Dr Baddiley?

Baddiley: There hasn't been much work on the nature of those patterns that one sees on a few walls. *Bacillus polymyxa* shows one of these regular square-form patterns and one can remove this with urea solutions. The pattern is made up from granules of glycoprotein (Goundry *et al.* 1967). What they are doing there I don't know, but they are not part of the murein.

References

ANDERSON, R. G., HUSSEY, H. & BADDILEY, J. (1972) The mechanism of wall synthesis in bacteria. The organization of enzymes and isoprenoid phosphates in the membrane. *Biochem. J.* **127**, 11–25

FORMANEK, H. & FORMANEK, S. (1970) Specific staining for electron microscopy of murein sacculi of bacterial cell walls. *Eur. J. Biochem.* **17**, 78–84

GOUNDRY, J., DAVISON, A. L., ARCHIBALD, A. R. & BADDILEY, J. (1967) The structure of the cell wall of *Bacillus polymyxa* (N.C.I.B. 4747). *Biochem. J.* **104**, 1c

HEPTINSTALL, S., ARCHIBALD, A. R. & BADDILEY, J. (1970) Teichoic acids and membrane function in bacteria. *Nature (Lond.)*, **225**, 519–521

HUGHES, A. H., STOW, M., HANCOCK, I. C. & BADDILEY, J. (1971) Function of teichoic acids and effect of novobiocin on control of Mg^{2+} at the bacterial membrane. *Nat. New Biol.* **229**, 53–55

KORNBERG, R. D. & MCCONNELL, H. M. (1971) Inside-outside transitions of phospholipids in vesicle membranes. *Biochemistry* **10**, 1111–1120

MARGOLIS, S. & LANGDON, R. G. (1966) Studies on human serum β_1-lipoprotein. *J. Biol. Chem.* **241**, 469

SEEMAN, P. (1972) *Chem. Phys. Lipids* **8**, 31

Molecular configuration and states of aggregation of biopolymers

P. J. FLORY

Stanford University, Stanford, California

Abstract Three types of spatial configurations characterize the molecules of biopolymers in various states of aggregation: (*i*) the random coil which occurs typically in elastic proteins and in contracted fibrous proteins such as collagen and myosin; (*ii*) helical forms of collagen, myosin and DNA, either dispersed in dilute solution or in fibrous structures; and (*iii*) the globular state. The dense packing and predominantly irregular (though unique) conformation in the globular state set it apart from the first two. The difficulties of achieving a high density without steric overlaps being imposed, whilst providing favourable tertiary interactions and respecting the preference of the peptide units for extended conformations, militate overwhelmingly against the existence of an acceptable spatial configuration for any sequence of peptide units assembled at random. Only a highly selective combination of residues may afford the possibility of a dense configuration of low energy. The invariance of the conformation of a globular protein throughout numerous evolutionary alterations in residue sequence can be comprehended on this basis. With respect to the fibrous state of aggregation, the propensity of polypeptides to adopt stable helical conformations in solution may provide the impetus for self-generation of organized structures (fibrils). The formation of liquid crystalline phases in solutions of synthetic polypeptides is illustrative of a process whereby ordered structures may be formed spontaneously.

The intricacies of the morphology and function of biological structures must surely be inscribed in the molecular constitution of the biopolymers from which they are constructed, but comprehension of the myriad connections implied by this assertion would be an insuperable task. The less ambitious objective of grasping an understanding of some of the broader aspects of the interrelationship between chemical structure and properties should, however, be within reach of attainment. While the more highly elaborated biological polymers—the proteins and polynucleotides, for example—present features not shared by the simpler macromolecules such as natural rubber, cellulose and the synthetics, the reverse is not true. That is, properties and behaviour characterizing the latter are exhibited in the former as well. The two categories are not separated by a remnant of the vitalist's barrier, and much can be learned by applying principles and theories established for one of them to the other. The present paper is premised on this remark. I shall attempt to indicate how some of the characteristics of macromolecules in biological systems may be related to their chemical structures.

The spatial configuration, or the conformation, of a macromolecule may be regarded as the intermediary between chemical constitution on the one hand and the macroscopic properties manifested by the polymer on the other. Indeed, without an adequate understanding of the factors determining the spatial configuration of a chain molecule, there can be little hope of achieving a rational understanding of the properties of the supermolecular structure of which it is a part. Widespread appreciation of this assertion is evidenced by the efforts devoted in recent years to experimental investigations of the configurations of biopolymers. Simultaneously, important advances have been made in the analysis of the spatial configurations and configurational statistics of linear macromolecules in general.

The spatial configurations prevalent among biopolymers are of three principal kinds. First, there is the random coil configuration, or array of configurations, in which the units of the chain are uncommitted to any definite conformation, but are subject only to the restrictions of bond geometry and hindrances to rotations about single bonds of the chain skeleton. The random coil is most familiar in dilute solutions, where it exists in virtual isolation from influences of other polymer molecules. It also occurs in the amorphous state, where the chain molecules are densely intertwined. It may confer high elastic deformability on the material. Examples are elastin and other elastic proteins, denatured fibrous and globular proteins, and single-stranded polynucleotides.

Regular conformations, for example the α-helices and β-extended conformation of polypeptides, comprise the second category. Such conformations are essential for the crystalline, or fibrous, state of organization.

The third category consists of the globular forms peculiar to certain native proteins. In this state the macromolecules are individually arranged in specific conformations, compact and unique for the particular protein.

Here I shall outline some of the principles which may serve to correlate the structural and configurational properties of polymer chains, and also examine the requirements for various states of aggregation.

THE SPATIAL CONFIGURATION

The spatial configuration of a long-chain molecule depends jointly on (*i*) the structural parameters consisting of bond lengths and bond angles dictated within narrow limits by the chemical bonding, and (*ii*) angles of rotation about single bonds of the chain skeleton which are susceptible to wider latitudes of variation. Although subject to minor variations which may sometimes be significant, the structural parameters (*i*) may usually be treated as fixed. Their values may be furnished by structural data on analogous small molecules. The rotations about skeletal bonds may therefore be considered the principal variables that determine the configuration.

FIG. 1. Macromolecular chain with a three-bond repeat unit. Skeletal bond rotation angles are denoted respectively by φ, ψ and ω, with subscripts *i* etc. denoting the serial indexes of units.

Consider, for example, the hypothetical macromolecule shown in part in Fig. 1 and made up of repeat units consisting of three skeletal atoms, or groups A, B, and C, and the three bonds joining them consecutively as indicated. The polypeptide molecule is such a chain; the α-carbon may be represented by A, the carbonyl carbon by B, and the NH group by C. An enormous variety of spatial configurations can be generated by assigning arbitrary values to the angles of rotation φ, ψ, and ω for every repeat unit *i*. The rotations are biased by rotational potentials, by steric interactions between non-bonded atoms or groups, and by coulombic interactions. Much of the configurational hyperspace is excluded on this account. In the polypeptide chain the angles ω_i are known to be restricted to values very near zero by the overwhelming preference of the amide partial double bond for the *trans* configuration. Thus, for each repeat unit, only the two rotation angles, φ and ψ, need to be treated as variables. The configuration partition function over these two angles is about 1.25 rad^2 according to calculations (Flory 1971). This corresponds to about 3.2% of the configuration space for these two rotations, or about 18% per bond. Obviously, the restrictions imposed by the interactions mentioned above severely limit the conformational range. Equally important is the particular range or ranges of conformational space that are accessible or favoured.

Important advances have been made in methods for dealing with the configurational characteristics of long-chain macromolecules in the last decade. In particular, statistical mechanical methods have been developed for evaluating averages of configuration-dependent properties over all configurations of the free chain (Flory 1969). These methods are general in scope and rigorous. The calculation of conformational energies is in a less satisfactory state. Available semi-empirical methods do, however, permit one to approximate the locations of preferred conformations and to gauge, roughly at least, the relative importances of competing conformations. Each type of chain presents peculiar geometrical features.

Special schemes may often be devised for simplifying the spatial representation of the chain molecule and the analysis of its configurations. Notable in this respect are the virtual bond schemes that have proved especially useful for the analysis of more complicated chains. The polypeptide chain, although not particularly complicated, is conveniently treated in terms of virtual bonds (Flory 1960; Ramachandran *et al.* 1963; Venkatachalam & Ramachandran 1967; Brant & Flory 1965; Scott & Scheraga 1966) representing the vectors extending from one α-carbon to the next (i.e. connecting consecutive atoms A in Fig. 1). Because of the prevalence of the amide group of the peptide unit in the *trans* configuration to the virtual exclusion of the *cis*, the virtual bond is legitimately considered to be of fixed length (3.80 Å) and embedded in the plane of the skeletal bonds of the peptide unit. This unit may be represented by the sequence ABCA in Fig. 1. The entire chain may be regarded as a sequence of virtual bonds thus defined. The orientation of one virtual bond with respect to the virtual bond and frame of atoms of the adjoining unit is determined by the rotations about the pair of chemical bonds flanking the intervening α-carbon (i.e. by φ_i and ψ_i for the i-th residue in Fig. 1). A configuration is uniquely specified by assigning angles φ_i, ψ_i for every residue. This scheme for generating configurations is applicable both to regular forms and to analysis of the multitude of random ones (Brant & Flory 1965; Brant *et al.* 1967).

Carbohydrate polymers, for example cellulose and amylose, have been treated in a corresponding manner (Rao *et al.* 1969; Yathindra & Rao 1970; Goebel *et al.* 1970; Brant & Dimpfl 1970). Recently, a similar scheme has been developed for the representation of polynucleotide chains (Olson & Flory 1972). In this case it is necessary to define two virtual bonds for the representation of each repeat unit embracing six chemical bonds along the chain backbone.

With the advent of the methods indicated, fairly reliable predictions can be made about the spatial configurations, both random and helical, on the basis of the chemical constitution and geometrical parameters which may be inferred from the chemical structure. The essential first step in relating chemical architecture to properties can thus be negotiated.

If every equivalent bond assumes the same rotation throughout the chain, for example, if in the case illustrated in Fig. 1 all φ_i take on the same value, all ψ_i are the same, and likewise all ω_i, then the conformation is a regular one, typically a helix. It is characterized by a rectilinear axis to which each repeat unit is equivalently related. Identical conformational assignments for equivalent bonds are the necessary and sufficient condition for the occurrence of a regular conformation which, trivial qualifications aside, may be designated as a helix of one sort or another. The most stable conformation is specified by the values of the rotations for the several kinds of bonds which, repeated throughout the chain, minimize the free energy. Seldom is the preference for this conformation sufficient to suppress all others in a free chain. In the absence of intermolecular constraints such as occur in a crystalline array, where each chain is forced to comply with its neighbours in adopting a geometrically symmetrical form, correlations of long range are destroyed by departures from the preferred conformation. These may consist of frequently occurring small deviations or of occasional departures of large magnitude.

In the chain of cellulose (Yathindra & Rao 1970), for example, the bonds (φ, ψ) of the ether linkage are confined overwhelmingly to one region of the conformation space favouring a relatively high extension of the chain, but variations within that region suppress correlations beyond a range of 20 to 40 units. The long polymeric chain accordingly is best described as a random coil, although it is an expanded one. Much the same holds for the polypeptide chain when in the random coil form (as opposed to the helix), although its degree of expansion is not as great as is typical for the chain of a cellulose derivative.

Isotactic vinyl chains bearing large substituents (e.g. isotactic polystyrene) provide further examples of macromolecules subject to severe steric constraints favouring a single conformation (e.g. a 3_1 helix). Either the occasional occurrence of a conformation of higher energy, or the presence of a small proportion of units having the alternative (racemic) stereochemical configuration, or both, serve to interrupt perpetuation of the helical conformation over long ranges.

As the examples cited illustrate, local conformational interactions alone seldom suffice to impose a regular conformation throughout a long polymer chain. In dilute solution or in the amorphous state, the polymer molecule ordinarily occurs, therefore, as a random coil. Rod-like form in the absence of special interactions is the exception. Apart from the crystalline state, its occurrence is confined to macromolecules in which adherence to the preferred conformation is stringently enforced either by severe, consistently recurring steric interactions or by strong secondary bonding. Examples of the former kind are afforded by the poly(*N*-alkyl isocyanates)

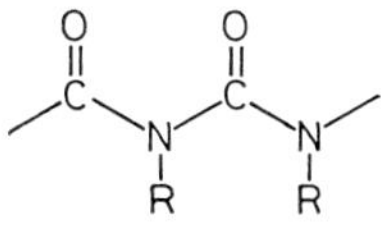

in which the opposing influences of partial double bonding in the main chain, which favour a planar structure, and steric interferences between pairs of substituents R ... R and O ... O which oppose it engender a helical conformation apparently of remarkable rigidity (Schmueli *et al.* 1969; Fetters & Yu 1971). Examples of chains that assume quasi-rigid conformations are poly-L-proline-II, polymers containing *p*-phenylene suitably joined, and, marginally, certain cellulose derivatives.

The principal examples of the second category are the α-helical polypeptides $-[NH-CHR-CO-]_x$, where R is an aliphatic side chain. Intramolecular hydrogen bonding plays a dominant role. A further example is the single-stranded polyribonucleotide chain under conditions strongly favouring base pairing. Multiple-stranded helices such as the famous Watson-Crick double helix and the collagen triple helix acquire enhanced rigidity from interchain (hydrogen) bonding, as is well known.

LIQUID CRYSTALLINE PHASES AND FIBROUS AGGREGATION

The accommodation of highly asymmetrical particles in a liquid medium poses special problems. Mutual correlation of the orientations of particles in proximity to one another becomes obligatory above a certain concentration that depends on their asymmetry; it is literally impossible for them to assume orientations at random at high concentrations. This much is self-evident. Deeper analysis shows that above a critical concentration such a system must separate into two phases (Onsager 1949; Isihara 1951; Flory 1956*a*): one, a dilute isotropic phase in which particle orientations are uncorrelated (i.e. random), and the other, a somewhat more concentrated anisotropic phase in which particle orientations are mutually correlated.

The statistical thermodynamic theory of solutions of rod-like particles has been developed in detail by resort to a simplified model (Flory 1956*a*). The mixture is predicted to separate into two phases above a solute volume fraction given approximately by $\varphi_2 \approx 8/x$ for large values of the axis ratio x. In the dilute phase of this concentration, the particles are orientated at random and are devoid of mutual correlations. In the coexisting phase of somewhat higher concentration the particles are aligned preferentially with respect to a common axis. The concentration of this tactoidal, or nematic, liquid-crystalline phase exceeds that of the more dilute phase only by a factor of about 1.5 (Flory 1956*a*). It is to be emphasized that the separation of the anisotropic phase is contingent solely upon the anisotropy of the particles. Intermolecular interactions (other than those of repulsion, which preclude overlaps of one particle with another) play at most a minor role, provided, however, that the interaction between solute and solvent is favourable (see below).

The early experiments of Robinson *et al.* (1958) revealed the occurrence of separation of solutions of α-helical polypeptides into two phases after the foregoing manner, the concentrations of the coexisting phases being roughly in accord with predictions (Flory 1961). More recent work by Nakajima *et al.* (1968) and by Wee & Miller (1971) demonstrate close agreement with theory.

A comparatively small attraction between large rigid particles should suffice to overcome the small entropy associated with the distribution in their spatial locations and orientations. This circumstance finds expression in the theory of such mixtures (Flory 1956*a*) by the prediction of a discontinuity in the phase diagram at a small positive value of the interaction parameter characterizing the enthalpy of mixing. The moderately dilute isotropic and anisotropic phases coexisting below this critical value are replaced above it by a very dilute phase and a dense, highly anisotropic phase in which the particles are in near-perfect alignment. The transformation is tantamount to aggregation of the solute particles. This prediction of theory is well confirmed for α-helical polypeptides by the experiments of Nakajima *et al.* (1968) and by those of Wee & Miller (1971).

With regard to the reverse process, namely dispersal of aggregated rod-like particles in a solvent, it follows that dissolution should occur only in a 'good' solvent for which the enthalpy of mixing is negative (i.e. favourable) or at most only slightly positive. Normally, London interactions between the large particles will favour their aggregation, and dispersion to a stable solution will be prevented. Insolubility of long, rod-like particles consequently should be the rule, their dispersal in a solvent the exception. In fact, stable solutions appear to be attained only under the aegis of a strong interaction between solvent and the particle, or its surface layer. This requisite may be provided by the mixing of side chains of a polypeptide α-helix with the solvent (Flory & Leonard 1965) or by an electric double layer; both oppose aggregation.

We thus arrive at the scheme depicted in Fig. 2 for describing the interrelationships between various configurations and states of aggregation for a system of long-chain macromolecules. In the dilute solution, the random coil configuration (I) is to be expected. If, however, severe conformational constraints are operative, then, in exceptional instances, a regular helical conformation (II) may be preferred even in dilute solution. In a polypeptide such as poly (γ-benzyl L-glutamate), for example, a change of solvent or temperature such as to favour intramolecular hydrogen bonding promotes the transition I → II. As pointed out above, avoidance of precipitation requires the solvent-polymer interaction to be favourable, i.e. interactions between two (or more) solute molecules must

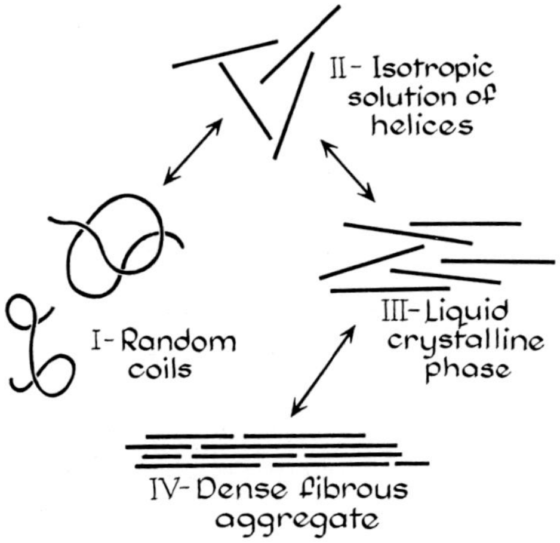

FIG. 2. Schematic representation of configurational transitions and of associated changes in states of aggregation.

be repulsive, or only slightly attractive at most. If the concentration is increased and the axis ratio is large, liquid crystalline domains constituting the tactoidal phase III may separate. Attractions between particles may then convert the dilute nematic domains to dense aggregates of macromolecules in the highly oriented fibrous state IV. Of course, the latter may be formed directly from II. Experiments on various systems serve to demonstrate the reversibility of each step in the scheme.

One may thus comprehend the spontaneous generation of an ordered supermolecular structure from random macromolecules. The underlying factors are in evidence in the primary chemical structure. Given a suitable linear assemblage of structural units, no external agent (surface, template or energy source) is required; the factors favouring the rigid conformation are latent in the chemical architecture, as discussed above. Moreover the changes in the physical state of aggregation are comprehensible in terms of basic principles implemented by theory which, though crude in some respects, appears to be adequate.

The processes I → II → IV call to mind the crystallization of a polymer, either from a dilute solution or from a melt. There are important differences however. In the processes shown schematically in Fig. 2, the generation of the ordered conformation precedes the aggregation to the crystalline array IV; the two processes are well separated so that the former can be completed before the latter sets in. In the crystallization of a polymer as it normally occurs, the adoption of the regular conformation that is essential for incorporation of the polymeric chain, or a section thereof, is induced by the crystallization process.

It occurs simultaneously with the deposition of the chain on the lattice of the crystal. Consequently, crystallization is invariably incomplete, and the crystalline domains extend over only a fraction of the length of the polymer molecule. Sections of the molecule falling between crystalline regions remain amorphous and disordered. The process depicted in Fig. 2 allows of a more efficient conversion to a state of high order. Well oriented fibres of α-helical polypeptides or of collagen may be formed in this manner. When the ordered state is rendered unstable by an increase in temperature or by a change in the surrounding liquid medium so that melting ensues, a fibre thus constituted undergoes shrinkage to a fraction of its length (Flory 1956*b*; Mandelkern 1964). The process may be reversed provided that the polymer chains are cross-linked or otherwise held together as a network. Examples are provided by collagen, muscle fibres, and cross-linked polyethylene (Flory 1956*b*; Mandelkern 1964, 1971; Harrington 1971). The melting of oriented crystalline fibres represents the underlying molecular mechanism responsible for contractility in fibrous proteins.

THE GLOBULAR STATE

The conformation of the polypeptide chain within a native globular protein partakes of some of the characteristics of both the crystalline and the amorphous states. Its differences with respect to these states are more noteworthy, however, than its superficial resemblances to them. As in a crystal, the chain is packed to high density, void space beyond that required to accommodate the side chains being negligible. Although the configuration of the chain is well defined (Perutz *et al.* 1965; Perutz 1965) and invariant from one molecule to another, even including molecules of corresponding proteins from widely different organisms, the individual chain residues occur in a variety of conformations (Watson 1969). There is no persistent conformation running throughout the chain, and the particle is accordingly devoid of the long-range order characteristic of a crystal. However, short α-helical segments are prominent in the myoglobins (Watson 1969); in α-chymotrypsin and elastase (Shotton & Watson 1970), short sections of the chain resemble the extended β-conformation, two such sections being paired with their directions antiparallel. The order in the α-helical and the extended β-sequences appears to be imperfect. These sequences may span of the order of five to twenty units. Within a given globular protein the directions of the axes of different α-helical sections or β-conformed domains show no obvious correlations. These semi-ordered runs of units are laced one to another by disordered sequences whose conformations are distributed more or less at random over the allowed ranges of conformation space (φ, ψ; see above) (Watson 1969; Brant & Schimmel 1967). They impart a random character to the overall conformation. The conformation cannot, however, be likened to that of a random coil for the reason, foremost amongst others, that the chain is densely packed rather than loosely spread over a volume many times its net volume.

This latter fact merits circumspection. We define the globular state as one in which the chain molecule is densely packed with itself, irrespective of the degree of ordering or disordering of the conformations of its various units. An acceptable conformation must meet the condition of confinement to a small region of space while avoiding multiple occupancy of any element of the space by different parts of the chain. These requirements impose a most severe restriction on the eligible conformations—so severe, in fact, as to cast doubt on the existence of even a single conformation meeting them. Analysis of the problem (Flory 1961, 1969) strongly suggests that a macromolecule such as a polypeptide with a preference for conformations in the extended β-region of the conformation space (φ, ψ) would offer a dearth of acceptable globular conformations. The difficulty may be relieved by departures of the conformations of some of the residues, at strategically selected positions, from those that are preferred according to energy criteria. Strategic placement of glycine or other residues offering alternative conformational possibilities might accomplish the same end. Additionally, parallel arrangements of neighbouring sequences of units may ameliorate the difficulties of packing the chain in a space-filling conformation.

In any event, analysis of the problem suggests a paucity of globular conformations for any long polypeptide chain. It is, in fact, problematical whether such a chain assembled at random from a representative set of amino acid residues would offer one or more globular conformations of low intramolecular energy. Add to the foregoing conditions for dense packing the further requirements of tertiary bonding and it becomes clear that the question to be posed is whether one such conformation exists for a given assemblage of residues, and not which of a supposed array of possible choices will be preferred.

The foregoing considerations raise the question of why globular conformations of proteins occur at all. How are the factors that should normally oppose the globular state overcome? The answer doubtless lies in the occurrence in globular proteins of a hydrophobic core made up of non-polar units and especially of hydrophobic side chains (Perutz *et al.* 1965; Watson 1969). These units are not confined to a given portion of the chain sequence but are distributed throughout the chain. They occur in both semi-ordered (e.g. helical) and disordered sequences. Their ordinal positions in the sequence of units appear to be more or less random if the three-dimensional structure is ignored. Yet when the chain is folded into its native conformation they occupy a contiguous domain, or core, shielded from contact with the surrounding aqueous solvent. Obviously, their locations in the chain sequence are vitally important. (Nearly all of the hydrophilic residues occur on the periphery of the particle where they may associate with the surrounding aqueous environment [Watson 1969]. Only two of the hydrophilic residues of myoglobin lie within the hydrophobic nucleus.)

It appears that the stability of the globular protein stems primarily from 'hydrophobic bonding' that opposes dispersal of the hydrophobic units (more specifically, of their side chains) in the aqueous environment. The α-helical and antiparallel β-sequences, and their associated hydrogen bonds, probably play a secondary role. A change in temperature or solvent may shift the thermodynamic balance in favour of the much more voluminous random coil in preference to the dense globular native conformation, whereupon denaturation occurs. Renaturation upon restoration of the original conditions demonstrates the reversibility of the process. The narrow range of the independent variable (e.g. temperature) within which the transformation takes place warrants designation of it as a transition, with deliberate allusion to a phase transition.

Ptitsyn *et al.* (1968) and Eizner (1969) have sought an analogy to formation of the globular state in the abrupt changes in spatial dimensions of certain synthetic polymers, for example poly(methacrylic acid) (Silberberg *et al.* 1957) and copolymers of maleic anhydride with alkyl vinyl ethers (Dubin & Strauss 1970), that occur when the solvent or the degree of neutralization of the polymer is suitably altered. The change in size of the polymer molecule may be quite large, as indicated, for example, by the change in viscosity of the solution. From a theoretical point of view, the lyophobic interactions within the domain of the macromolecule may be likened to a negative excluded volume interaction, i.e. to an interaction that favours intramolecular encounters between units remote in sequence along the chain instead of opposing them as the term 'excluded volume' normally implies. If the effective interaction volume for a pair of units is appreciably negative, then, as Ptitsyn *et al.* (1968) and Eizner (1969) have shown by extension of excluded volume theory into this regime, a discontinuous collapse of the spatial configuration may be predicted. If proper account is taken of the thermodynamics of mixing at high concentrations, it will be found, however, that the resulting comparatively dense conformation falls far short of collapse to a size equal to the net volume of the polymer. The macromolecule should remain swollen with an appreciable quantity of solvent well beyond the 'transition' marked by a discontinuous change in volume. This prediction appears to be supported by the experiments cited above.

In contrast to the behaviour in this respect of a polymer consisting of undifferentiated units, the density of packing of the chain in a globular protein is such as virtually to exclude all solvent from the interior of the particle. The hydrophobic character of the core obviously makes it hostile to the presence of water molecules, but, according to the arguments recited above, the formation of dense hydrophobic nucleus is conditional on the occurrence of an acceptable assignment of the residue sequence.

In sum, the inevitable difficulties of packing a long polypeptide chain to high density are enhanced in a globular protein by the further requirement that the various units and their side chains be juxtaposed in a manner providing for favourable tertiary interactions, especially the hydrophobic interactions within the core, but also including interactions between polar groups. A conformation meeting all these demands must be a rarity. If such a conformation is found, its strict conservation throughout biological evolution becomes a matter of virtual necessity.

Further to this point, Lim & Ptitsyn (1970) have examined the sum of the volumes of the residue side chains occupying the hydrophobic core in myoglobins and haemoglobins of different origins. Variances of this sum for 14 examples that differ by at least two hydrophobic residues were shown to be much smaller than would be expected if substitution of one hydrophobic residue occurred non-selectively from amongst all the various available hydrophobic residues. More detailed comparisons between the globins from different sources revealed a marked correlation between substitutions of one residue by another in a given neighbourhood. If a large side chain at a given site in one example is replaced by a smaller one at the same site in another example, a compensating substitution of a small side chain by a larger one is almost always found to have occurred in the adjoining space. Thus, the native conformation has survived many evolutionary substitutions of residues, but not without considerable pains in the selection amongst the various amino acids (Lim & Ptitsyn 1970). It is a reasonable conclusion that conservation of the conformation has been enforced by the unavailability of alternatives short of those requiring a drastic revision of the entire sequence.

References

BRANT, D. A. & DIMPFL, W. L. (1970) *Macromolecules* **3**, 655–664
BRANT, D. A. & FLORY, P. J. (1965) *J. Am. Chem. Soc.* **87**, 2791–2800
BRANT, D. A. & SCHIMMEL, P. R. (1967) *Proc. Natl Acad. Sci. U.S.A.* **58**, 428–435
BRANT, D. A., MILLER, W. G. & FLORY, P. J. (1967) *J. Mol. Biol.* **23**, 47–65

DUBIN, P. L. & STRAUSS, V. P. (1970) *J. Phys. Chem.* **74**, 2842–2847
EIZNER, YU. YE. (1969) *Vysokomol. Sowjet* **A11**, 364–371 [Translation: (1969) *Polym. Sci. U.S.S.R.* **11**, 409–417]
FETTERS, L. J. & YU, H. (1971) *Macromolecules* **4**, 385–389
FLORY, P. J. (1956*a*) *Proc. R. Soc. Lond.* **A234**, 73–89
FLORY, P. J. (1956*b*) *Science (Wash. D.C.)* **124**, 53–60
FLORY, P. J. (1960) in *Protein Structure and Function (Brookhaven Symp. Biol.)* **13**, 89–103
FLORY, P. J. (1961) *J. Polym. Sci.* **49**, 105–128
FLORY, P. J. (1969) *Statistical Mechanics of Chain Molecules*, Wiley-Interscience, New York
FLORY, P. J. (1971) *Pure Appl. Chem.* **26**, 309–326
FLORY, P. J. & LEONARD W. J. (1965) *J. Am. Chem. Soc.* **87**, 2102–2108
GOEBEL, C. V., DIMPFL, W. L. & BRANT, D. A. (1970) *Macromolecules* **3**, 644–654
HARRINGTON, W. F. (1971) *Proc. Natl Acad. Sci. U.S.A.* **68**, 685–689
ISIHARA, I. (1951) *J. Chem. Phys.* **19**, 1142–1147
LIM, V. I. & PTITSYN, O. B. (1970) *Mol. Biol.* **4**, 372–382
MANDELKERN, L. (1964) *Crystallization of Polymers*, ch. 7, McGraw-Hill, New York
MANDELKERN, L. (1971) Muscle as a fibrous protein system. In *Contractile Proteins and Muscle* (LAKI, K., ed.), Dekker, New York
NAKAJIMA, A., HAYASHI, T. & OHMORI, M. (1968) *Biopolymers* **6**, 973–982
OLSON, W. K. & FLORY, P. J. (1972) *Biopolymers* **11**, 1–66
ONSAGER, L. (1949) *Ann. N.Y. Acad. Sci.* **51**, 627–659
PERUTZ, M. F. (1965) *J. Mol. Biol.* **13**, 646–668
PERUTZ, M. F., KENDREW, J. C. & WATSON, H. C. (1965) *J. Mol. Biol.* **13**, 669–678
PTITSYN, O. B., KRON, A. K. & EIZNER, YU. YE. (1968) *J. Polym. Sci. Part C* **16**, 3509–3517
RAMACHANDRAN, G. N., RAMAKRISHNAN, C. & SASISEKHARAN, V. (1963) *J. Mol. Biol.* **7**, 95–99
RAO, V. S. R., YATHINDRA, N. & SUNDARARAJAN, P. R. (1969) *Biopolymers* **8**, 325–333
ROBINSON, C., WARD, J. C. & BEEVERS, R. B. (1958) *Disc. Faraday Soc.* **25**, 29–42
SCHMUELI, U., TRAUB, W. & ROSENHECK, K. (1969) *J. Polym. Sci. Part A-2* **7**, 515–524
SCOTT, R. A. & SCHERAGA, H. A. (1966) *J. Chem. Phys.* **45**, 2091–2101
SHOTTON, D. M. & WATSON, H. C. (1970) *Nature (Lond.)* **225**, 811–816
SILBERBERG, A., ELIASSAF, J. & KATCHALSKY, A. (1957) *J. Polym. Sci.* **23**, 259–284
VENKATACHALAM, C. M. & RAMACHANDRAN, G. N. (1967) in *Conformation of Biopolymers*, pp. 83–105, Academic Press, New York
WATSON, H. C. (1969) *Prog. Stereochem.* **4**, 299–333
WEE, E. L. & MILLER, W. G. (1971) *J. Phys. Chem.* **75**, 1446–1452
YATHINDRA, N. & RAO, V. S. R. (1970) *Biopolymers* **9**, 783–790

Discussion

Levitt: Professor Flory, you suggested that it is very difficult to find sequences of amino acids that are able to fold into a compact three-dimensional structure. If this restriction is really severe, one may find that different protein conformations are built according to a few conformational themes. Of course, in many cases different proteins have evolved from a common ancestor and so they continue to have similar conformations. Already there are several different proteins known to have similar conformations, for example, myoglobin and haemoglobin, chymotrypsin and elastase, and lysozyme and lactalbumin. Conceivably, there may be a small number of conformational themes followed by all protein conformations, say less than 100 themes.

Flory: For chains consisting of some 200 residues, for example, it is conceivably so, but the number of themes, as you call them, is not predictable.

Changeux: Some of the membrane proteins are obviously in a highly hydrophobic phase, constituted by the hydrocarbon chain of the lipid bilayer. Could some of these proteins have an inverted structure, that is hydrophilic groups inside the molecules and hydrophobic groups outside?

Flory: I don't know. I noted with interest earlier that you said some of the bilayer proteins may have the inverted arrangement in a hydrophobic environment, with the hydrophilic residues forming the interior.

Changeux: Is such a situation known in synthetic polymers?

Flory: I did not discuss in detail the subject of the possible occurrence of globular forms amongst synthetic polymers. So far as I am aware, none duplicates the compact structure of a globular protein. Ptitsyn (Ptitsyn *et al.* 1968) has considered the possibility of analogous conformations in polymers with strong self-interactions. His examples are not synthetic polymers so fashioned as to have a hydrophobic core. Rather, he has considered polymers that are subject to self-attractions operative uniformly throughout the chains. The attractions may be viewed as the converse of the excluded volume situation. That is, instead of having an excluded volume disfavouring contacts between chain units, there is a negative excluded volume that favours them. The equations of excluded volume theory may be used, but in the range of negative values. Ptitsyn (Ptitsyn *et al.* 1968) pointed out that there could be a critical point for a transition from the random coil to comparatively compact conformations under some circumstances. But on closer inspection, the theory is not unambiguous on this point, in my opinion. There are a few isolated examples of synthetic polymers for which fairly dense particles are indicated in solution, but even these are considerably swollen with solvent. They do not meet the terms of the globular particle of a protein. There is thus no close prototype for the globular protein, so far as I am aware.

Changeux: You mentioned the effect of shape as a basis for order in lipids. Is the shape factor important in establishing bilayer structure?

Flory: Onsager (1949) considered solutions of highly asymmetric particles—discs as well as rods. He was the first to show that such systems should exhibit phase transitions. The separation of an ordered phase from a solution of discs or rods solely in consequence of the geometric asymmetry of the particles might be one step towards the generation of a membrane.

Chapman: Can anyone envisage inverting a protein which has the particular amino acid sequence that gives rise in water to hydrophobic residues inside and hydrophilic residues outside, so that at the same time it retains an appreciable amount of α-helical structure?

Changeux: The sequence of the membrane proteins concerned might have been selected to have an inverted structure in a hydrophobic environment.

Eigen: How many proteins are known to sit in the hydrophobic region?

Baddiley: One is known so far.

Chapman: There is a great deal of information that suggests that proteins are situated in various positions in a membrane (Steck *et al.* 1972). X-ray and various experimental evidence suggests that proteins can indeed sit in the hydrophobic region of the membrane.

Eigen: Phase transitions in lipid bilayers have been studied quite extensively by Hermann Träuble (unpublished work). He showed that very sharp cooperative transformations occur in lipid vesicle systems at around (and above) room temperature. They occur within milliseconds (down to fractions of milliseconds), depending on the composition of the membrane. These phase transitions are very sensitive to the chain length of the hydrocarbon tail and especially to a possible content of cholesterol. The transitions occur between a highly ordered and a liquid crystalline phase. Of course, in the liquid crystalline phase it is much easier for a protein to move. The whole transport phenomenon changes drastically with these phase transitions. Since they are very sharp one can shift them by changing the external conditions, e.g. by electric fields. In other words, with an electric pulse one can trigger a certain conformational change of the whole system, which then can suddenly set free some enzymic activity. One can do the same by bringing about small changes in ionic strength. I am sure these structures are of great significance in processes involving electric membrane excitation.

Katchalski: Are there any indications in biological systems that there are liquid crystals in the protoplasm or anywhere else apart from the membrane?

Luzzati: What do you mean by 'liquid crystal'?

Eigen: There is a highly ordered phase where all the hydrocarbon tails have the same orientation and with increasing temperature they start to rotate and form 'defects', apparently in a cooperative manner.

Luzzati: In fact the hydrocarbon chains always rotate.

I do not think that the meaning of 'liquid crystal' is just a matter of semantics, since a clear definition was given by Friedel (1922). You seem to imply that any object displaying order and partial disorder is so widespread that we do not need a special expression for it. I think that calling a membrane a 'phase' and/or a 'liquid crystal' is just misusing well-established notions. Perhaps myelin sheath, or retinal rods, are 'phases' and 'liquid crystals', but not isolated membranes.

As a rule biopolymers display both long-range and short-range order, whereas lipids (and to some extent membranes) have long-range order and short-range disorder. This is, in my opinion, a fundamental difference between lipids and biopolymers like proteins and nucleic acids.

Flory: Strictly speaking the ordered phase containing rod-like particles is a tactoidal or nematic phase, and not a liquid crystal in the crystallographic sense. However, there is sufficient order in the tactoidal phase to give a lateral spacing in the X-ray diffraction. The point is that one can go continuously from a dilute tactoidal phase to a more concentrated highly ordered phase. Eventually it must show order in two dimensions in the transverse plane. When do you call it a liquid crystal and when do you call it a partially ordered phase?

I mentioned earlier that, in theory at least, solutions of platelets may exhibit phase separation. The transition here involved is not analogous to a bilayer transition, however. The bilayer transition may correspond in some degree to the phase transitions occurring in solutions of rod-like particles, but not the platelets or discs. I know of no analogue or counterpart to the latter in the membrane layer.

Changeux: I asked about that to see whether one could distinguish the contribution of shape and the contribution of bonding.

Flory: Conceptually the bonding factor is very different from the shape factor. It is more difficult to distinguish these experimentally, which is in answer to your question. One may note that these shape-induced transitions I have talked about are accompanied by a very small change in enthalpy, and it follows that the change in entropy associated with the transition must be very small as well.

References

FRIEDEL, G. (1922) *Annls Phys.* **18**, 273
ONSAGER, L. (1949) *Ann. N.Y. Acad. Sci.* **52**, 627
PTITSYN, O. B., KRON, A. K. & EIZNER, YU. YE. (1968) *J. Polym. Sci. Part C* **16**, 3509
STECK, T. L., FAIRBANKS, G. & WALLACH, D. F. H. (1972) *Biochemistry* **10**, 2617

Formation of the native structure of proteins: inferences from the kinetics of denaturation and renaturation

CHARLES TANFORD

Department of Biochemistry, Duke University Medical Center, Durham, North Carolina

Abstract Kinetic equations applicable to any network of unimolecular reactions show that the observable change in any physical property can always be expressed in terms of a sum of exponential factors. Complete solutions have been obtained both for the amplitudes and decay constants of these factors for a variety of possible sequences of reaction steps, and the solutions have been applied to experimental data for the reversible denaturation of several proteins by guanidine hydrochloride. The product of this reaction is a structureless random coil, and the reversal of the reaction thus provides a model for the folding of structureless polypeptide chains in any situation, including that which occurs *in vivo* after a polypeptide chain has first been synthesized. The results for two proteins, cytochrome *c* and β-lactoglobulin, show that a strictly sequential mechanism for the folding process can be unambiguously rejected. The predominant initial product in the renaturation of these proteins is a 'dead-end' incorrect structure, which has to be unfolded again before progress to the correct native structure can occur. The overall rate of formation of the native state depends on the rate of formation of unique intermediates that do not contain structural elements in conflict with the native structure. In cytochrome *c* and β-lactoglobulin the rate of formation of these particular intermediates is evidently slower than the rate of formation of partly folded states that do not satisfy this criterion. It is suggested that this situation is likely to occur frequently, and that incorrectly folded states will represent an important feature of the mechanism of folding for most proteins.

The purpose of the work described in this paper is to learn something about the mechanism by which structureless polypeptide chains assume their native conformations. This is done by study of the kinetics of the reversible denaturation of proteins by guanidine hydrochloride: the product of this reaction has been shown to be a structureless, randomly coiled polypeptide chain for most proteins studied (Tanford *et al.* 1966*b*; Tanford 1968), so that the reversal of denaturation represents in effect the reaction of interest. The conditions under which the renaturation reaction is studied differ from the conditions under which newly synthesized polypeptide chains adopt their native structures in an *in vivo* system. The two processes are nevertheless sufficiently similar to allow consideration of the renaturation reaction as a model for the *in vivo* process. To interpret kinetic data on this process we have made use of the general theory of the kinetics of networks of unimolecular reactions (e.g. Frost & Pearson 1953; Benson 1960). This limits us to systems in which bimolecular reaction steps do not influence the results and thus precludes at present any consideration of proteins in which subunit interactions are an important feature of the native structure.

THEORY

The various possible conformational states of a particular protein may be designated as A_1, A_2, $\ldots A_n$. One of them (usually A_1) will be the native state (N), and one of them (usually A_n) will be the fully denatured randomly coiled state (D). Other states are reaction intermediates, the number of them depending on the complexity of the reaction. Only those states that accumulate to a significant extent during the time course of the reaction affect the kinetics: unstable transient states do not.

The microscopic rate constants for unimolecular conversions $A_i \rightarrow A_j$ will be designated as k_{ij}. The values of k_{ij} are of course a function of the external variables: in our study the concentration of denaturant is the only variable affecting the system. The microscopic rate constants do not, however, depend on the state of the protein. At a particular concentration of denaturant the same values apply to the kinetics of unfolding of the native protein and to the kinetics of refolding of the random coil.

The solutions to the set of differential equations that govern such a process (Frost & Pearson 1953; Benson 1960) show that the time-dependent concentration $C_i(t)$ of any species A_i is governed by a set of composite rate constants λ_s, each of which is a unique function of the microscopic rate constants k_{ij}. The theoretical number of distinct λ_s values is $n-1$, where n is the number of distinct conformational states that constitute a significant fraction of the total population at some stage of the process. The equation for any $C_i(t)$ is

$$C_i(t) = C_i(eq) + \sum_{s=1}^{n-1} a_{is} e^{-\lambda_s t} \qquad (1)$$

where $C_i(eq)$ is the concentration of species A_i at equilibrium, i.e. when no further change with time occurs. The same set of λ_s applies to each species, but the time-independent coefficients a_{is} differ for each species. They depend not only on the k_{ij} values but also on the concentration of each species at $t = 0$. Methods for the evaluation of the λ_s and a_{is} in terms of the k_{ij} and (for a_{is}) initial conditions are given in the references cited.

In the study of the kinetics of protein denaturation one cannot measure the concentrations of individual conformational states, and we are confined to the determination of some physical property of the system, such as optical density, that depends on the conformational state. If y_i is the value of this property in state A_i, then the average value measured at time t, $y(t)$, will be given by

$$y(t) = \sum_{i=1}^{n} y_i \, C_i(t)/C_o \tag{2}$$

where $C_o = \Sigma_i \, C_i(t)$ is the total concentration of protein in all its states, which remains constant in a system undergoing only unimolecular transitions. Since each C_i varies with time according to the same set of exponential decay terms, $y(t)$ must be governed by the same set. The resulting expression is most conveniently written in the form

$$\frac{y(t) - y(eq)}{y(0) - y(eq)} = \sum_{s=1}^{n-1} P_s e^{-\lambda_s t} \tag{3}$$

where $y(0)$ and $y(eq)$ are the values of y at $t = 0$ and at equilibrium, respectively, and the coefficients P_s are proportional to $\Sigma_i \, y_i a_{is}$, i.e. they depend on the characteristic optical or other properties of the states A_i, as well as on the initial conditions in an experiment and the values of k_{ij}.

The values of y_i for the native and fully denatured states can be experimentally determined, but those for other states must be considered as unknowns. It is convenient to define an auxiliary parameter α_i relating each y_i to the known values, y_N and y_D, in the native and randomly coiled states,

$$\alpha_i = \frac{y_i - y_N}{y_D - y_N} \tag{4}$$

The parameter α_i is zero for the native state, unity for state D, and between 0 and 1 for a state with intermediate properties. The values for each coefficient P_s of equation 3 can then be expressed in terms of α_i and a_{is} values. Because the left-hand side of equation 3 is unity at $t = 0$,

$$\sum_{s=1}^{n-1} P_s = 1 \tag{5}$$

EXPERIMENTAL PROCEDURE

The state of equilibrium of the system, as a function of denaturant concentration, must be known before kinetic experiments are carried out. For many small proteins the process cannot be separated into stages displaying intermediate states but occurs as a single cooperative transition, as illustrated by Fig. 1. From such data one can determine an apparent equilibrium constant,

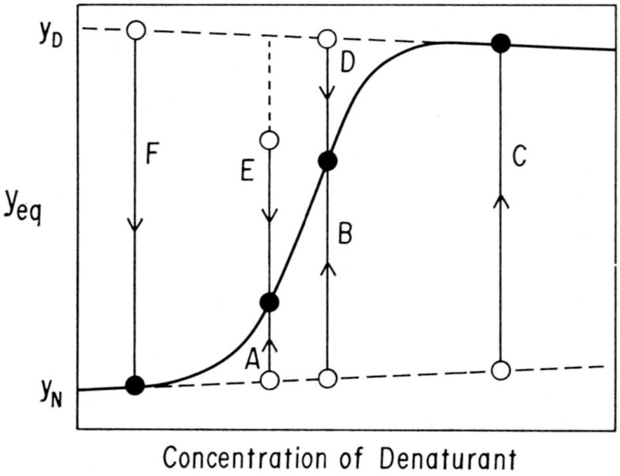

FIG. 1. Kinetic experiments in relation to an equilibrium curve for the denaturation of proteins by guanidine hydrochloride, as followed by measurement of some optical property y. The equilibrium values of y are indicated by the heavy line, and values for the native and fully denatured states by the dotted lines. The vertical lines are representative of typical kinetic experiments: open circles represent initial points obtained by extrapolation of kinetic data to zero time, filled circles represent final values of y attained.

K_{app}, as a function of denaturant concentration. This equilibrium constant is related to the thermodynamic equilibrium constants for formation of each species A_i from the native state, $K_{Ni} = C_i(eq)/C_N(eq)$, as well as to the characteristic value y_i of the measured property for each state, again conveniently expressed by means of equation 4. The relation between these parameters (Tanford 1968) is

$$\frac{y_{eq} - y_N}{y_D - y_{eq}} \equiv K_{app} = \frac{\sum_{i=2}^{n} \alpha_i K_{Ni}}{1 + \sum_{i=2}^{n} (1 - \alpha_i) K_{Ni}} \tag{6}$$

The rate of approach of y to its equilibrium value is determined at many concentrations of denaturant. The starting material can be the native state (lines A, B and C of Fig. 1) or the randomly coiled state (lines D, E and F of Fig. 1), and sometimes it is also useful to begin with an equilibrium mixture at the transition mid-point and measure the rate of approach to N or D outside the transition zone. As will be seen below, an important outcome of the application of the theoretical treatment to such data is to show that it is essential to study the reaction in the transition zone in both directions at identical denaturant concentrations, as illustrated by lines B and D or A and E in Fig. 1.

For many proteins the reaction is very rapid, and a stopped-flow system must be used to make the measurements. For β-lactoglobulin even the stopped-flow system did not provide sufficiently fast mixing and part of the reaction occurred too rapidly for observation, as illustrated by line E of Fig. 1. Measurements are always extrapolated to $t = 0$ and continued to at least 90–95% completion of the reaction. In most of our experiments the observed total change in y, $y(eq) - y(0)$, agrees with expected values based on the equilibrium data, as illustrated by all the lines of Fig. 1 except line E.

For each experiment, the logarithm of the left-hand side of equation 3 is plotted versus time, as shown in Fig. 2. If a straight line is obtained (curve A) it means that the system is governed by a single exponential decay term, i.e. equation 3 becomes

$$\frac{y(t) - y(eq)}{y(0) - y(eq)} = e^{-\lambda t} \tag{7}$$

If a straight line is not obtained, the data are fitted to equation 3 with the minimum number of exponential terms, and values of P_s and λ_s are determined. Values of P_s may all be positive (curve B of Fig. 2), or one or more of them may be negative (curve C of Fig. 2), but their sum is always unity, obeying equation 5. A negative value of P_s generally indicates an induction period in the kinetics, as shown by curve C. In a situation such as is illustrated by line E of Fig. 1, the logarithmic plot may be based on the true value of $y(0)$, which in this case would be y_D, as illustrated by curve D of Fig. 2. The right-hand side of equation 3 would then contain one exponential term for which λ_s would not be determinable.

If a reaction leading to the same equilibrium state is studied in both directions, as illustrated by lines B and D of Fig. 1, the theory requires that the same values of λ_s apply in each direction, for the λ_s values are independent of the initial state of the system. The P_s values, however, do depend on the initial state, because the a_{is} terms of equation 1 do so. We have used the symbol P_{sF} to represent the P_s values for kinetic studies in the forward direction, that is starting with the native state, and the symbol P_{sR} for the reverse direction, starting with the random coil. The theory predicts that values of P_{sF} and P_{sR}

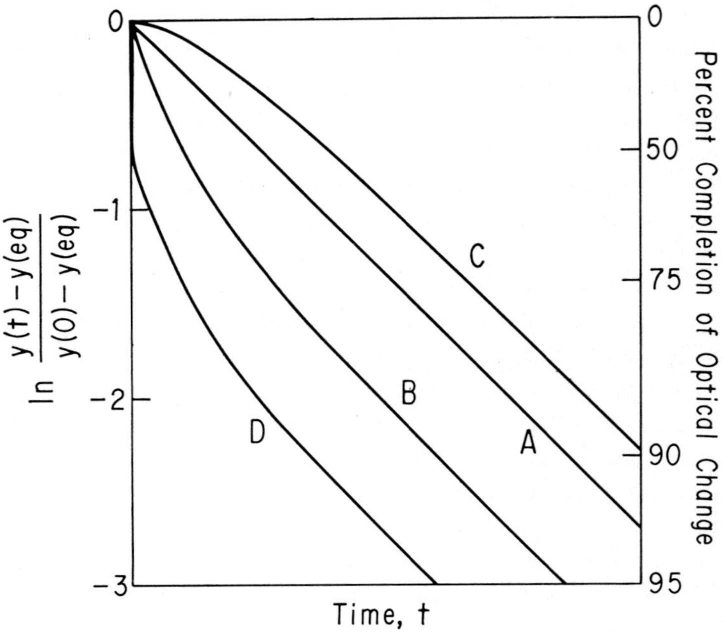

FIG. 2. Possible varieties of kinetic plots, used to calculate the parameters of equation 3. Curve A: single exponential term. Curve B: two exponential terms, $\lambda_1 = 5\lambda_2$, $P_1 = 0.5$, $P_2 = 0.5$. Curve C: two exponential terms, $\lambda_1 = 3\lambda_2$, $P_1 = -0.5$, $P_2 = 1.5$. Curve D: three exponential terms, λ_1 too fast to measure, $\lambda_2 = 5\lambda_3$, $P_1 = 0.5$, $P_2 = 0.25$, $P_3 = 0.25$.

for the same step can differ greatly, and our experimental studies have confirmed this: we nearly always obtain the same λ_s values in both directions (exceptions will be discussed), but the associated values of P_{sF} and P_{sR} generally differ.

Sometimes an exponential term on the right-hand side of equation 3 is required to fit the data in one direction, but not in the other, the λ_s values for other terms being the same. Since the principle of microscopic reversibility requires that the mechanism be the same in both directions, this is interpreted as indicating that either P_{sF} or P_{sR} is zero for one of the terms on the right-hand side of equation 3. It is easy to see qualitatively how this can occur. For example, a fast step followed by a slow step will be seen kinetically as two successive steps. In the reverse direction the slow step will be rate-determining for the entire reaction and the fast step will not contribute to the observed kinetics.

It should be noted that the number of terms required experimentally on the right-hand side of equation 3 determines the minimum number of species that must be involved in the mechanism of the reaction: for a mechanism involving n species, regardless of the details of the mechanism, $n-1$ different values of

λ_s are obtained. As stated in the preceding paragraph, one can only be sure of observing all necessary terms if a reaction leading to a given state is studied in both forward and reverse direction. Another possibility is that two distinct λ_s values will not differ sufficiently to be experimentally detectable as distinct. In many cases this problem can be recognized if data are obtained at a sufficient number of denaturant concentrations: since each λ_s will have a different dependence on denaturant concentration, a composite exponential term may be observed to split into two terms as the denaturant concentration is changed.

USE OF DIFFERENT PHYSICAL PARAMETERS

An important conclusion reached from the theoretical treatment is that the additional information to be obtained by using different physical parameters to follow the progress of the reaction is limited. The same values of λ_s apply, regardless of the physical probe that is used, for λ_s depends only on the rate constants k_{ij}. The coefficients P_s of the decay terms are functions of the parameters y_i or α_i as well as of the k_{ij}, and different values will in principle be obtained with each different physical probe employed. However, each reaction step makes some contribution to each exponential term: even if two intermediates A_i and A_j differ considerably in one physical property but not at all in another, it is not likely that this will lead to disappearance of one of the exponential decay terms (i.e. $P_s = 0$) if the second physical property is employed. An exception to this generalization would arise only if the interconversion of A_i and A_j (in both directions) is much faster than any other step in the reaction mechanism that involves either A_i or A_j. In that case one of the λ_s values would be nearly equal to $k_{ij} + k_{ji}$ for this step alone, and the corresponding P_s would be zero with the physical probe by which A_i and A_j are indistinguishable.

In the stopped-flow apparatus we use we are limited to use of light absorption as a physical probe. Different wavelengths have been employed, but we have never observed a significant difference in results, i.e. virtual disappearance of an entire exponential term at one wavelength, in any experiments done in the transition zone for the reaction. In one instance such a difference was observed in kinetic experiments under extreme conditions (such as line C of Fig. 1).

ANALYSIS: TWO-STATE TRANSITIONS

If the logarithmic plot of Fig. 2 is linear and equation 7 is obeyed, the kinetics is that of a two-state process:

$$N \underset{k_{21}}{\overset{k_{12}}{\rightleftharpoons}} D$$

If this result is obtained in a variety of conditions, in both directions with the same λ where data in both directions can be obtained, it means that no intermediate states accumulate to a sufficient extent to affect the kinetics, and therefore means that no information about the reaction mechanism can be extracted from the results. The single value of λ at any denaturant concentration is equal to $k_{12} + k_{21}$ and $K_{app} = K_{12} = k_{12}/k_{21}$, so that both rate constants and their dependence on external factors can be determined. From this one can deduce some of the properties of the overall activated state for the transition, but this is meagre information compared to what can be learned when the kinetics are complex.

The unfolding and refolding of lysozyme by guanidine hydrochloride has proved to be a two-state process over a wide range of pH values and denaturant concentrations (Tanford *et al.* 1966*a*; Aune & Tanford 1969). The same is true for the unfolding and refolding of ribonuclease by guanidine hydrochloride (Salahuddin & Tanford 1970), though experiments in this system were done over a narrower range of conditions. Both these proteins contain four disulphide bonds, and the consequent restriction on the randomly coiled state and on other states that may be formed from it is likely to be a factor in the inability to observe reaction intermediates.

ANALYSIS: COMPLEX KINETICS

If two terms are required on the right-hand side of equation 3, at least three states are involved in the mechanism. More than three states would lead to a two-term equation if some λ_s values happen to be nearly equal, so that they would be experimentally indistinguishable. More than three states would be indicated by the experimental data if the two values of λ_s for a forward and reverse experiment at the same denaturant concentration are not the same in the two directions: this would mean that equation 3 is actually a three-term equation, with one P_s value very small in one direction, and another very small in the other direction.

When three states are involved, two of them must be the native and fully denatured states N and D, and the third state must be a reaction intermediate X. The physical properties of this intermediate are not known, and α_x (equation 4) becomes one of the unknown parameters of the system. We have used the general theory described earlier to obtain relations between the experimental parameters and the unknown variables (microscopic rate constants and α_x) for all possible three-state mechanisms. We have also derived auxiliary equations that greatly facilitate the determination of the unknowns from the experimental data directly, rather than on a trial and error basis. The details will be presented elsewhere (Ikai & Tanford 1972). The most important result obtained from this theoretical analysis is that certain mechanisms are associated with obligatory relations between the experimental variables, which can be used as diagnostic tools. If such an obligatory relation is not obeyed the mechanism to which it refers can be unambiguously rejected. All such obligatory relations involve results for forward and reverse kinetics at the same concentration of denaturant: they cannot be applied where data are available in one direction only.

For example, if X is a true intermediate on the pathway between N and D, so that the kinetic mechanism would be $N \rightleftharpoons X \rightleftharpoons D$, an obligatory relationship is (Ikai & Tanford 1971),

$$K_{app} \frac{P_{1F}}{P_{1R}} < 0 \tag{8}$$

i.e. with K_{app} positive, either P_{1F} or P_{1R} must be negative. (We have used the convention of numbering λ_s values in order of magnitude, such that in the general case, $\lambda_1 > \lambda_2 > \lambda_3$ .... Equation 8 for a two-term equation thus applies to the coefficients of the exponential terms with the larger value of λ_s.) On the other hand, if X represents a conformation quite dissimilar from the native state, which would have to be unfolded again before one could go to the native state, the kinetic mechanism would be $N \rightleftharpoons D \rightleftharpoons X$, and for this mechanism

$$K_{app} \frac{P_{1F}}{P_{1R}} > 0 \tag{9}$$

An additional relationship for the mechanism $N \rightleftharpoons D \rightleftharpoons X$ is

$$\frac{P_{1F}}{P_{1R}} = \frac{\lambda_2 + P_{1F}(\lambda_1 - \lambda_2)}{\lambda_1 + K_{app} P_{2F}(\lambda_1 - \lambda_2)} \tag{10}$$

If three terms are required on the right-hand side of equation 3 ($\lambda_s = \lambda_1, \lambda_2, \lambda_3$, with $\lambda_1 > \lambda_2 > \lambda_3$), two intermediates, X_1 and X_2, are required, with corresponding unknown α values, which will be called α_1 and α_2. Complete solutions of the equations for possible mechanisms of this type have also been obtained (Ikai & Tanford 1972) and diagnostic equations can again be obtained. For example, for the mechanism $N \rightleftharpoons D \rightleftharpoons X_1 \rightleftharpoons X_2$,

$$\frac{\lambda_1 P_{1F}}{P_{1R} + K_{app}\, P_{1F}} = \frac{\lambda_2 P_{2F}}{P_{2R} + K_{app}\, P_{2F}} = \frac{\lambda_3 P_{3F}}{P_{3R} + K_{app}\, P_{3F}} \tag{11}$$

The method can be extended to mechanisms involving any number of intermediates, but the equations increase in complexity.

If the diagnostic test for a mechanism is obeyed, the values for unknown parameters can be determined. For any linear mechanism

$$A_1 \rightleftharpoons A_2 \rightleftharpoons \ldots \rightleftharpoons A_n$$

this can be done unambiguously, provided that kinetic experiments have been done in both directions under the same conditions, and explicit values for each P_{sF} and P_{sR} are available. For a mechanism involving n species there would then be $3n-4$ experimental parameters: $n-1$ values of λ_s, $n-2$ independent values of P_{sF} and $n-2$ independent values of P_{sR} and the value of K_{app} determined from equilibrium measurements. (One P_{sF} and one P_{sR} are not independent by virtue of equation 5.) There are also $3n-4$ unknowns: $2n-2$ microscopic rate constants and $n-2$ unknown α_i values.

For a non-linear mechanism the number of unknowns exceeds the number of experimental parameters, and this would also be true for a linear mechanism if explicit values for one or more of the λ_s, P_{sF} or P_{sR} cannot be obtained, as, for instance, if one of the λ_s values is too large to be determined. Even then, however, some of the k_{ij} can often be determined and severe restrictions on possible values of the others and on the values of α_i can be obtained.

These values for the characteristic parameters applicable to a particular mechanism provide additional criteria for interpretation and may often be used to reject a mechanism. For example, if ultraviolet light absorption of the aromatic residues is used to follow the progress of the reaction, the values of α_i can be physically interpreted as a measure of the degree of exposure of aromatic side chains to the solvent. This immediately sets strict limits on possible values for α_i, and allows rejection of any mechanism requiring unacceptable values, such as a value corresponding to $> 100\%$ exposure. Furthermore, it is generally true that reactions leading to greater exposure of aromatic residues are favoured by an increase in concentration of guanidine hydrochloride (Tanford 1970), so that one can make the plausible assumption that in a reaction step $A_i \rightleftharpoons A_j$ the value of k_{ij} may not decrease significantly with an increase in denaturant concentration if $\alpha_j > \alpha_i$. This leads to additional grounds for rejection of a mechanism on the basis of comparison between values of the unknown parameters required by the experimental data at different denaturant concentrations.

RESULTS FOR CYTOCHROME *C*

A detailed experimental study and analysis of the results has been carried out for cytochrome *c*. A preliminary report has been published (Ikai & Tanford 1971) and a full report is in preparation (Ikai *et al.* 1972). The transition from native to randomly coiled protein at 25 °C occurs between 2 and 3M-guanidine hydrochloride and kinetic measurements in both forward and reverse direction were made at 2.28, 2.50 and 2.61M denaturant. In each case biphasic kinetics were obtained, i.e. two exponential decay terms were required on the right-hand side of equation 3. At 2.28M, however, significantly different values were obtained for λ_1 in the forward and reverse directions, indicating that a three-species mechanism would not suffice to explain the results.

The great power of the mathematical analysis described above is illustrated by the fact that the requirement for a mechanism involving a minimum of four species can be deduced from the data at 2.50 and 2.61M denaturant alone, even though no overt indication of the need for more than three species is evident. The values for the experimental parameters are shown in Table 1. The sequential

TABLE 1

Experimental parameters for cytochrome *c*[a]

Concentration of GuHCl (molar)	λ_1 s^{-1}	λ_2 s^{-1}	P_{1F}	P_{2F}	P_{1R}	P_{2R}	K_{app}
2.28	0.9 (F) / 1.4 (R)	0.06	0.50	0.50	0.65	0.35	0.35
2.50	1.2	0.06	0.65	0.35	0.85	0.15	0.95
2.61	1.35	0.10	0.80	0.20	0.85	0.15	1.6

(*a*) Parameters of equation 3, with two exponential terms. At 2.28M denaturant different values of λ_1 were obtained in the forward and reverse directions. Experimental uncertainty for all parameters is about ± 10%: for the smaller P_{sR} values it is about ± 0.05–0.10.

mechanism $N \rightleftharpoons X \rightleftharpoons D$ can be immediately rejected by equation 8 because P_{1F} and P_{1R} are both positive. The mechanism $N \rightleftharpoons D \rightleftharpoons X$ comes closer to being valid, but equation 10 is not obeyed, and the disagreement greatly exceeds the experimental error. In fact, any three-species mechanism can be rejected on the basis of diagnostic relations for the most general such mechanism:

Even without the result at 2.28M one would thus have to invoke a four-species mechanism and explain the failure to observe three distinct λ_s values on the basis that two of them have nearly the same value.

When the analysis is extended to four-species mechanisms, it is found that most such mechanisms can again be rejected, and the only acceptable mechanism proves to be

$$N \underset{k_{21}}{\overset{k_{12}}{\rightleftharpoons}} X_1 \underset{k_{42}}{\overset{k_{24}}{\rightleftharpoons}} D \underset{k_{34}}{\overset{k_{43}}{\rightleftharpoons}} X_2$$

in which one intermediate (X_1), with a large value of α_1, is on the pathway between N and D, while the second intermediate (X_2), which must have a relatively small value for α_2, is a dead-end product, directly interconvertible with D, but not with N or X_1. Another requirement is that the rate constant k_{43} for the reaction $D \rightarrow X_2$ must be larger than the rate constant k_{42} for the reaction $D \rightarrow X_1$. The major contribution to the optical properties arises in this case from the haem group, and the α values indicate that the haem group in X_1 is in an environment much like that of the random coil, whereas in X_2 it is in a hydrophobic environment, i.e. X_2 is presumably a highly folded form of the polypeptide chain.

Unequivocal determination of the parameters of the mechanism is not possible because two of the three exponential decay terms are not experimentally separable, but trial-and-error procedures indicate that the permissible variations are not large. One set of parameters which satisfies the experimental data at all three denaturant concentrations, as well as the requirements given earlier for the dependence of the k_{ij} values on denaturant concentration, is given in Table 2. The experimental data calculated from these parameters agree closely with those observed experimentally, and, in particular, account for the difference between the observed λ_1 values in the two directions at 2.28M and the failure to detect such a difference at the higher denaturant concentrations.

The central portion of Table 2 shows the calculated parameters for the first two of the three terms on the right-hand side of equation 3 that are generated by this mechanism with the given kinetic parameters. It is seen that at 2.28M-guanidine hydrochloride, P_{1F} is essentially zero, so that only the second term will contribute to the result in the forward direction. In the reverse direction, on the other hand, P_{1R} is much larger than P_{2R}. The contribution from the second term, while not zero, could not be distinguished (primarily because of the relatively small difference between λ_1 and λ_2, as compared to the difference of both from λ_3) by the limited resolution of the experimental and curve-fitting procedures. Thus only a single term will again be observed experimentally, but with a λ value closer to λ_1 than to λ_2. At the other two concentrations of

TABLE 2

Kinetic parameters and calculated results

Concentration of GuHCl (molar)	*Rate constants (s^{-1})* k_{12}	k_{21}	k_{24}	k_{42}	k_{34}	k_{43}	α_1	α_2
2.28	0.22	0.78	0.08	0.12	0.35	1.00	0.75	0.30
2.50	0.50	0.65	0.12	0.06	0.40	0.80	0.75	0.30
2.61	0.73	0.50	0.18	0.05	0.60	0.70	0.75	0.30

Calculated parameters for first two terms of three-term equation

	λ_1 s^{-1}	λ_2 s^{-1}	P_{1F}	P_{2F}	P_{1R}	P_{2R}
2.28	1.46	1.05	−0.04	0.61	0.55	0.19
2.50	1.28	1.18	−0.22	0.89	0.27	0.63
2.61	1.36	1.28	−0.58	1.30	0.41	0.50

Calculated parameters for two-term equation

	$\lambda_{(1+2)}$	λ_3	$P_{(1+2)F}$	P_{3F}	$P_{(1+2)R}$	P_{3R}	K_{app}
2.28	1.05 (F) 1.35 (R)	0.045	0.57	0.43	0.74	0.26	0.39
2.50	1.2	0.065	0.67	0.33	0.90	0.10	0.91
2.61	1.3	0.12	0.72	0.28	0.91	0.09	1.45

denaturant the values of λ_1 and λ_2 are so close to each other that only a single term will be observed in either direction, with essentially the same value of λ, regardless of the values of P_1 and P_2. The kinetic parameters thus predict that only two terms will be required in equation 3 to represent the experimental data at all three denaturant concentrations. The first term will be a composite of the initially calculated terms for λ_1 and λ_2, and the second term will correspond to λ_3 of the initial calculation. The final calculated parameters (including K_{app}) are shown at the bottom of Table 2, and they are seen to be in excellent agreement with the experimental data shown in Table 1.

RESULTS FOR β-LACTOGLOBULIN

Native β-lactoglobulin is a protein with two identical polypeptide chains, but the interaction between them is very weak, and dissociation to monomers with essentially unaltered conformation occurs at a very low concentration of guanidine hydrochloride. The reversible unfolding to a random coil occurs at much higher concentrations of denaturant and is thus a unimolecular process to which the present treatment is applicable.

The dominant feature of the experimental results in the transition zone is that a large part of the optical change in the reverse direction occurs too rapidly to be measured (line E of Fig. 1). The forward reaction under the same conditions, on the other hand, is a simple first-order process, in which the entire expected optical change is seen. In effect this is the same situation as is observed for cytochrome *c* at 2.28M-guanidine hydrochloride. The coefficient P_{1F} of the first term on the right-hand side of equation 3 is clearly negligibly small, whereas P_{1R} is large. For β-lactoglobulin, however, λ_1 is very much larger than all other λ_s values, in fact too large to be measured even by the stopped-flow procedure. The mechanism required to fit the data is correspondingly similar to that deduced for cytochrome *c*. A four-species mechanism is required with two intermediates X_1 and X_2. However, because the forward reaction is a simple first-order process, both intermediates must be placed on a dead-end pathway. Analysis is in the process of being completed (Takagi *et al.* 1972). It has not yet been determined which of the two simplest mechanisms

(i) $$N \rightleftharpoons D \rightleftharpoons X_1 \rightleftharpoons X_2$$

(ii) $$N \rightleftharpoons D \begin{matrix} \nearrow\!\!\swarrow X_1 \\ \searrow\!\!\nwarrow X_2 \end{matrix}$$

provides the best agreement with experiment. In either case, to account for the very large value of λ_1, the rate of the reaction $D \rightarrow X_1$ has to be much faster than the rate of the reaction $D \rightarrow N$.

DISCUSSION

The most important aspect of these results is the necessary inclusion, both for cytochrome *c* and β-lactoglobulin, of one or more states that lie on a dead-end pathway and not on the direct pathway between N and D. Such states must be physically interpreted as incorrectly folded forms of the polypeptide chain, in which residues are brought close together that in the native structure need to be far apart. From such states direct progress to the native state would be impossible: it can occur only after disruption of the initially formed structure, i.e. by return to the randomly coiled state or something closely resembling it. In both cytochrome *c* and β-lactoglobulin the rates of formation of these incorrectly folded states from the disordered polypeptide chain have been shown to be faster than the rates of formation of the native state, or of intermediates leading directly to the native state.

The existence of this dominant feature of the mechanism was reached on the basis of a formal analysis of the kinetic results, but the necessity for it is qualitatively evident from the results alone. Thus for cytochrome *c*, the *major* part of the optical change, in both forward and reverse directions under otherwise identical conditions, is associated with the fastest and thus *earliest* part of the reaction. In other words, starting from the native state, a product optically similar to the denatured state is formed rapidly, while, starting from the denatured state, a product optically similar to the native state is formed rapidly. At the same time the overall equilibrium constant must be about unity. It is difficult to imagine any mechanism that could account for these circumstances other than a mechanism with one of the initial products off the direct pathway from N to D. The major features of the reaction are thus necessarily described by mechanistic schemes of one of two alternative types:

$$\mathrm{N} \overset{\text{fast}}{\rightleftharpoons} \mathrm{X}_1 \text{ (optically like D)} \overset{\text{slow}}{\rightleftharpoons} \mathrm{D} \overset{\text{fast}}{\rightleftharpoons} \mathrm{X}_2 \text{ (optically like N)}$$

$$\mathrm{X}_1 \text{ (optically like D)} \overset{\text{fast}}{\rightleftharpoons} \mathrm{N} \overset{\text{slow}}{\rightleftharpoons} \mathrm{X}_2 \text{ (optically like N)} \overset{\text{fast}}{\rightleftharpoons} \mathrm{D}$$

The first of these corresponds to the mechanism we have suggested. The second can be excluded on physical grounds. While it is easy to visualize a folded form of the polypeptide chain (X_2) that cannot be directly converted to the native state, it seems highly improbable that there can be a largely unfolded form (X_1) that cannot be directly converted to D.

It is reasonable to suggest the hypothesis that a mechanism of this type, involving an incorrectly folded form of the polypeptide chain as initial product, applies quite generally to the formation of the native structure of a protein from a disordered polypeptide chain, both *in vivo* and *in vitro*. It is likely that any disordered polypeptide chain can make a multitude of intrachain contacts leading to a large number of possible initial products of the folding process. The great majority of them are likely to be unproductive in that they bring together amino acid residues that are far apart in the native structure: to attain the native state from any such product the initial step would first have to be reversed. Only a small fraction of the possible initial steps is likely to be so structured that direct progress to the native structure is possible. The relative rates of formation of the various possible initial products will be influenced by factors, such as the proximity of portions of the chain that are brought together, that bear no relation to the thermodynamic factors that determine the most stable ultimate product of the folding process. There is therefore no *a priori* reason to believe that the most rapidly formed initial product will be a product on the direct pathway to the native state. Such coincidence would be expected only fortuitously.

A special aspect of the *in vivo* process is that the amino-terminal portion of the polypeptide chain is available for initial folding before the synthesis of the entire chain is complete. This could increase or decrease the probability of incorrectly folded intermediates being formed, depending on the kinds of intermediates that can be formed from the incomplete chain and their relative rates of formation. There is no compelling reason to believe that either alternative will in general be favoured.

It should be pointed out in this connection that the guanidine-denatured forms of cytochrome *c* and β-lactoglobulin do not correspond exactly to structureless polypeptide chains as they exist *in vivo* after biosynthesis. In cytochrome *c* the covalently-linked haem group remains attached to the denatured polypeptide chain, forming a link between residues 14 and 17 (Dickerson *et al.* 1971), and a possible directing influence on the folding process. In guanidine-denatured β-lactoglobulin there are two disulphide bonds, which, however, link only short stretches of the polypeptide chain, so that their directing influence is likely to be small. In both cases the constraints imposed by the additional structural elements *favour* formation of an initial product leading directly to the native state, and the observation that an incorrectly folded state in fact appears to be an initial product strengthens the hypothesis that formation of an incorrectly folded state will occur when no constraints are present.

It is significant that the two proteins that have been shown to have simple first-order kinetics in denaturation and renaturation by guanidine hydrochloride—lysozyme and ribonuclease—each have four disulphide bonds, linking distal parts of the polypeptide chains. These disulphide bonds are intact in the denatured product and provide severe constraints on the possible initial folding steps that the denatured protein can undergo. The probability of formation of an incorrectly folded state is thus greatly diminished. Evidence for the transient formation of incorrectly folded forms of lysozyme in the renaturation of the denatured protein when disulphide bonds have been reduced has been presented by Yutani *et al.* (1968) and by Saxena & Wetlaufer (1970).

ACKNOWLEDGEMENTS

The original theoretical and experimental studies on which this paper is based were carried out in collaboration with Dr Atsushi Ikai, and his major contribution to the work is hereby acknowledged. Research support was received from the National Science Foundation and the National Institutes of Health, United States Public Health Service. The author is supported by a Research Career Award from the latter institution.

References

AUNE, K. C. & TANFORD, C. (1969) Thermodynamics of the denaturation of lysozyme by guanidine hydrochloride. *Biochemistry* **8**, 4579

BENSON, S. W. (1960) in *The Foundations of Chemical Kinetics*, McGraw-Hill, New York

DICKERSON, R. E., TAKANO, T., EISENBERG, D., KALLAI, O., SAMSON, L., COOPER, A. & MARGOLIASH, E. (1971) Ferricytochrome *c*. *J. Biol. Chem.* **246**, 1511

FROST, A. A. & PEARSON, R. G. (1953) in *Kinetics and Mechanism*, Wiley, New York

IKAI, A. & TANFORD, C. (1971) Kinetic evidence for incorrectly folded intermediate states in the refolding of denatured proteins. *Nature (Lond.)* **230**, 100

IKAI, A. C. & TANFORD, C. (1972) Kinetics of protein denaturation and renaturation. I. Mathematical analysis. *J. Mol. Biol.* in press

IKAI, A., FISH, W. W. & TANFORD, C. (1972) Kinetics of protein denaturation and renaturation. II. Results for cytochrome *c*. *J. Mol. Biol.* in press

SALAHUDDIN, A. & TANFORD, C. (1970) Thermodynamics of the denaturation of ribonuclease by guanidine hydrochloride. *Biochemistry* **9**, 1342

SAXENA, V. P. & WETLAUFER, D. B. (1970) Formation of three-dimensional structure in proteins. *Biochemistry* **9**, 5015

TAKAGI, T., IKAI, A. & TANFORD, C. (1972) In preparation.

TANFORD, C. (1968) Protein denaturation. Parts A and B. *Adv. Protein Chem.* **23**, 121

TANFORD, C. (1970) Protein denaturation. Part C. *Adv. Protein Chem.* **24**, 1

TANFORD, C., PAIN, R. H. & OTCHIN, N. W. (1966*a*) Equilibrium and kinetics of the unfolding of lysozyme by guanidine hydrochloride. *J. Mol. Biol.* **15**, 489

TANFORD, C., KAWAHARA, K. & LAPANJE, S. (1966*b*) Proteins in 6M-guanidine hydrochloride. *J. Biol. Chem.* **241**, 1921

YUTANI, K., YUTANI, A., IMANISHI, A. & ISEMURA, T. (1968) The mechanism of refolding of the reduced random coil form of lysozyme. *J. Biochem. (Tokyo)* **64**, 449

Discussion

Eigen: The analysis of relaxation spectra using amplitudes and time constants may not be as simple and straightforward as you have shown, Dr Tanford. First, the number of states one counts depends entirely on the time resolution.

If there are further fast processes which you could not resolve it looks as if there was only one final state, which actually represents a whole sum of states. Secondly, even if you can resolve all the steps, you might still have degeneracies in the system. For instance, a three-state system involves a quadratic equation but the roots may be so small that one sees only one of the two solutions. With a multiple-step system you might get many degeneracies, which means there are still many different states, but the superimposition of time constants is such that you can't resolve them. You need a very sharp resolution in order to distinguish two exponential functions with very similar time constants. Therefore, one has to be careful in drawing conclusions from such a kinetic analysis; it may still be possible that other techniques exist by which one may identify further states.

Tanford: I agree with you about degeneracy, but there cannot be any unresolvable fast steps with significant amplitude because the total change in optical density from the initial to the final state is known from equilibrium measurements, and we can normally account for all of it in terms of the resolvable decay terms. The only exception would be in the situation illustrated by line E of Fig. 1 (p. 128).

Eigen: An example was provided by chymotrypsin, where an increased time resolution brought about more fine structure of the relaxation spectrum and changed the interpretation of previous results (R. L. Baldwin, unpublished results). However, I agree with what you said about the examples you studied.

Engel: The mechanism you proposed for cytochrome *c* was also suggested for the renaturation of collagen (Beier & Engel 1966) on the basis of a direct display of the products formed. In this instance it was found that the relative amounts of side products, X_2, decreased when the transition temperature was approached. At temperatures which are only a few degrees below denaturation temperature essentially only native molecules are re-formed, whereas at very low temperatures only side products appear. This is explained by the finding that side products, in which interactions are saturated in a fairly unspecific way, are formed much faster than collagen molecules, but these products are also less stable. The question I would like to raise in this context is whether you are observing different relaxation times and amplitudes when you perform your kinetic experiments at different temperatures or different concentrations of denaturants. Another problem which remains unsolved, at least for collagen, is that renaturation *in vitro* is much slower than collagen synthesis under all experimental conditions.

Tanford: The concentration of guanidine hydrochloride is the only external variable in our studies. At each concentration one gets an entirely different set of λ and P parameters because, in a highly cooperative transition, the rate constants and the equilibrium constants must depend steeply on denaturant concentration.

Eigen: Are the λ values all the same?

Tanford: The λ values at any given concentration of denaturant are the same, regardless of what property of the system one is measuring and regardless of whether one is going in the forward or reverse direction. The λ values have to differ as the denaturant concentration is changed because the denaturant concentration is the parameter that shifts the equilibrium from the native to the denatured state. This parameter plays the same role that temperature would play in a thermally induced transition.

Levitt: Can you give a figure for the degree of order in the states X_1 and X_2?

Tanford: In the cytochrome *c* transition, $\alpha_1 \simeq 0.75$ and $\alpha_2 \simeq 0.30$. This means that in terms of optical density X_1 lies about 75% from the native state towards the random coil, and X_2 only about 30%.

Changeux: I am surprised that you think the concurrent use of different physicochemical techniques might not be of help to study this phenomenon. The problem of interest is the identification of the specific pathways of unfolding. Some amino acid residues might be involved in one pathway and others in another pathway. Are these pathways used during the synthesis of protein *in vivo*, or do you think they are completely irrelevant to what is actually happening during the biological unfolding of the proteins?

Tanford: I think that the principle applies to the biological situation. The experimental data will have to be a great deal better before we can tell whether the results apply in detail. In general, we have found that as a system gets closer to water, new intermediate states appear, so that in fact a four-state mechanism on our time scale will no longer satisfy the data for cytochrome *c* when we get down to about 1 M-guanidine hydrochloride. Since the initial protein solution in renaturation experiments contains 6 M-guanidine hydrochloride, the stopped-flow apparatus does not permit study of the reaction at guanidine hydrochloride concentrations much below 1 M, i.e. the actual physiological conditions are not accessible to us. It seems questionable to me whether it is really worth knowing the exact pathway of folding *in vivo* if one arrives at a unique equilibrium regardless of the pathway.

Changeux: I think it matters.

Phillips: Zabin's recent work on β-galactosidase has provided some experimental evidence that this enzyme is folded on the ribosome as it is being made (Hamlin & Zabin 1972). The fact that some proteins can be unfolded and refolded *in vitro* under extremely non-biological conditions does not establish that they, or proteins generally, fold in an analogous way in the cell to a minimum energy conformation.

Tanford: I don't know about β-galactosidase but most enzymes that I know about can be fully renatured *in vitro* after denaturation by guanidine hydrochloride or by other denaturants, such as dodecyl sulphate, which form structured rather than randomly coiled denatured states. Renaturation from guanidine hydrochloride often requires only a few seconds.

Phillips: Refolding in seconds isn't the sort of time constant I have heard discussed before, except for molecules like lysozyme and ribonuclease which are held together with disulphide bridges.

Tanford: With both β-lactoglobulin and cytochrome *c* it occurs in seconds.

Revel: It might be very relevant to know what the correct folding is for a virgin protein which is just being synthesized. My main concern is the time it takes. Even in a test tube, when one makes an enzyme like lysozyme from phage T4, starting from the messenger, it should take about ten seconds to make a whole polypeptide chain with 160 amino acids. The important question is how long it will take for the polypeptide chain to fold and for us to see activity in the test tube. Those studying protein synthesis are not taking this into account. If I understood Dr Engel correctly, it takes collagen about ten times longer to fold than to be synthesized.

Tanford: In the formation of native lysozyme, both refolding and disulphide bond formation are necessary. Our results can be considered as representative of the total time required for attainment of the native structure only for those proteins that do not contain disulphide bonds.

Revel: How long would it take for a newly synthesized protein to fold?

Tanford: Sometimes seconds, sometimes minutes.

Eigen: In nature there are many proteins and they are all different. Some change their conformations within milliseconds, others take minutes or even longer. In experiments with nucleic acids where conformational transitions were found between AU–, AA– and triple helices, even real hysteresis phenomena were observed (Katchalsky & Neumann 1972). There are also proteins where one can see such hysteresis phenomena. For those proteins which refold in seconds it doesn't matter how they reach their equilibrium conformation. One can put them into any configuration and in a reasonable time they return to the equilibrium position. But there may be cases of slow change or hysteresis where the sequence of events is of importance.

Katchalski: As far as I am aware, even long linear fragments of the polypeptide chain of a globular protein often do not attain in aqueous media the conformation which they assume in the intact protein. However, if the folding of the growing polypeptide chain occurs while it is still attached to the ribosome, it is possible that the folding at the ribosome surface proceeds by a different route than in an aqueous medium.

Phillips: I think one could extend this sort of pattern to folding on the ribosome. At an intermediate stage, a variety of conformations would be present, one of which would be selected out as the length of completed chain increased.

Silman: It is possible that the correct folding occurs in the procollagen molecule and then when you get it you would chop off the unnecessary parts enzymically, *in vivo*. This would eliminate the problem of having incorrect folding in the collagen molecule itself.

Engel: Several people have speculated that the additional cysteine-containing peptide piece which is present in procollagen helps in the folding processes. We have done renaturation experiments with procollagen which was prepared in Prockop's laboratory (Jimenez *et al.* 1971). There was no change in renaturation rate when the material was treated with 2-mercaptoethanol in order to split possible S–S bridges. Furthermore, renaturation was not enhanced compared with normal soluble collagen.

Flory: Since this is supposed to be a symposium relating to both the esoteric biopolymers and the more mundane synthetic polymers and structural polymers of any source, natural or synthetic, I can't resist mentioning that there seem to be many parallels here with the process of polymer crystallization, and perhaps also with polymer melting. Protein denaturation was considered as a chemical reaction years ago, then as a phase transition and now, properly, as a combination of the two. If a polymer is supercooled or supersaturated to a large degree, highly imperfect crystals tend to be formed. It is exceedingly difficult to convert these imperfect, semicrystalline forms directly into a more stable, better organized, crystalline state. It is essential to melt them first, and then to re-form them (slowly) into a more stable state of better order, under carefully controlled conditions.

You mentioned the importance of cyclic connections, or cross-linkages, in assisting the molecule to find its way back to the native state, Dr Tanford. This has its counterpart in the cross-linking of highly oriented fibres of polyethylene or of collagen, where the melting and shrinkage is spontaneously reversible to an extent that depends on the perfection of the original structure. Here the recrystallization and the re-formation of the 'native' structure is favoured by the insertion of cross-linkages in the ordered state.

Baddiley: In the biosynthesis of a lipid-soluble protein would it be necessary for a virgin polypeptide chain to go immediately into a lipid environment to acquire the correct conformation, or could this be achieved via an aqueous phase, where presumably other conformations would come in?

Tanford: As far as I know it could happen either way.

References

BEIER, G. & ENGEL, J. (1966) The renaturation of soluble collagen products formed at different temperatures. *Biochemistry* **5**, 2744

HAMLIN, J. & ZABIN, I. (1972) β-Galactosidase: immunological activity of ribosome-bound, growing polypeptide chains. *Proc. Natl Acad. Sci. U.S.A.* **69**, 412–416

JIMENEZ, S. A., DEHM, P. & PROCKOP, D. J. (1971) Further evidence for a transport form of collagen. Its extrusion and extracellular conversion to tropocollagen in embryonic tendon. *FEBS (Fed. Eur. Biochem. Soc.) Lett.* **17**, 245

KATCHALSKY, A. & NEUMANN., I (1972) *Int. J. Neurosci.* **3**, 175–182

Folding of nucleic acids

MICHAEL LEVITT

MRC Laboratory of Molecular Biology, Cambridge

Abstract This paper describes and discusses nucleic acid conformations, the energy contributions that stabilize them, and the ways the stable conformations are formed. Preferred conformations of nucleotides and of double, triple and super helices in nucleic acids are described briefly. Next the enthalpy and entropy of nucleic acid order/disorder transitions are reviewed. It is concluded that (*a*) the double helix is stabilized primarily by interbase hydrogen bonds, and (*b*) the available thermodynamic parameters for single-strand stacking are probably incorrect. The conformation of single strands in solution is discussed briefly. The kinetics of helix formation is described; the transition state of three base pairs that has been proposed supports the idea that the strength of hydrogen bonds depends on the access of water molecules to the base pair. Hairpin formation and more complex secondary structures of nucleic acids are described and explained in terms of rate of formation, stability, and control of secondary structure. Transfer RNA has tertiary structure as well as secondary structure. Schematic diagrams are used to show how experimental evidence fixes different parts of the tRNA tertiary structure. The detailed interactions in the proposed model are supported by the sequence of cell-wall glycine tRNA, which can be charged with an amino acid without participating in protein synthesis. Sequential steps in the folding and unfolding of the propcsed tRNA conformation are outlined. Finally, the abilities of nucleic acids and proteins to form stable tertiary structures are compared and the evolutionary consequences are considered.

Nucleic acids and proteins have been studied intensively over the past 25 years. Although similar techniques have been used on each of these important classes of macromolecule, available information for each is very different. X-ray diffraction methods have elucidated conformations of both nucleic acids and proteins but, whereas the known conformations of different proteins are very different, those of nucleic acids are all simple variants of the Watson-Crick double helix. It is easier, however, to study nucleic acid conformations in solution, for they consist of only four repeating units while proteins consist of 20 repeating units. Available information from proteins and nucleic acids could be complementary: protein conformations indicate the construction principles of stable three-dimensional structures; nucleic acid folding processes indicate ways in which stable structures may be attained in solution.

Nucleic acids are built up from three structural components: a base, a sugar ring and a phosphodiester linkage. The regular backbone of the molecule is formed from sugar rings connected by phosphodiester linkages; the side-chains attached to each sugar ring consist of the different bases. Five types of bases

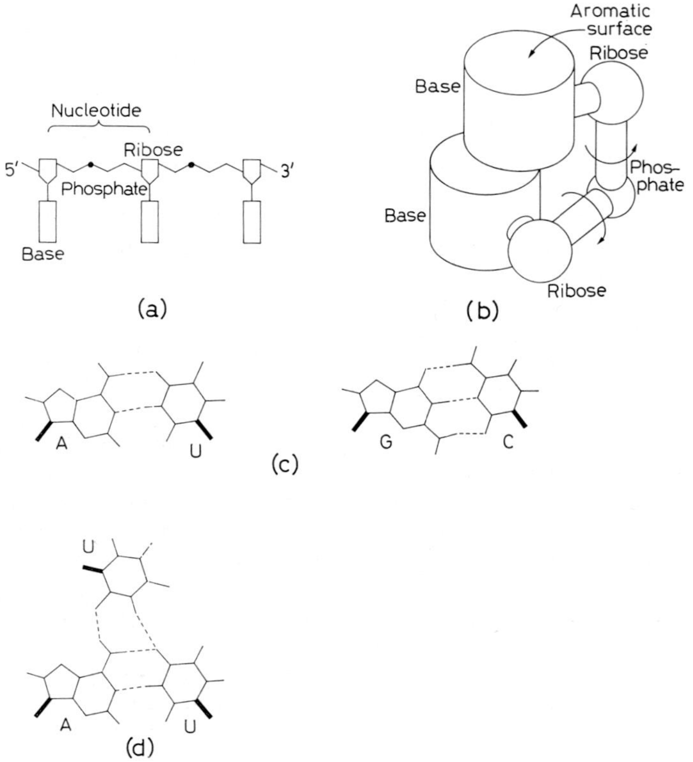

FIG. 1. (*a*) A schematic diagram of nucleic acid primary structure. The nucleic acid chain has a definite polarity as the two ends are chemically distinct. (*b*) A schematic model of a dinucleotide showing the flat aromatic surfaces of the bases, the rigidity of the backbone and the main degrees of freedom. Nucleotide conformations are very restricted (see Sundaralingam 1969): there are only two sugar-ring conformations, three phosphodiester conformations, and one or two base orientations, depending on the type of base. Hence, there are only six possible conformations for a pyrimidine nucleotide and twelve possible conformations for a purine nucleotide. One-third of these give a right-handed helix like that found in double-helical DNA or RNA. (*c*) The Watson-Crick complementary base pairs A·U and G·C. (*d*) A possible base triplet that could be incorporated into the poly(A·2U) triple helix.

occur most commonly: uracil (U), thymine (T), cytidine (C), adenine (A), and guanine (G). The first three, U, T, and C (pyrimidines), are formed from a six-membered aromatic ring; the other two, A and G (purines), are formed from a five-membered ring fused to a six-membered ring (see Fig. 1*a*). Deoxyribonucleic acid (DNA) differs from ribonucleic acid (RNA) in only two respects: (*a*) the base T occurs in DNA in place of U in RNA; (*b*) the sugar ring of DNA lacks a hydroxyl group present in RNA.

X-ray crystallography can be used to study the conformation of the nucleic acid repeating unit: the crystal structures of more than 13 nucleotides, 25 nucleosides, and eight phosphodiesters have been determined. An analysis of these data (Sundaralingam 1969) indicates that the nucleotide has very few allowed conformations. The main degrees of freedom of a nucleotide seem to be rotation about the glycosyl bond and the two O—P bonds in the phosphodiester linkage (see Fig. 1*b*), but the ranges of possible values for these torsion angles are very restricted.

Aqueous solutions of bases or nucleotides are far from ideal for the bases tend to associate by stacking (cf. Ts'o 1968). What causes this association? Non-polar aromatic groups in proteins are thought to interact in water through hydrophobic forces (Kauzmann 1959). This interaction is explained as follows. Water molecules in contact with aromatic surfaces are more ordered than other water molecules. When the aromatic groups come together, less surface is exposed and consequently less water is ordered. For aromatic side-chains the driving force for stacking comes from the increased disorder of water molecules in the stacked state, i.e. from a change in entropy rather than enthalpy. By analogy the entropy of the surrounding water molecules could be responsible for base stacking in nucleic acids. Recently Bugg *et al.* (1971) have studied the stacking of bases in crystals of nucleotides, nucleosides and bases. They conclude that as bases usually stack but do not overlap fully, dipole/induced–dipole forces cause stacking in crystals. The dipole moment of the polar substituent of one base induces a dipole moment in the π electron cloud of the neighbouring base, the bases are brought together, and the enthalpy of the system drops. As the distance between fully overlapped bases is always greater than that between partially overlapped bases, this apparent attraction could be an artifact of the close-packing of the crystals.

Besides stacking, bases can also associate through hydrogen bonds. In fact, two of the four commonly occurring bases A, U, C, and G can form a base pair connected by two hydrogen bonds in 29 ways (Donohue & Trueblood 1960). Only G and C, however, can form a base pair connected by three hydrogen bonds. The bases A and U can form a base pair connected by two hydrogen bonds with the same shape and dimensions as the strong G·C base-pair. These G·C and A·U base pairs, which were used by Watson & Crick (1953) when they defined a molecular model for DNA, are known as complementary base pairs.

Two long chains of DNA with complementary sequences can be connected through complementary base pairs. The well-known conformation of this double-stranded helical complex was determined almost 20 years ago by a combination of model-building and X-ray crystallography (Watson & Crick 1953; Wilkins *et al.* 1953; Franklin & Gosling 1953). Since then more detailed crystallographic methods have revealed a second conformation for DNA as well as a conformation for RNA (see Arnott *et al.* 1969). The different conformations fall into two classes (see Fig. 2): (*a*) the base pairs are perpendicular

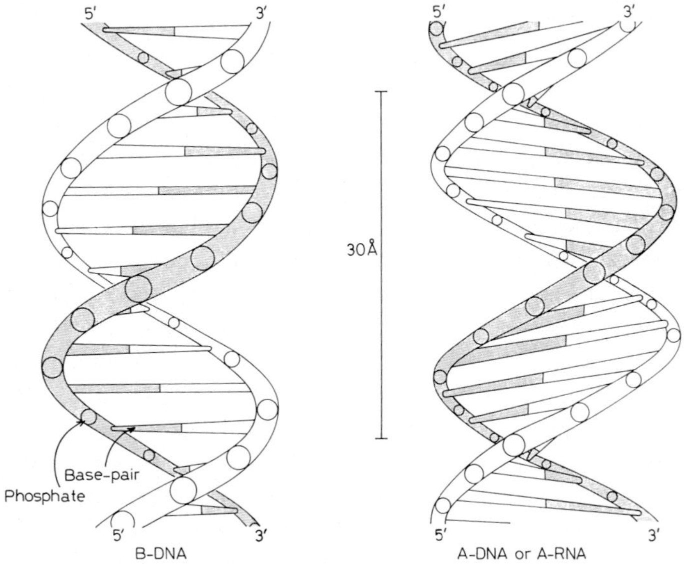

FIG. 2. Schematic drawing of the two forms of the nucleic acid double helix formed between complementary strands. Note that the two complementary strands are antiparallel, for the 5′ terminal of one strand is next to the 3′ terminal of the other strand. A-DNA is more stable than B-DNA only when there are no free cations in solution. RNA is always found in the A form. The two helical forms differ in the tilt of the base pairs, the degree of base pair overlap, the backbone conformation and the minimum interstrand phosphate separation.

to the axis of the helix (B form), and (*b*) the base pairs are tilted at about 20° from the perpendicular position (A form). Whereas DNA has the A form in crystalline fibres of low ionic strength and the B form in fibres of high ionic strength, all RNA double-helical conformations are very similar to the A-form DNA. The stereochemistry of the nucleic acid double helix is surprisingly good:

the conformation of the nucleotide backbone is very similar to the preferred conformation found in crystals of isolated nucleotides, the bases stack but do not overlap fully, the hydrogen bonds between the bases are inaccessible to water, and the backbone phosphates are as far apart as they can be.

Synthetic RNA and DNA polymers with suitable sequences also form triple helices. Although no triple-helical conformation has been determined by X-ray methods, a third strand can be model-built onto the nucleic acid double helix. The third strand is accommodated in the deep groove of the double helix, making hydrogen bonds with the sides of the existing base pairs. Phosphate groups of the three strands have to come very close together in this structure.

More complex helical conformations have been observed in circular double-stranded viral DNA, in which the DNA helix winds onto itself, forming a super-helix or a coiled coil (see Vinograd *et al.* 1968). Super-helix formation can be understood as follows. Circular DNA consists of two intertwined closed circles of single-stranded DNA. The number of times one strand winds round the other strand is fixed when the circle is closed; any changes in this number produce a twisting force that causes super-helix formation. The number of turns of super-helix equals the number of extra turns of one strand of DNA about the other; the sense of the super-helix is the same as that of the extra turns in the DNA helix. For example, if A-DNA is closed into a circle, and the pitch of the double helix then changes to give B-DNA, several turns of right-handed super-helix will form. (A fixed number of nucleotides can form fewer turns of A-DNA than B-DNA, and the DNA double helix is right-handed.)

Information about nucleic acid conformations from X-ray crystallography can be summarized as follows: (*a*) the nucleotide backbone is stiff and has very few allowed conformations, and (*b*) complementary chains of nucleic acid form rigid strongly stabilized double-helical structures.

What forces stabilize nucleic acid conformations? It is accepted that the Gibbs free energy has a minimum value when any molecule has an equilibrium conformation. The free energy change, ΔG, that accompanies a change in molecular conformation has two components: $\Delta G = \Delta H - T\Delta S$, where ΔH, the enthalpy change, indicates how well the atoms interact in the two conformations, and ΔS, the entropy change, indicates the relative degree of order of the two conformations. When a biological macromolecule folds into a precisely ordered conformation the entropy decreases, causing a corresponding increase in the free energy. This increase in free energy must be offset by an even larger decrease in the enthalpy. In other words, to compensate for the increased order, the atoms of the ordered conformation must interact more favourably than in the disordered conformation.

We can learn about these interaction energies by measuring changes in enthalpy when the conformation of the molecule changes. The direct approach is to heat a solution of ordered nucleic acid molecules and measure the extra heat absorbed when the conformation changes. Such calorimetric experiments are not easy, for the energy absorbed when the nucleic acid structure breaks down is small compared with the heat absorbed by the solution. Nevertheless, enthalpies of several nucleic acid transitions have been determined by this method. A less direct method of measuring enthalpy changes is to follow the change in order as the temperature is raised (see Felsenfeld & Miles 1967). As the characteristic ultraviolet absorption spectra of bases change when the bases stack or form hydrogen bonds, it is possible to monitor the amount of ordered structure in the solution. If one knows the spectrum of a base in the fully ordered state and also in the fully disordered state, one can calculate the fraction of bases represented by any intermediate spectrum. Fig. 3 shows a

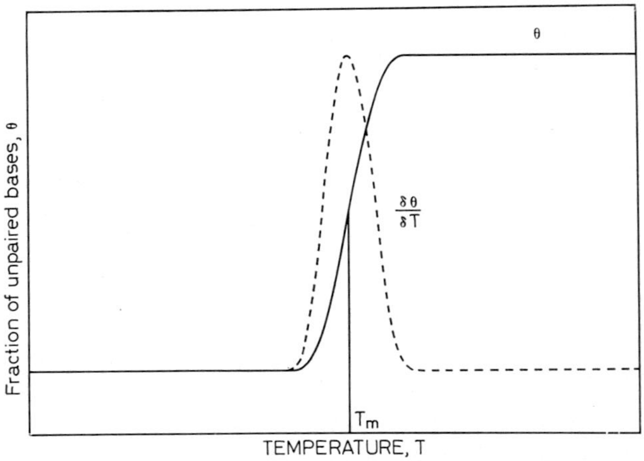

FIG. 3. A schematic melting curve of a nucleic acid double helix. θ, the fraction of unpaired bases, is proportional to the ultraviolet absorption of the solution. The differentiated melting curve, $\delta\theta/\delta T$, gives a clearer indication of the transition, particularly when two or more transitions overlap. T_m, the melting temperature, is defined as the temperature where $\theta = \frac{1}{2}$, i.e. paired and unpaired bases occur equally.

schematic melting curve of a nucleic acid double helix. At low temperatures all base pairs are formed, while at high temperatures no base pairs are formed. At any intermediate temperature the fraction of unpaired bases depends on (*a*) the free energy of base-pair formation, ΔG, and (*b*) the nature of the partially paired intermediates present in the transition from fully paired to fully unpaired structures. At the melting temperature, T_m, paired and unpaired bases occur equally. T_m depends only on the free energy of base-pair formation and not on the intermediates, for at T_m, $\Delta G = 0$ and $T_m = \Delta H/\Delta S$. Furthermore, if one assumes a theoretical model for the partially paired intermediates, it is possible to calculate ΔH and ΔS from the melting curve. It is usual to assume that no intermediates are present, for then ΔH and ΔS can be calculated most easily.

When the DNA double helix was first discovered, hydrogen bonds between bases on complementary strands were supposed to stabilize the structure against the electrostatic repulsion between the negatively charged phosphates. Since then opinion has shifted considerably. Many people now believe that stacking of base pairs provides the major stabilizing energy in the double helix. Consideration of the enthalpies and entropies of nucleic acid denaturation may help to decide this question. Table 1 gives the transition enthalpies and entropies per ordered structural unit, i.e. a nucleotide in single strands, a base pair in double strands and a base triplet in triple strands. Most of the enthalpies were measured calorimetrically, and the entropies have been calculated from the melting temperature using the formula given above. Before considering the consequences of these results, I must mention that cation concentration affects both ΔH and ΔS. Both these quantities increase with higher ionic strength; the

TABLE 1

Changes of enthalpy and entropy in order/disorder transitions of nucleic acids

Transition	ΔH *(kcal/mol)*	ΔS *(eu)*	*Reference*
RNA single strands			
ApA unstacking	6 to 13*	20 to 40	Cf. Davis & Tinoco (1968)
Poly(A) unstacking	5 to 10*	20 to 40	Cf. Felsenfeld & Miles (1967)
RNA double strands			
Poly(A·U) → Poly(A) + Poly(U)	8.0	24	Krakauer & Sturtevant (1968)
Poly(G·C) → Poly(G) + Poly(C)	10.2	24	Estimate from T_m and $\Delta S_{A.U}$
DNA double strands			
Poly(A·T) → Poly(A) + Poly(T)	7.2	24	Klump & Ackermann (1971)
Poly(G·C) → Poly(G) + Poly(C)	9.0	24	
RNA triple strands			
Poly(A·2U) → Poly(A) + 2Poly(U)	12.7	37	Krakauer & Sturtevant (1968)
Poly(A·2U) → Poly(A·U) + Poly(U)	4.3	13	
Poly(A·2U) → A + 2Poly(U)†	12.8	42	Scruggs & Ross (1970)
Poly(AA·2U) → AA + 2Poly(U)†	15.8	48	

* Not measured calorimetrically, but taken from a two-state fit to the melting curve.
† A and AA in these complexes are monomers of adenosine and 2-aminoadenosine respectively.
ApA: Adenosine dinucleoside phosphate.

changes are quite appreciable (for poly(A·U) up to 10% change in ΔH for a tenfold change in cation concentration), and ΔH is affected more than ΔS. Furthermore different cations have different effects; for example ΔH of a given poly(A·U) transition is less when potassium cations are present in place of sodium cations (Krakauer & Sturtevant 1968). Although we do not really understand these effects several theories have been proposed: (*a*) certain workers have considered the shielding of the repulsion between phosphates to be due to cations in the solution (Kotin 1963; Schildkraut & Lifson 1965); (*b*) others have considered the effect of preferential binding of ions to the ordered conformation (Krakauer & Sturtevant 1968). Nevertheless these complications will not affect the trends shown in Table 1, for all enthalpies that are to be compared were measured at the same cation concentrations.

Hydrogen bonds appear to contribute to the stabilizing enthalpy of a base pair in a double helix: a G·C base pair is about 2 kcal/mol more stable than an A·U base pair; AA (2-aminoadenosine) forms a triple base with two U's that is 3 kcal/mol more stable than the same triple base with normal adenosine. (AA can make three hydrogen bonds with U, whereas A can only make two.) At first this result is surprising. How can a hydrogen bond contribute stabilizing enthalpy when the base could just as well hydrogen-bond to water? I believe that hydrogen bonds to water may be a little weaker than interbase hydrogen bonds, and that interbase hydrogen bonds can be stabilizing if inaccessible to water molecules. Support for this point of view comes from the calorimetrically measured enthalpies of protein denaturation (Privalov *et al.* 1971), where each internal hydrogen bond seems to contribute between 1 and 2 kcal/mol of stabilizing enthalpy. Not all hydrogen bonds in a base pair will be equally strong; for example the middle hydrogen bond in a G·C base pair might be stronger because it is more inaccessible to water.

If hydrogen bonds appear to contribute most of the enthalpy of base-pair formation, why has the idea of stabilizing by stacking been so popular? Two possible reasons are: (*a*) Double-helix formation is known to be a highly cooperative process, for an isolated base pair is much less stable than a base pair inside a helical region. As the isolated base pair makes no stacking interactions, while the internal base pair makes two stacking interactions, stabilization by stacking will account for this cooperative behaviour. Both the isolated and internal base pair seem to have the same number of inter-base hydrogen bonds and should therefore be equally stable. However, I have suggested that only hydrogen bonds protected from water contribute stabilizing enthalpy, and helix formation could be cooperative even if hydrogen bonds provided stabilization. (*b*) Single-stranded nucleic acids appear to be strongly stabilized by stacking enthalpy. Ultraviolet spectroscopy detects an apparent order/disorder transition when poly(A) is heated. Different groups have obtained different enthalpy and entropy changes for this transition (see Table 1). The largest enthalpy changes seem very unreasonable: dipole/induced-dipole interactions could hardly contribute this much enthalpy; hydrophobic stacking interactions will reduce the entropy but not affect the enthalpy. The large entropy change of between 20 and 40 entropy units (e.u.) seems even more unreasonable. When a rigid double helix breaks up, the entropy change per nucleotide is only 12 e.u. How could a stacked single strand be more ordered than a single strand in a rigid double helix? I believe that the real enthalpy and entropy changes of a nucleotide in an ordered single strand are likely to be less than 3 kcal/mol and 9 e.u. respectively.

Why has the order/disorder transition of single strands been so problematic? Davis & Tinoco (1968) suggested that the apparent two-state order/disorder transition of ApA may be an artifact: as spectroscopic methods only report on the average associated order of the bases, the decrease in order could be explained by an oscillating dimer model in which the relative oscillation of the bases increases with temperature. Recent experiments by Achter & Felsenfeld (1971) indicate that single strands of apurinic acid, a DNA from which most A's and G's have been removed, has the same considerable rigidity in solution as poly(U) or poly(A) at high temperatures. This rigidity can only be due to the restricted rotation about the backbone single bonds. A plausible model for the conformation of single-stranded nucleic acid is as follows: the sugar-phosphate backbone rigidity causes nucleotides to have a conformation like that of the single strands in double helices where the bases stack (see Fig. 1*b*). Stacking does not contribute stabilizing enthalpy, but may stabilize the ordered conformation by increasing the entropy. As the temperature is raised most bases do not unstack but slide back and forth over their neighbours with increasing amplitude. This causes the apparent thermal transition measured by perturbation of base absorption spectra. At low temperatures (near 0 °C) almost all nucleotides will be in a helical conformation and the molecule will be very extended; but as the temperature is raised some nucleotides will take up non-helical conformations and the molecule will become less extended (see Eisenberg & Felsenfeld 1967). Low-angle X-ray diffraction studies on single strands of poly(C) support this model for a single-stranded nucleic acid in solution (Gulik *et al.* 1970). Nevertheless, the real nature of single-strand stacking is not fully resolved. My criticism of the thermodynamic parameters derived from the melting curve of single strands may be too naive. A peculiar ordering of water molecules could possibly cause the large enthalpy changes associated with the optically measured unstacking.

The stability of DNA and RNA helices is different (Table 1): a pair of nucleotides in RNA interact more favourably than the same pair in DNA. DNA double helices are B-form, while RNA double helices are A-form. A-form RNA is, therefore, more stable than B-form DNA and much more stable than A-form DNA. As the A-forms of RNA and DNA are virtually identical (Arnott *et al.* 1969), the extra hydroxyl group on the sugar ring of RNA must provide stabilization of A-RNA over A-DNA. As yet there is no evidence that the extra 2′ hydroxyl group stabilizes the RNA double helix by forming a hydrogen bond.

The entropy changes given in Table 1 indicate that the entropy decreases by about 12 e.u. when one nucleotide is ordered into a double or triple helix. This entropy change has two components: (*a*) the configurational entropy due to the restricted freedom of a nucleotide in a helix, and (*b*) the hydrophobic bond entropy due to the increased freedom of water when the base is stacked in a double helix. As different nucleotides cause the same total entropy change when they form a double helix, both the configurational and stacking entropy could be similar for different types of nucleotides.

The contributions to the free energy of formation of nucleic acid double helices can be summarized as follows: (*a*) each hydrogen bond in a base pair that is inaccessible to water reduces the enthalpy by 2 to 3 kcal/mol. Other interactions, such as chain rigidity, dipole/induced-dipole stacking forces, etc., contribute 4 to 2 kcal/mol. (*b*) Each ordered nucleotide reduces the entropy of the double helix by 12 e.u., which is made up of the configurational entropy of the nucleotide and the entropy of water that is disordered by stacking hydrophobic bases. Thus, most of the enthalpy of a base pair comes from the hydrogen bonds.

Next I should like to consider the steps in the formation of nucleic acid double helices. To do this one has to consider the intermediates and transition states that occur in going from unpaired single strands to paired double strands. For long chains one has to consider very many partially base-paired intermediates (Lifson & Zimm 1963) and double-strand formation is very complicated (see Fig. 4*a*). For short chains no intermediates are present; we only have to consider the fully paired state, the fully unpaired state and one partially paired transition state (see Fig. 4*b*). The transition state grows as fast as it dissociates for it is at a free energy maximum. Two groups, Pörschke & Eigen (1971) and Craig *et al.* (1971), have used the temperature jump method developed by Eigen & De Maeyer (1963) to investigate the transition state. The rate of helix association and dissociation was measured over a range of temperatures for oligo(A) and oligo(U) helices of different lengths. As the rate of association depends on the free energy difference between the unpaired state and the transition state, while the rate of dissociation depends on the free energy difference between the fully paired state and the transition state, it is possible

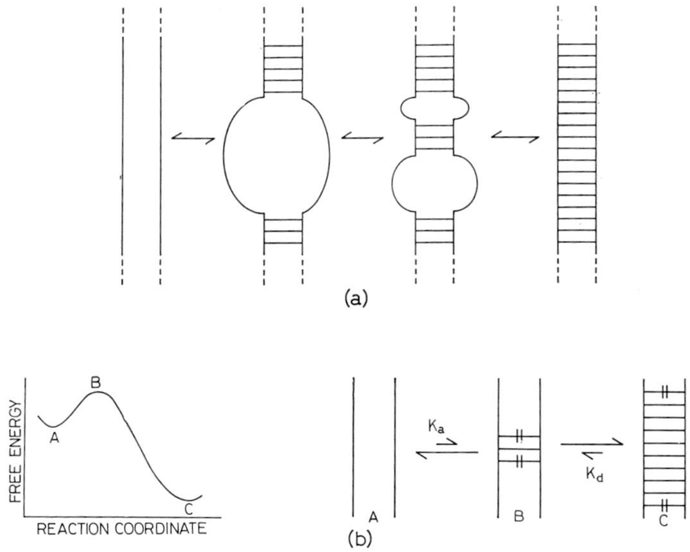

FIG. 4. (*a*) Schematic representation of intermediates in the melting of double helices of long nucleic acids. The mean length of regions of paired bases between loops depends on the cooperativity of helix formation. (*b*) Melting of an oligonucleotide double helix, showing the proposed conformation of the transition state B. The last base pair in a stack is assumed to contribute much less enthalpy than the other base pairs.

to calculate the enthalpy and entropy of both the transition state and the fully paired state. The enthalpy and entropy changes of the transition state, which do not depend on the length of the fully paired helix, are approximately 9 kcal/mol and 50 e.u. respectively. This suggests that the transition state is a stack of three base pairs, for a smaller double helix would not have sufficient enthalpy. This fits in well with the idea that stabilizing enthalpy comes from hydrogen bonds in base pairs inside a helical stack, for a stack of one or two base pairs would have very little stabilizing enthalpy. The enthalpy of a short double helix increases with the number of base pairs, though it seems that the last two bases contribute very little enthalpy (Martin *et al.* 1971). Helix formation depends on the concentration of the single strands, whereas helix dissociation does not. Because base pairs are added very rapidly to the transition state, the rate of helix formation depends on the rate at which the transition state is formed.

Base pairs can also form between complementary regions of a single strand of nucleic acid. In the process the chain must double back on itself so that a hairpin is formed (Fig. 5). The rate of formation of a short base-paired region is limited by the rate of formation of the transition state, which depends in

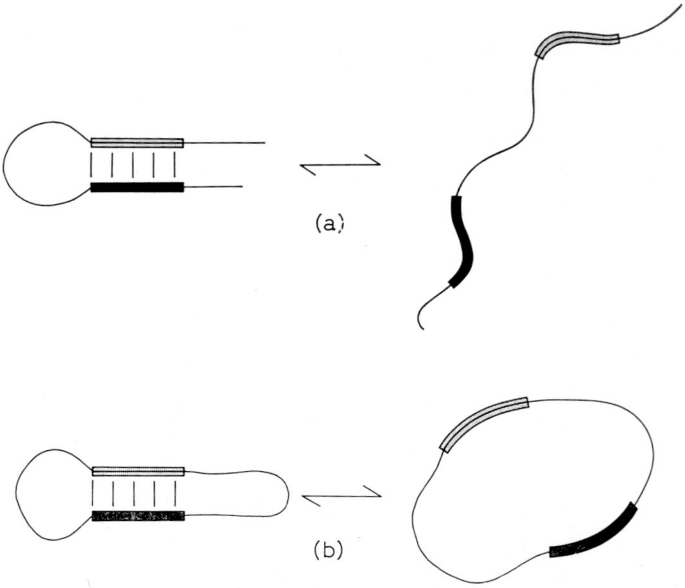

FIG. 5. (*a*) Melting of a base-paired hairpin in a single chain of nucleic acid. Even when the base-pairing is disrupted the two complementary sequences are quite close together in solution. (*b*) Melting of a base-paired hairpin in a closed circle of single-stranded nucleic acid. The disrupted circular single strand has greater order (lower entropy) than a linear single strand, and melting is less likely.

turn on the probability that the two complementary parts of the chain are close together in solution. Complementary regions separated by a few nucleotides along the chain come together often, while regions separated by many nucleotides come together much less frequently. Base-paired regions enclosing small loops always form faster than those enclosing big loops, irrespective of the number of base pairs in the region. Longer stretches of base-pairing are, of course, more stable than short stretches. Scheffler *et al.* (1970) have found that hairpins formed in closed circles of nucleic acid are more stable than identical hairpins formed in open chains. When a hairpin in an open chain melts, the chain becomes very free and the entropy increases; but when a hairpin in a closed chain melts, the chain cannot become so free for it is closed into a circle and the entropy increases less. Hence, joining the ends of a nucleic acid chain strengthens hairpins in the chain.

Fig. 6 shows the most plausible secondary structures (regions of complementary base-pairing) of various ribonucleic acid sequences that have been determined over the past five years. A common feature is the formation of small hairpins which are linked together into a circle by base-pairing of two

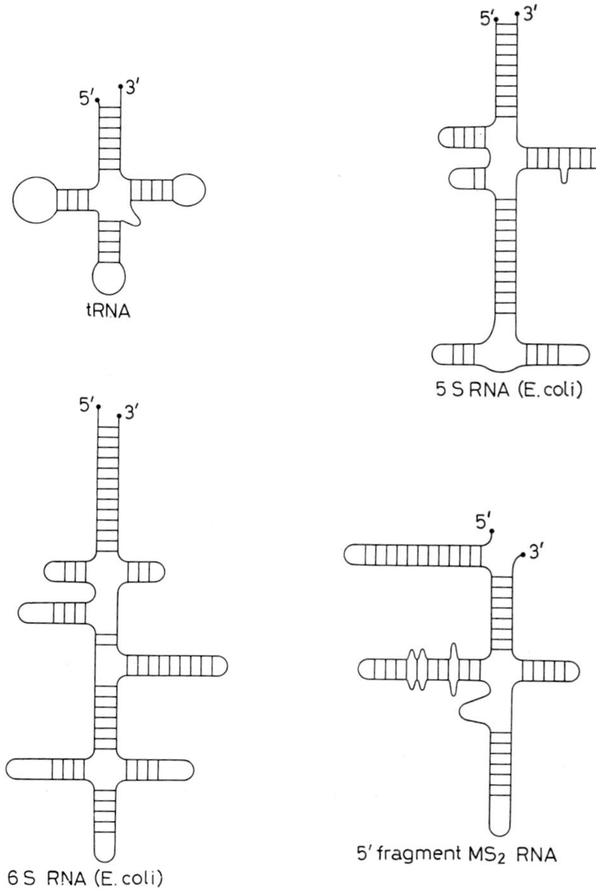

FIG. 6. Secondary structure in single-stranded RNA. The diagrams are only schematic and do not show which bases are involved in loops or double-helical secondary structure. Secondary structures different from those shown are possible for all these molecules, though long uninterrupted stretches of base-pairing are probably genuine. The tRNA secondary structure is commonly known as a clover-leaf because of its shape. 5S RNA from *E. coli* was sequenced by Brownlee *et al.* (1967), 6S RNA, also from *E. coli*, by Brownlee (1971), and the fragment of MS2 viral RNA by Wachter *et al.* (1971).

more distant parts of the chain. Hairpins with small loops occur because they form faster than hairpins with big loops. The arrangement of groups of hairpins into closed circles can also be explained. Consider the melting of one hairpin in such a closed circle. Provided that the base-pairing holding the hairpins in a circle is more stable than the base-pairing closing the hairpin, the hairpin will have to melt into a closed circle and will be stabilized. All the hairpins linked together in a circle are stabilized in this way. This is a more efficient way to stabilize secondary structure than by making each hairpin more stable. A second reason for the common occurrence of closed rings could be control of secondary structure. If the strong base-pairing closing the chain into a circle were to be disrupted in some way, all the enclosed hairpins would be less stable. In this way the secondary structure of viral RNA could be disrupted after infection merely by opening up the base pairs that tie the chain ends together.

Besides base-paired secondary structure, some single-stranded nucleic acids also have tertiary structure. Transfer RNA (tRNA) is the best-studied example of a nucleic acid with a well-defined tertiary structure, i.e. ordering of the double-helical base-paired regions and the single strands between them into a precise, compact conformation. Experiments by Fresco *et al.* (1966) indicate that tRNA must have tertiary structure for its biological activity, and that this tertiary structure is only formed in the presence of divalent cations like magnesium (Mg^{2+}).

Nucleotides seem to have been specifically designed to form very stable double helices; it is difficult to imagine how loops and helices can interact to form a precise tertiary structure. Fortunately, tRNA has been studied extensively by a great variety of techniques. Although X-ray diffraction methods have not yet revealed the conformation of tRNA, several indirect clues suggest how the tertiary structure is formed. Fig. 7 shows schematically how indirect experimental information can define a three-dimensional model for tRNA. The clover-leaf pattern for the secondary structure is well established, as the 33 tRNA species that have been sequenced to date can all form the same regions of complementary base-pairing (Fig. 7*a*). The most definite information about tRNA tertiary structure comes from a photoreaction that covalently links two nucleotides distant in the sequence (Yaniv *et al.* 1969; Favre *et al.* 1971). Fig. 7*b* shows how two arms of the clover-leaf must be positioned to fit this result. Next, we have the only coordinated change of sequence not incorporated into the clover-leaf secondary structure. Whenever the nucleotide at position 15 is A, that at position 49 is U, and whenever 15 is G, 49 is C. No other combination occurs, yet each alternative occurs frequently (to date 24 tRNA sequences have G·C, and 9 have A·U). This coordinated change strongly suggests that the nucleotides concerned should be close in the tertiary structure, possibly forming a standard Watson-Crick base pair. Fig. 7*c* shows the necessary positioning of a third arm of the clover-leaf. Other available data are more general and consequently less useful. The shape of unit cells in crystals of tRNA (Kim *et al.* 1971) and results of low-angle X-ray scattering of tRNA solutions (Ninio *et al.* 1969) suggest that the tRNA molecule is long and thin. Hence, the anticodon arm is arranged as shown in Fig. 7*d*. Finally, the presence of extra base pairs

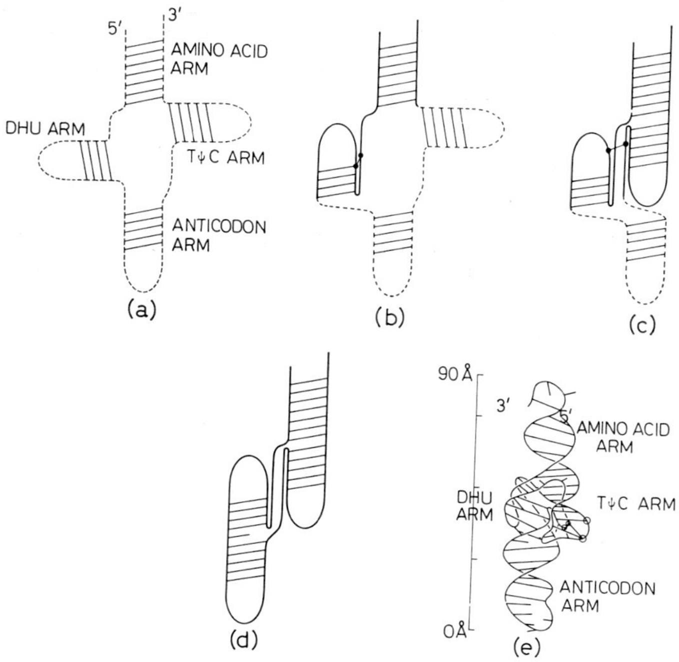

FIG. 7. A schematic diagram showing how indirect evidence can be used to define a tertiary structure of transfer RNA. (*a*) Clover-leaf secondary structure supported by 33 independent tRNA sequences. (*b*) Positioning of two arms necessitated by the photoreaction observed by Yaniv *et al.* (1969). (*c*) Positioning of a third arm to bring the only nucleotides whose sequence change is coordinated (other than those in the clover-leaf) close together. (*d*) Positioning of the fourth arm to give a long thin molecule. (*e*) Formation of base pairs between loops at the centre of the molecule to give a precise tertiary structure. The circled bases are constant in all tRNA species except $tRNA^{Gly}$ from *Staph. epidermidis*. DHU: dihydro U.

indicated by tritium exchange experiments (Englander *et al.* 1972) and the unreactivity of many of the nucleotides in the loops (cf. Levitt 1969) suggests that bases in the loops interact through base pairs to form a compact structure (Fig. 7*e*). Thus, one obtains the model for tRNA tertiary structure that I proposed some time ago (Levitt 1969). When a detailed model of the interactions that stabilize the central part of the tRNA molecule was built, bases were connected by hydrogen bonds and then stacked to make these hydrogen bonds inaccessible to water molecules. Naturally, this could not be done as neatly or symmetrically as in the regular double helix. In several places three or more phosphate groups come very close together, which indicates that divalent cations like Mg^{2+} might be needed to neutralize electrostatic repulsion. Experimental data that fix the conformation of the anticodon loop or the position of the anticodon arm relative to the rest of the molecule are weak. Even when the central part of the molecule is stabilized by detailed interactions between the loops, the anticodon arm is left fairly free. I feel that any changes of conformation of tRNA that might occur during protein synthesis would involve the anticodon arm of the molecule.

I do not want to discuss the proposed tRNA model in detail, but I would like to mention support for the model that came from an unexpected quarter. Transfer RNA^{Gly} from *Staphylococcus epidermidis* is used for cell wall synthesis, but cannot work in protein synthesis (Stewart *et al.* 1971). As this tRNA is charged by the normal glycine synthetase, and as normal glycine tRNA molecules used for protein synthesis can also be used for cell wall synthesis, only that part of the tRNA that interacts with ribosomes should be abnormal in the cell-wall glycine tRNA. Roberts (1972) has sequenced this tRNA and finds several sequence changes in all three loops. However, if this cell-wall tRNA is folded into the tertiary structure I proposed, all the changes occur in one part of the three-dimensional structure (see Fig. 7*e*). This strongly supports the idea that one side of tRNA is mainly concerned with amino-acid charging, etc., while the other side is concerned with codon recognition and binding to the ribosome.

Römer *et al.* (1970*a*, *b*) have shown that the different steps in the melting of tRNA in the absence of Mg^{2+} correspond to the melting of individual hairpins of the tRNA clover-leaf secondary structure. Addition of increasing quantities of Mg^{2+} raises the melting temperature of each resolvable step until the different transitions become coupled in a single rapid cooperative melting of the tertiary and secondary structure (see Fig. 8). I believe that the proposed model for tRNA will fold in solution as follows: firstly, complementary sequences close together along the chain form hairpins with small loops, which are kinetically favoured over hairpins with large loops. Next the beginning and end of the molecule are joined by a region of strong base-pairing which stabilizes all the previously formed hairpins. Finally, nucleotides in loops interact through base pairs to form a stable tertiary structure. The tertiary structure stabilizes each hairpin. Once the tertiary structure melts, the hairpins of the clover-leaf secondary structure are no longer stable. They melt immediately, giving a single sharp transition (see Fig. 8).

Biological systems are characterized by macromolecules with precise tertiary structure. It is interesting to compare the abilities of proteins and nucleic acids to form such precise three-dimensional shapes. As nucleotides have few allowed backbone conformations, and as the bases impose severe packing problems, nucleic acids can probably form only a limited number of different tertiary structures. Proteins, by contrast, are much more versatile, for they can have

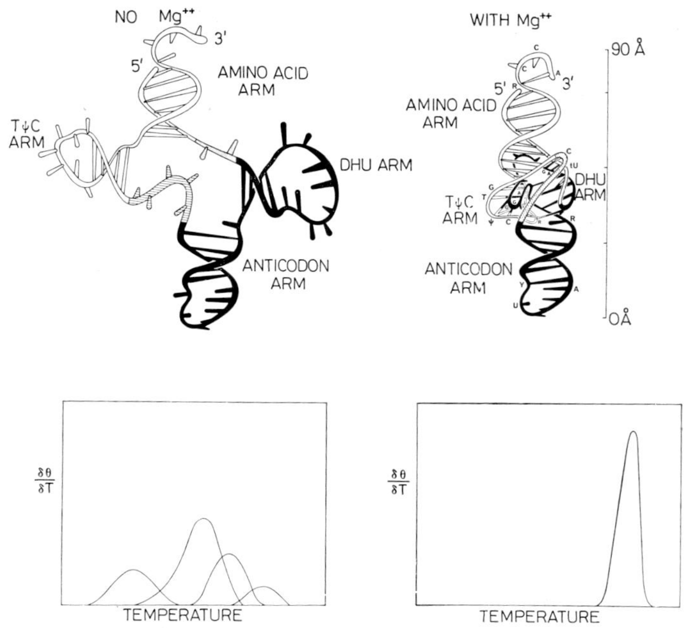

FIG. 8. A schematic diagram contrasting the properties of tRNA in the absence and presence of Mg^{2+}. Each peak in the differential melting curve of tRNA without Mg^{2+} corresponds to independent melting of ordered structure.

many types of complicated tertiary structure. The four types of nucleotides are much more alike than the 20 types of amino acids, and a random change in the chemical sequence probably has a more disturbing effect on protein conformations than on nucleic acid conformations. Crick (1968) proposed that nucleic acid may have performed many enzymic tasks in primitive cells. The features of possible tertiary structures of nucleic acids strongly support this idea; a random sequence of 80 nucleotides could possibly form a precise tertiary structure, whereas a similar sequence of amino acids could not. The lower sensitivity of nucleic acid tertiary structure to sequence errors would also be a great advantage in primitive cells. As cells evolved and protein synthesis became more precise, proteins would have taken over functions previously performed by nucleic acids. Transfer RNA could be considered as a relic from the past, for it is the one example where a nucleic acid behaves like a protein (Crick 1966). Recently Crick (1971) suggested that the tertiary structure of nucleic acid may be important in controlling DNA transcription in higher organisms. He proposes that chromosomal DNA has some tertiary structure so that it can generate a sufficient variety of recognition sites.

It seems increasingly clear that the secondary and tertiary structure of nucleic acids plays a far greater role in molecular biology than was ever expected from molecules ostensibly designed to transmit information coded in a chemical sequence.

References

ACHTER, E. K. & FELSENFELD, G. (1971) The conformation of single-strand polynucleotides in solution: sedimentation studies of apurinic acid. *Biopolymers* **10**, 1625–1634

ARNOTT, S., DOVER, S. D. & WONACOTT, A. J. (1969) Least-squares refinement of the crystal and molecular structures of DNA and RNA from X-ray data and standard bond lengths and angles. *Acta Crystallogr. Sect. B* **25**, 2192–2206

BROWNLEE, G. G. (1971) Sequence of 6S RNA of *E. coli*. *Nat. New Biol.* **229**, 147–149

BROWNLEE, G. G., SANGER, F. & BARRELL, B. G. (1967) Nucleotide sequence of 5S-ribosomal RNA from *Escherichia coli*. *Nature (Lond.)* **215**, 735–736

BUGG, C. E., THOMAS, J. M., SUNDARALINGAM, M. & RAO, S. T. (1971) Stereochemistry of nucleic acids and their constituents. X. Solid-state base-stacking in nucleic acid constituents and polynucleotides. *Biopolymers* **10**, 175–219

CRAIG, M. E., CROTHERS, D. M. & DOTY, P. (1971) Relaxation kinetics of dimer formations by self-complementary oligonucleotides. *J. Mol. Biol.* **62**, 383–401

CRICK, F. H. C. (1966) The genetic code—yesterday, today and tomorrow. *Cold Spring Harbor Symp. Quant. Biol.* **31**, 3–9

CRICK, F. H. C. (1968) The origin of the genetic code. *J. Mol. Biol.* **38**, 367–379

CRICK, F. H. C. (1971) General model for the chromosomes of higher organisms. *Nature (Lond.)* **234**, 25–27

DAVIS, R. C. & TINOCO, I. (1968) Temperature-dependent properties of dinucleoside phosphates. *Biopolymers* **6**, 223–242

DONOHUE, J. & TRUEBLOOD, K. M. (1960) Base pairing in DNA. *J. Mol. Biol.* **2**, 363–371

EIGEN, M. & DE MAEYER, L. (1963) Relaxation methods. In *Technique of Organic Chemistry* (FREISS, S. L., LEWIS, E. S. & WEISSBERGER, A., eds.) Vol. 8, part 2, pp. 895–1054. Interscience, New York

EISENBERG, H. & FELSENFELD, G. (1967) Studies of the temperature-dependent conformation and phase separation of polyriboadenylic acid solutions at neutral pH. *J. Mol. Biol.* **30**, 17–37

ENGLANDER, J. J., KALLENBACH, N. R. & ENGLANDER, S. W. (1972) Hydrogen exchange study of some polynucleotides and transfer RNA. *J. Mol. Biol.* **63**, 153–169

FAVRE, A., MICHELSON, A. M. & YANIV, M. (1971) Photochemistry of 4-thiouridine in *Escherichia coli* transfer RNA_1^{Val}. *J. Mol. Biol.* **58**, 367–379

FELSENFELD, G. & MILES, H. T. (1967) The physical and chemical properties of nucleic acids. *Annu. Rev. Biochem.* **36**, 407–448

FRANKLIN, R. E. & GOSLING, R. G. (1953) Molecular configuration in sodium thymonucleate. *Nature (Lond.)* **171**, 740–741

FRESCO, J. R., ADAMS, A., ASCIONE, R., HENLEY, D. & LINDAHL, T. (1966) Tertiary structure in transfer ribonucleic acids. *Cold Spring Harbor Symp. Quant. Biol.* **31**, 527–537

GULIK, A., INOUE, H. & LUZZATI, V. (1970) Conformation of single-stranded polynucleotides:

small-angle X-ray scattering and spectroscopic study of polyribocytidylic acid in water and in water-alcohol solutions. *J. Mol. Biol.* **53**, 221–238

KAUZMANN, W. (1959) Some factors in the interpretation of protein denaturation. *Adv. Protein Chem.* **14**, 1–63

KIM, S. H., QUIGLEY, G., SUDDATH, F. L. & RICH, A. (1971) High-resolution X-ray diffraction patterns of crystalline transfer RNA that show helical regions. *Proc. Natl Acad. Sci. U.S.A.* **68**, 841–845

KLUMP, H. & ACKERMANN, T. (1971) Experimental thermodynamics of the helix-random coil transition. IV. Influence of the base composition of DNA on the transition enthalpy. *Biopolymers* **10**, 513–522

KOTIN, L. (1963) On the effect of ionic strength on the melting temperature of DNA. *J. Mol. Biol.* **7**, 309–311

KRAKAUER, H. & STURTEVANT, J. M. (1968) Heats of the helix-coil transition of the poly A—poly U complexes. *Biopolymers* **6**, 491–512

LEVITT, M. (1969) Detailed molecular model for transfer ribonucleic acid. *Nature (Lond.)* **224**, 759–763

LIFSON, S. & ZIMM, B. H. (1963) Simplified theory of the helix-coil transition in DNA based on a grand partition function. *Biopolymers* **1**, 15–23

MARTIN, F. H., UHLENBECK, O. C. & DOTY, P. (1971) Self-complementary oligoribonucleotides: adenylic acid–uridylic acid block copolymers. *J. Mol. Biol.* **57**, 201–215

NINIO, J., FAVRE, A. & YANIV, M. (1969) Molecular model for transfer RNA. *Nature (Lond.)* **223**, 1333–1335

PÖRSCHKE, D. & EIGEN, M. (1971) Co-operative non-enzymic base recognition. III. Kinetics of the helix-coil transition of the oligoribouridylic-oligoriboadenylic acid system and of oligoriboadenylic acid alone at acidic pH. *J. Mol. Biol.* **62**, 361–381

PRIVALOV, P. L., KHECHINASHVILI, N. N. & ATANASOV, B. P. (1971) Thermodynamic analysis of thermal transitions in globular proteins. I. Calorimetric study of chymotrypsinogen, ribonuclease and myoglobin. *Biopolymers* **10**, 1865–1890

ROBERTS, R. J. (1972) Structures of two glycyl-tRNAs from *Staphylococcus epidermidis*. *Nat. New Biol.* **237**, 44–45

RÖMER, R., RIESNER, D. & MAASS, G. (1970*a*) Resolution of five conformational transitions in phenylalanine-specific tRNA from yeast. *FEBS (Fed. Eur. Biochem. Soc.) Lett.* **10**, 352–357

RÖMER, R., RIESNER, D., COUTTS, S. M. & MAASS, G. (1970*b*) The coupling of conformational transitions in alanine-specific transfer ribonucleic acid from yeast studied by a modified differential absorption technique. *Eur. J. Biochem.* **15**, 77–84

SCHEFFLER, I. E., ELSON, E. L. & BALDWIN, R. L. (1970) Helix formation by d(TA) oligomers. II. Analysis of the helix-coil transitions of linear and closed oligomers. *J. Mol. Biol.* **48**, 145–171

SCHILDKRAUT, C. & LIFSON, S. (1965) Dependence of the melting temperature of DNA on salt concentration. *Biopolymers* **3**, 195–208

SCRUGGS, R. L. & ROSS, P. D. (1970) A calorimetric study of monomer-polymer complexes formed by polyribouridylic acid and some adenine derivatives. *J. Mol. Biol.* **47**, 29–40

STEWART, T. S., ROBERTS, R. J. & STROMINGER, J. L. (1971) Novel species of tRNA. *Nature (Lond.)* **230**, 36–38

SUNDARALINGAM, M. (1969) Stereochemistry of nucleic acids and their constituents. IV. Allowed and preferred conformations of nucleosides, nucleoside mono-, di, tri-, tetra-phosphates, nucleic acids and polynucleotides. *Biopolymers* **7**, 821–860

TS'O, P. O. P. (1968) The physicochemical basis of transitions of nucleic acids. In *Molecular Associations in Biology* (Pullmann, B., ed.), pp. 39–75, Academic Press, New York

VINOGRAD, J., LEBOWITZ, J. & WATSON, R. (1968) Early and late helix-coil transitions in closed circular DNA. The number of superhelical turns in polyoma DNA. *J. Mol. Biol.* **33**, 173–197

WACHTER, R. DE, MERREGAERT, J., VANDENBERGE, A., CONTRERAS, R. & FIERS, W. (1971) Studies on bacteriophage MS2. The untranslated 5′ terminal nucleotide sequence preceding the first cistron. *Eur. J. Biochem.* **22**, 400–414

WATSON, J. D. & CRICK, F. H. C. (1953) A structure for deoxyribose nucleic acid. *Nature (Lond.)* **171**, 737–738

WILKINS, M. H. F., STOKES, A. R. & WILSON, H. R. (1953) Molecular structure of deoxypentose nucleic acids. *Nature (Lond.)* **171**, 738–740

YANIV, M., FAVRE, A. & BARRELL, B. G. (1969) Structure of transfer RNA. Evidence for interaction between two non-adjacent nucleotide residues in $tRNA^{Val}$ from *Escherichia coli*. *Nature (Lond.)* **223**, 1331–1333

Discussion

Tanford: The hydrophobic effect should be accompanied not only by an entropy change but also by a change in heat capacity. All hydrophobic effects that I know about have a ΔH that is temperature-dependent. The heat of solution of benzene in water, for example, changes by 0.42 kJ (0.1 kcal) mol^{-1} K^{-1}. Have you observed this effect and, if not, why?

Levitt: The specific heat, $\frac{\delta H}{\delta T}$, of polynucleotides is much closer to zero than that of polypeptides.

Eigen: It depends on the nature of the nucleic acid what ΔH values you get. When a double helix is melted into single strands, these still involve stacking of bases and the degree of stacking depends on the temperature of the transition and also on the type of structure. For instance, you mentioned the work of Römer *et al.* (1970) on differential melting curves. They analysed these curves and tried to attribute ΔH parameters to the different loops of the tRNA molecule. It turned out that the magnitude of ΔH was not a simple function of the number of base pairs involved. It depends not only on the number of A·U and G·C pairs but also on the degree of 'stacking' in the melted region. One could, of course, try to measure independently the degree of stacking and correct all values for the transition from the completely 'unstacked' to a completely double helical state.

Levitt: I don't dispute that single-stranded polynucleotides are stacked. What I find difficult to believe is that the change of enthalpy on unstacking is as high as calculated from a two-state interpretation of optically measured melting curves. Unfortunately, the single-strand melting occurs over such a wide range that it is impossible to measure the change in enthalpy by calorimetry.

Eigen: No, but you can combine calorimetric data with optical and sound absorption data and then get a pretty safe figure for these values (cf. Pörschke & Eggers 1972).

Klug: On these very short polynucleotides don't you get a rather broad curve, not a sharp transition, which it may be rather dangerous to fit by a two-state model?

Eigen: You can never get sufficient data from just one melting experiment. We tried to solve the problem in the following way (Eigen & Pörschke 1970; Pörschke & Eigen 1971; Pörschke 1971). We prepared oligomers of A_3 up to A_{18} (or U_3 to U_{18}) and then compared the melting and relaxation behaviour. For each base pair we used a different stability and rate constant of formation. The analysis was carried out with the help of a computer. The results show that the stability of the 'first' base pair (nucleation) differs strongly from that of any following pair (propagation). Beyond three pairs differences in the parameters (stability constants, ΔH, ΔS and rate constants) were not detectable within the limits of experimental error. Similar results could also be obtained with certain G and C oligomers. The ΔH values which are evaluated for the propagating pair formation are in the range of 29.4 to 42 kJ mol^{-1} (7 to 10 kcal mol^{-1}) (referring to one base pair). If we correct for the single-strand stacking—which can be determined independently—they come up to around 54.6 kJ mol^{-1} (13 kcal mol^{-1}). The stability constant for G·C at room temperature is about ten times as high as for A·U (related to the 'cooperative' pair). In these values the ΔH contribution from stacking is appreciable.

Klug: When you say you extrapolate to a single strand how is that state defined?

Eigen: You can heat the single strands alone (i.e. oligo (A) in the absence of oligo(U)) and evaluate the data (e.g. relaxation amplitudes and times). The phenomenon turns out to be non-cooperative, so that it can be described by one equilibrium constant and corresponding ΔH and ΔS parameters. From there on you can calculate everything. In the paper by Pörschke & Eggers (1972) I think a reasonable evaluation with self-consistent parameters was done.

Katchalski: What is the state of X-ray analysis of crystals of tRNA?

Tanford: I have been told by Drs S. H. Kim and A. Rich that they expect to have the complete structure in the very near future.

Flory: Dr Olson has studied the problem of the conformation of polynucleotide chains in great detail and with results that should be of interest (Olson & Flory 1972). She was able to interpret the random coil and to correlate predictions with some of the conformations for helices. Polynucleotide chains are not highly hindered. There is a considerable degree of latitude of rotation around the P—O bonds and about the preceding two bonds of the sequence.

Of course, one of the bonds occurs in the pentose ring and the ring imposes a constraint. Also, the C—O bond following the ring is highly hindered sterically. In order to achieve agreement with experimental work, it proved necessary to take account of electrostatic interactions. These interactions lead to preferences within the allowed conformational domains. In the random coil, at least, this may be fairly important. Whether the ring conformation is 3′-*endo* or 2′-*endo* has a very marked effect on the calculations of average properties. The results also depend sensitively on rotation about the following C—O bond.

Luzzati: I wish to point out that in all the cases in which we looked for a random coil in nucleic acids we found an organized structure, which in the range of 20 nm has the properties of a stiff rod (Luzzati *et al.* 1967; Gulik *et al.* 1970); this is probably the kind of stacked conformation that you describe.

Flory: I addressed my remarks to single-stranded polynucleotide chains—mainly RNA—and these *are* random coils. Their dimensions have been determined as random coils, both for poly(rA) and for poly(rU).

Luzzati: Poly(rA) is probably a single-stranded helix, with stacked bases.

Flory: Eisenberg & Felsenfeld (1967; see also Inners & Felsenfeld 1970, and Achter & Felsenfeld 1971) found that at room temperature poly(A) is stacked to a considerable degree. At higher temperatures the stacking diminishes; it is almost exclusively a function of temperature. By a plausible extrapolation they were able to ascertain the characteristics for the unstacked material. If you extrapolate to a state of no stacking you need not worry about stacking. You can debate whether or not they have done it correctly but for poly(U) we have no such problem—poly(U) is virtually free of stacking.

Luzzati: I am rather sceptical about the interpretation of the spectroscopic measurements in terms of conformation, at least for nucleic acids (see Gulik *et al.* 1970).

Flory: Felsenfeld and co-workers (see above) have measured radii of gyration by light scattering, they have determined chain dimensions from viscosity, and they have taken sedimentation data. All are in excellent agreement, not only in showing poly(A) and poly(U) to be random coils, but in quantitating their dimensions.

Luzzati: A polymer can behave as a random coil as far as light scattering and sedimentation are concerned, and yet display a quite well ordered short-range conformation.

Flory: A rod-like segment of 20 nm is irreconcilable with experiments and with theory. Single-stranded polynucleotides behave as random coils, not as stiff rods. This is well established.

Levitt: There is tentative evidence (Anderson & Smith 1972) that suggests that some tRNA molecules may fold up in the manner suggested by Professor

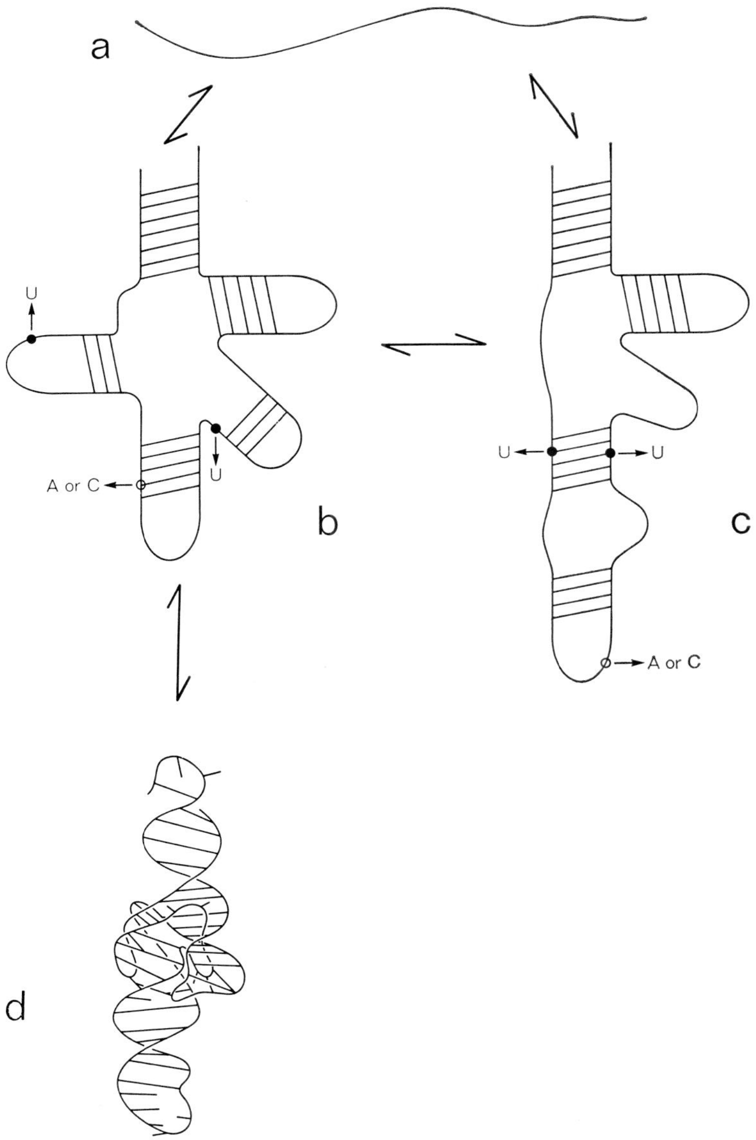

FIG. 1. (Levitt). Proposed folding of tyrosine suppressor tRNA.

(a) Unfolded molecule.
(b) Correctly folded clover-leaf secondary structure.
(c) Incorrectly folded extended secondary structure.
(d) Correctly folded tertiary structure.
○ Site of the primary mutation.
● Site of the revertant sequence changes.

Tanford for cytochrome *c*, i.e. with intermediate partially folded states not on the direct path between the unfolded and correctly folded conformation. (The tentative interpretation of Anderson & Smith's results is my contribution.)

Certain mutants of the bacterium *E. coli* have a defect in a gene whose expression is essential for their survival under some growth conditions. These bacteria can then only grow if they also have a special tRNA, known as a suppressor tRNA, which permits synthesis of the protein specified by the defective gene. These bacteria, therefore, depend on the functionality of the suppressor tRNA for their survival. For this reason it is possible to alter this tRNA by mutation so that the *E. coli* cannot survive, and then to mutate the tRNA gene at a second site so that the *E. coli* are again able to survive. The first set of mutations makes the tRNA non-functional; the second set, known as revertants, restores its functionality. Smith and his co-workers have produced many different mutants and revertants of the suppressor tRNA gene; they have also found the changes in the tRNA sequence in each case. One particular set of mutants and revertants was puzzling (see Fig. 1). The primary mutation changes the chemical sequence in such a way as to disrupt a base pair in the anticodon arm of the molecule; in the revertant the sequence changes (other than the trivial restoration of the disrupted base pair) occur in several different positions in the sequence. In addition, the sequence changes of the revertants are non-specific: either of the two different mutations that disrupt the particular base pair in the anticodon arm is compensated by any one of the sequence changes of the revertant. It seems, therefore, that the sequence changes of the revertants cannot be compensating for the mutant by stabilizing the tertiary structure in the region disrupted by that mutation.

The results can be explained using partially folded intermediates according to the scheme shown in Fig. 1. The sequence change of the initial mutation disrupts a base pair, which weakens the clover-leaf secondary structure (Fig. 1*b*), but leaves the alternative extended secondary structure (Fig. 1*c*) unaffected. Much more of the tRNA now goes into the extended secondary structure, which is subsequently destroyed by ribonuclease in the cell. The net result is that much less of the functional tRNA is formed, and the bacterium cannot survive. The revertant sequence changes all weaken the extended secondary structure, but leave the clover-leaf unaffected (no base pairs are disrupted). More of the tRNA molecules now go into the clover-leaf secondary structure which then folds into the functional tertiary structure, and the bacterium can survive again.

Revel: Isn't this type of scheme misleading for the biologists? We know that there are maturation processes in the synthesis of tRNA (base modifications, shortening of the RNA chain). Is it right to consider folding of tRNA only in terms of nucleic acid interactions without introducing the specific enzymes which are required for maturation?

Levitt: Quite possibly; I have tried to give the simplest explanation of a very puzzling situation. One can always evoke enzymes that recognize specific nucleotide sequences and are unable to process a modified tRNA. Coupled with Professor Tanford's results on proteins, these results on nucleic acids are suggestive of a new consideration in macromolecular folding: that it is important for incorrectly folded forms to be avoided and for the ability to fold into the correct conformation to be retained.

References

ACHTER, E. K. & FELSENFELD, G. (1971) The conformation of single-strand polynucleotides in solution: sedimentation studies of apurinic acid. *Biopolymers* **10**, 1625–1634

ANDERSON, K. W. & SMITH, J. D. (1972) *J. Mol. Biol.* **69**, 349–356

EIGEN, M. & PÖRSCHKE, D. (1970) *J. Mol. Biol.* **53**, 123

EISENBERG, H. & FELSENFELD, G. (1967) Studies of the temperature-dependent conformation and phase separation of polyriboadenylic acid solutions at neutral pH. *J. Mol. Biol.* **30**, 17–37

GULIK, A., INOUE, H. & LUZZATI, V. (1970) Conformation of single-stranded polynucleotides: small-angle X-ray scattering and spectroscopic study of poly C in water and in water-alcohol solutions. *J. Mol. Biol.* **53**, 221

INNERS, L. D. & FELSENFELD, G. (1970) *J. Mol. Biol.* **50**, 373–389

LUZZATI, V., WITZ, J. & MATHIS, A. (1967) in *Genetic Elements* (SHUGAR, D., ed.), p. 41, Academic Press, London

OLSON, W. K. & FLORY, P. J. (1972) *Biopolymers* **11**, 1–66

PÖRSCHKE, D. (1971) *Biopolymers* **10**, 1989

PÖRSCHKE, D. & EGGERS, F. (1972) *Eur. J. Biochem.* **26**, 490

PÖRSCHKE, D. & EIGEN, M. (1971) *J. Mol. Biol.* **62**, 361

RÖMER, R., RIESNER, D. & MAASS, G. (1970) Resolution of five conformational transitions in phenylalanine-specific tRNA from yeast. *FEBS (Fed. Eur. Biochem. Soc.) Lett.* **10**, 352–357

Interactions between synthetic polymer molecules

P. CORRADINI and V. PETRACCONE

Istituto Chimico della Università di Napoli

Abstract An 'idealized' model of packing of the macromolecules of a polymer in the crystalline state and some kinds of deviations from this structure are discussed.

In the crystalline state, the chain conformation is always very near one of the minima of the internal energy surface for the isolated macromolecule, with the condition of periodical repetition along an axis. The mode of packing of the macromolecules—unit-cell constants and atomic coordinates—can be foreseen with reasonable accuracy in some cases. Various types of disorder may be present in the mode of packing of macromolecules; cases of constitutional (i.e. isotactic copolymers), configurational (a classical example is polyvinyl alcohol) and conformational isomorphism have been observed. The conformational isomorphism may refer to different conformations which may be assumed by structural units of the main chain [i.e. poly(1,4-*trans*-butadiene), high temperature modification] or by lateral groups [i.e. poly(*S*-3-methyl-1-pentene)].

Calculations of the conformational energy involved in some kinds of 'defects' such as folds and kinks in polyethylene show their feasibility.

In this paper, we shall discuss some aspects of the ways in which the macromolecules of a synthetic polymer of regular constitution and configuration form three-dimensional structures.

We shall discuss in more detail the compatibility of long-range, regular, side-by-side packing of linear macromolecules with various kinds of conformational defects, the regularity of the side-by-side packing of macromolecules being established by X-ray diffraction spectroscopy.

In general, the conformation of the chain of a polymer in the crystalline state is defined by a succession of structural units, geometrically equivalent in respect to an axis (Natta & Corradini 1960); the possible line repetition groups have been defined (Corradini 1960). For polymer chains with non-equivalent directions and asymmetric carbon atoms the only allowed symmetry of the chain is helical.

In the framework of the allowed symmetries, that is under the restrictions imposed by the need of having an axis of repetition, the conformation of the chain will tend to approach one of minimum conformational energy for the isolated chain. Of course, crystal packing effects may influence the choice among conformations of nearly equal energy for the isolated chain (Corradini 1969).

Many calculations of conformational energy of the above kind have been performed in recent years (De Santis *et al.* 1963; Scott & Scheraga 1966).

Attempts have also been made to arrive at reasonable predictions of the mode of packing of macromolecules in a crystalline lattice. Where only van der Waals forces are operating, the lattice binding energy per equivalent unit in the lattice can be written (Kitaigorodskij 1965):

$$E = \frac{1}{2} \sum_{i,k} U(r_{i,k})$$

The summation takes place over all the interatomic vectors $r_{i,k}$ connecting atoms of one unit with all the atoms of the other units. $U(r)$ is the potential of interaction between non-bonded atoms.

The results obtained a few years ago by Corradini & Avitabile (1968) in the study of isotactic polyacetaldehyde

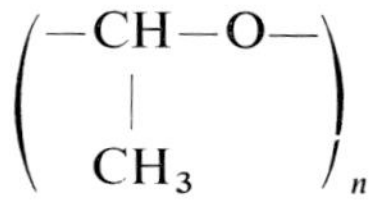

are given below.

From calculations of minimum potential energy, Borisova & Birshtein (1963) showed that the minimum energy conformations of the isolated chain repeating along an axis correspond to a fourfold helix, with a succession of approximately *trans* and *gauche* bonds along the chain.

On the assumption that the fourfold symmetry of the chain is maintained in the crystal lattice, Corradini & Avitabile have minimized the van der Waals energy with respect to (1) the unit cell constants of the various compatible tetragonal space groups, (2) the orientation of the macromolecules with respect to the chain axis and (3) the cylindrical coordinate z of one of the atoms of the helix. The results are shown in Fig. 1.

The non-bonded energy is a minimum for the $I4_1/a$ space group, which is the one experimentally observed; also the computed specific volume and the predicted mode of packing of the macromolecules are in good agreement with the experimental results.

One point of interest in this energy calculation is that the computed specific

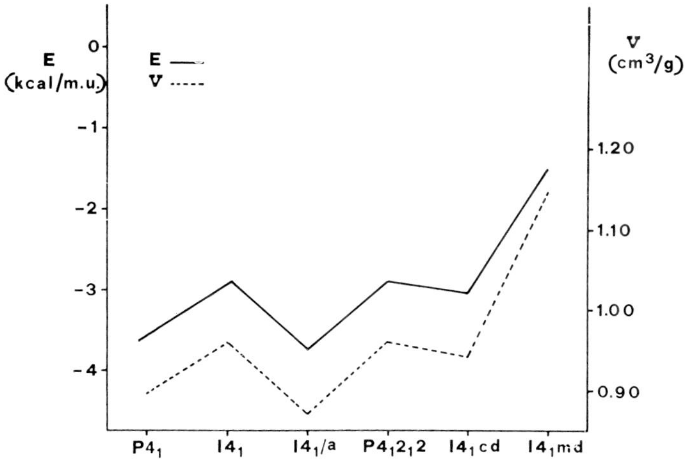

FIG. 1. The packing energy (E) and specific volume (V) for various space groups considered for isotactic polyacetaldehyde (the experimental space group is $I4_1/a$, $V_{exp} = 0.87\ cm^3/g$).

volumes for the various space groups follow the same order as the values of the non-bonded energies, in accordance with the close-packing principle put forward by Kitaigorodskij (1965).

This whole discussion refers, of course, to an 'idealized' crystal of infinite periodicity; no disorder in the crystal lattice is taken into consideration.

Sources of disorder may be of constitutional, configurational or conformational character. In a polymer, examples of constitutional faults with respect to the idealized model include inversions in the head-to-tail succession of monomeric units, the presence of units derived from a different monomer and, in any case, the ends of the chain; an example of a configurational fault is the possible presence, in an isotactic polymer, of racemic dyads in place of *meso*-dyads.

It is to be noted that crystallinity may be preserved even in the extreme case of almost complete randomness in the succession of constitutional units: polymers of aldehydes (Corradini 1968) with randomly distributed acetaldehyde and propionaldehyde units are crystalline over the whole range of possible composition (Fig. 2), provided that the copolymers are isotactic.

We now discuss in greater detail the possible occurrence of conformational disorder in the chain of a crystalline polymer.

We shall use the term 'conformational isomorphism' to refer to the more or less random occurrence in the same lattice site of different, but almost isoenergetic, conformers of the same portion of a molecule. A classification of the

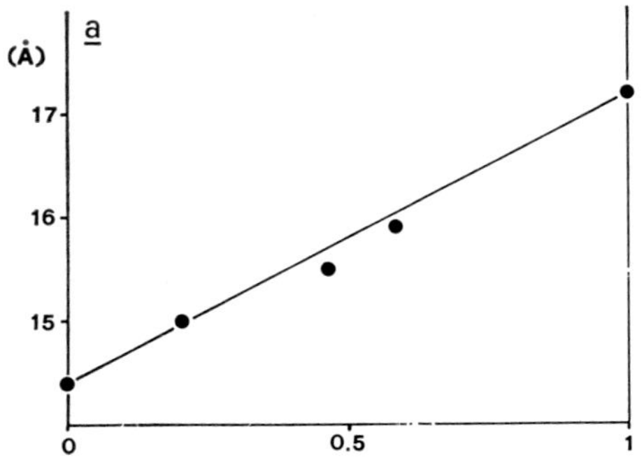

FIG. 2. Variation of the *a* constant in the crystalline copolymers of acetaldehyde and propionaldehyde as a function of composition (mole fraction of propionaldehyde units).

possible cases of conformational isomorphism in polymers can be made on the basis of the following two possibilities (Corradini, 1971, unpublished):

(1) The atoms of the chain may assume, in the same crystal and more or less at random, different (although nearly isosteric) conformations;*

(2) If the geometry of the main chain is fixed, the atoms of the lateral groups may assume different conformations in the same crystal, more or less at random.

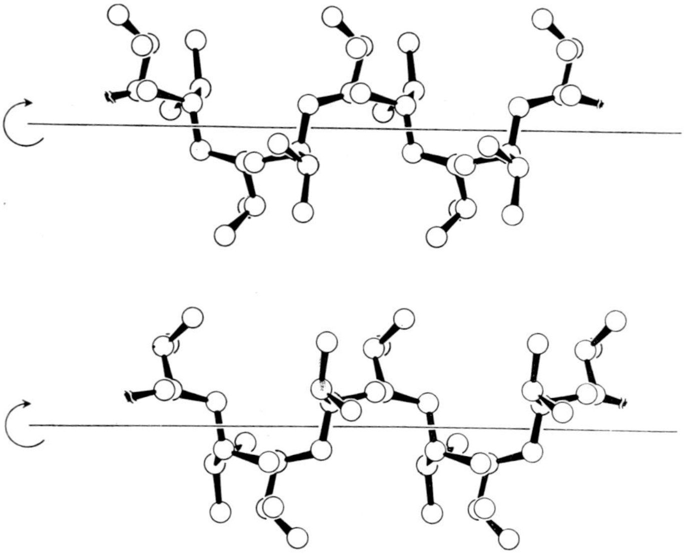

FIG. 3. Side views of two isosteric 'anticlined' helices with the same direction of isotactic poly(3-methyl-1-butene). The two helices are 'anticlined', differing in the up or down orientation of the pending lateral groups.

* Or they may adopt two non-equivalent but isosteric orientations of the same conformation. This is quite common (Natta *et al.* 1961; Allegra & Bassi 1969) in helical polymers such as isotactic polymers (Fig. 3).

We may also speak in a broader sense of conformational isomorphism in point and line defects in polymers, whenever such defects are due to different conformations of portions of the macromolecules being tolerated by the crystal lattice.

As we shall see, the occurrence of cases of conformational isomorphism of the first kind shows that a polymer chain can remain straight, as would happen if it were constrained to run inside the walls of a tight cylinder, while its conformational freedom remains at least of the same order of magnitude as it would have in a melt.

We have recently studied two examples of conformational isomorphism of the chain: the high temperature polymorphic form of poly(1,4-*trans*-butadiene) and natural rubber [poly(1,4-*cis*-isoprene)].

Poly(1,4-*trans*-butadiene) has two crystalline modifications. The low temperature form has a regular chain conformation:* (*trans*, S_+, T, S_-). The identity period is 4.85 Å.

Poly(1,4-*trans*-butadiene) (Corradini 1969) undergoes a reversible transition in the crystalline state at 75 °C, accompanied by a marked increase in the specific volume (+9 %) and a contraction of the identity period along the chain axis (−5 %). The new identity period is 4.65 Å. The chains are packed like cylinders in a pseudo-hexagonal lattice.

The entropy of transition has been calculated to be 6.3 entropy units (e.u.) per monomeric unit, more than twice the entropy of melting (2.7 e.u. per monomeric unit) which occurs at 145 °C, with a further volume increase of about 9 %. It is clear from the analogies that can be drawn with other polymers such as guttapercha that at least half the entropy changes quoted above must be attributed to the conformational freedom gained by the chains when transitions occur. Theoretical considerations and experimental results on low molecular weight compounds (Corradini *et al.* 1969) have shown that *skew* (S) and *cis* (C) conformations of the single bonds adjacent to the double bonds, as well as *trans* (T) and *gauche* (G) conformations of the CH_2-CH_2 bond, are practically isoenergetic. The conformational calculations on unperturbed polybutadiene chains performed by Abe & Flory (1971) confirm this.

An approximate calculation of the partition function for the free chain of poly(1,4-*trans*-butadiene), shows that each chain has 17^n available (isoenergetic) conformations, where *n* is the number of monomeric units in the chain.**

* The symbols T, S, G and C refer here and in the following pages to *trans*, *skew*, *gauche* and *cis* conformations of single bonds.

** We assume the simple approximation of unitary statistical weights σ and ρ (i.e., assuming that the three minima for the rotation around all the single bonds are isoenergetic) and zero statistical weights β for the conformations in which unfavourable interactions between non-bonded atoms arise. The statistical weights σ, ρ and β are defined by Abe & Flory (1971).

Surprisingly, if we prohibit only the G_+ and G_- conformations about the CH_2-CH_2 bond, we obtain a set of nearly straight chain irregular conformations:

$$(S_+ \text{ or } S_- \text{ or } C, \textit{trans}, S_+ \text{ or } S_- \text{ or } C, T)_n$$

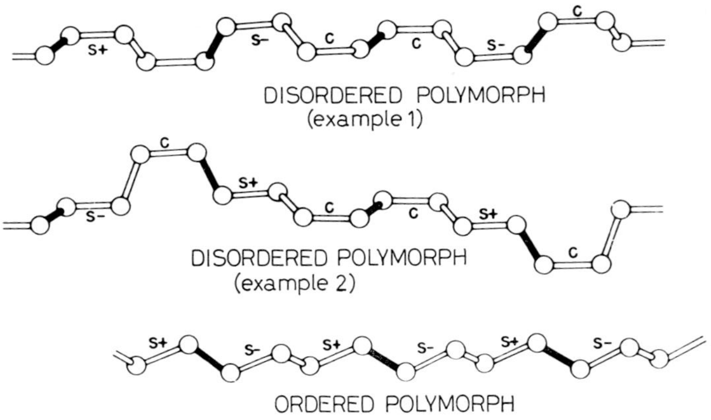

FIG. 4. Two examples of the possible statistical structures of the high temperature form of poly(1,4-*trans*-butadiene), compared with the ordered structure of the low temperature form.

Models of an irregular chain of this kind are shown in Fig. 4. We have shown that this is indeed the model of the high temperature form of poly(1,4-*trans*-butadiene). For this model, in the simplified approximation above (setting $\sigma = 0$), the number of available conformations is about 9^n.

A statistical chain conformation of the kind illustrated above, with the low activation energy which can be predicted from the local conformational changes, explains the shortening of the identity period in a quantitative way, the sudden drop of the second moment which occurs in the broad line n.m.r. spectrum, the fact that in the i.r. spectrum the so-called 'crystallinity bands' disappear, and the high entropy of transition.

Another interesting case of conformational disorder along a chain in the crystalline state is provided by poly(1,4-*cis*-isoprene):

$$\left(\begin{array}{ccccccccccc} & \overset{CH_3}{|} & & & & \overset{CH_3}{|} & & & & \overset{CH_3}{|} & \\ -CH= & C- & CH_2- & CH_2- & CH= & C- & CH_2- & CH_2- & CH= & C- & \end{array}\right)_n$$

$$\begin{array}{cccccc} (S_+ & T & S_-) & (S_+ & T & S_-) \\ (S_- & T & S_+) & (S_- & T & S_+) \end{array}$$

Here the chain has the shape of a ribbon with hollows and bulges provided by the methyl groups. However, although the chain always maintains the same

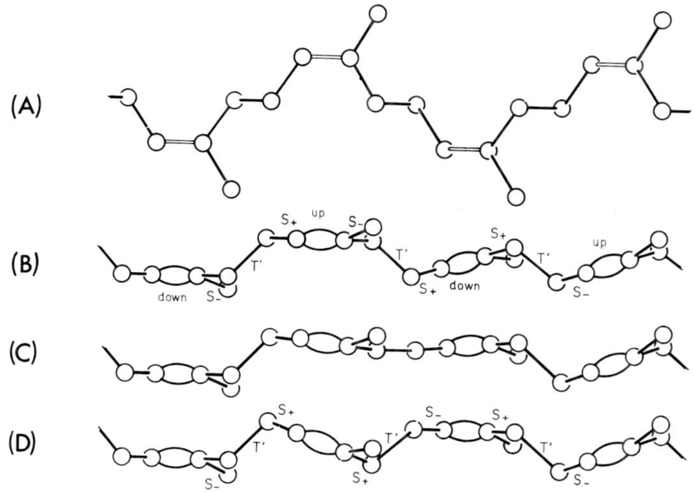

FIG. 5. Model of the conformation of the chain of poly(1,4-*cis*-isoprene) as viewed along the *b* axis (A) and along the *a* axis (B, C and D).

B and D are nearly isosteric but differ because of the exchange of an S_-TS_+ conformational sequence with an S_+TS_- conformational sequence. C represents a possible intermediate conformation.

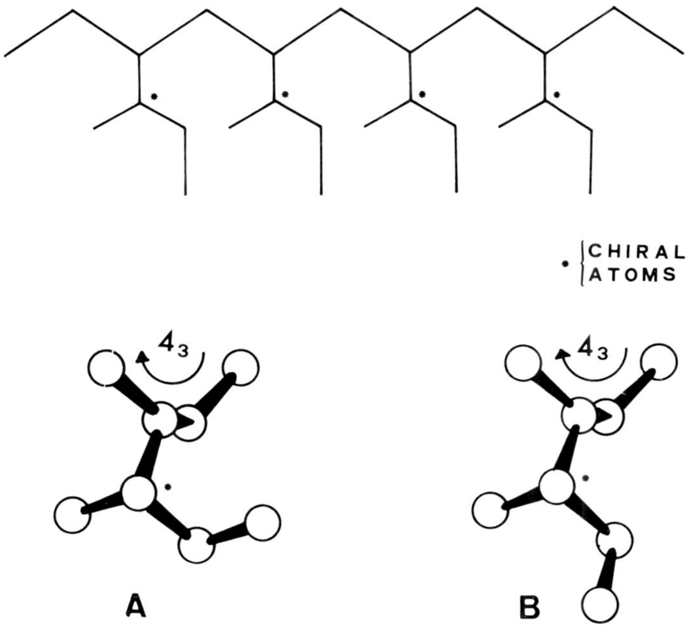

FIG. 6. Possible conformations of the monomeric units of isotactic poly(*S*-3-methylpentene) along the helix in the crystalline state.

approximate shape, we have to deal again with fluctuating statistics for the conformations. This is in accordance with the low entropy of fusion of poly-(*cis*-isoprene).

A concerted motion, which can be illustrated as shown in Fig. 5, provides a perpetual low-activation energy transition of each monomeric unit from one to the other available conformation. Nyburg (1954), who first described the existence of disorder in crystalline rubber and attributed to it a different nature, did not consider such a model because the possibility of slight deformations in bond angle, coupled with large variations in internal rotation angle, was not taken into account at the time.

We shall now refer briefly to a case of conformational isomorphism of lateral groups (type 2, p. 176). The crystal structure of isotactic poly(*S*-3-methyl-pentene) (Petraccone *et al.* 1972*b*) is characterized by the presence of fourfold helices of only one sense (left-handed). The conformations of opposing lateral groups of different chains are different; let us call them A and B (Fig. 6). Pairs A—B or B—A are randomly distributed along each column (Fig. 7). The

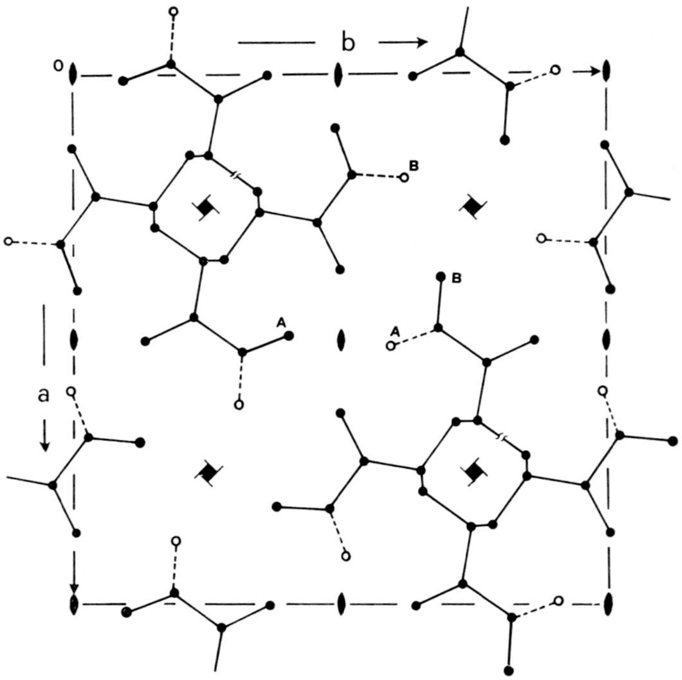

FIG. 7. Projection of the statistical model of the structure of isotactic poly(*S*-3-methyl-pentene) along the *c* axis.

The most efficient low-energy packing corresponds to relative conformations of the side methyl groups, marked in the same way (all in dotted or all in full line).

calculated optical activity in the solid state is in agreement with our findings.

Finally, it is also possible to speak in a broad sense of conformational isomorphism for point and line defects, in so far as they are tolerated by the crystal lattice, whenever they are due to different conformations of portions of the macromolecules (Fig. 8). Here again a correct approach is to make conformational energy calculations and to study model compounds.

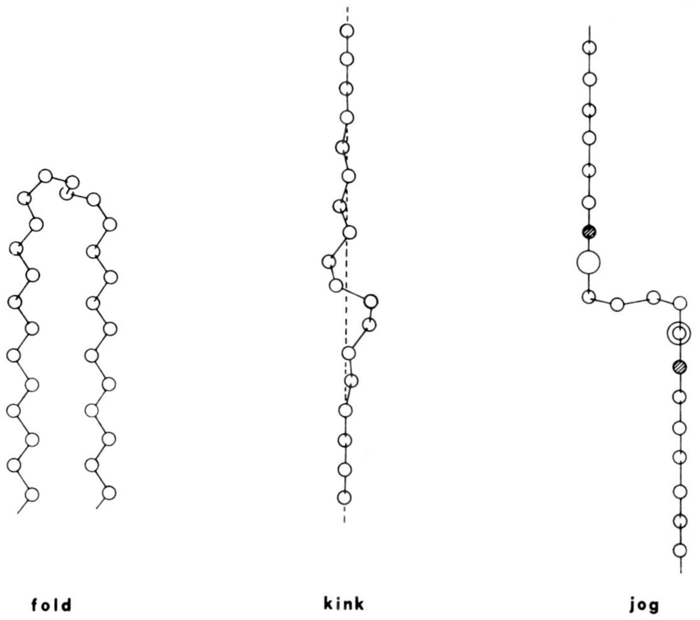

FIG. 8. Possible crystallographic point defects in chain molecules: model of a fold and of a kink in polyethylene, and of a jog in nylon 6–6.

In calculating conformational energies for such systems, it is important to take into account the possibility of coupling small deformations in bond angle with large variations of the internal rotation angles (Lifson & Warshel 1968).

The kinks which are possible in the polyethylene structure are of interest, because they are mechanically responsible for the transport of matter along the chain of a polymer crystal. Calculations by McMahon *et al.* (1967) have led to unrealistically high values of the energy involved, because bond angle deformations were not taken into account within the constraints posed by the problem. Moreover, at least some of the possible models of kinks escaped consideration.

A possible model of a kink of small energy (Petraccone *et al.* 1972*a*), fitting

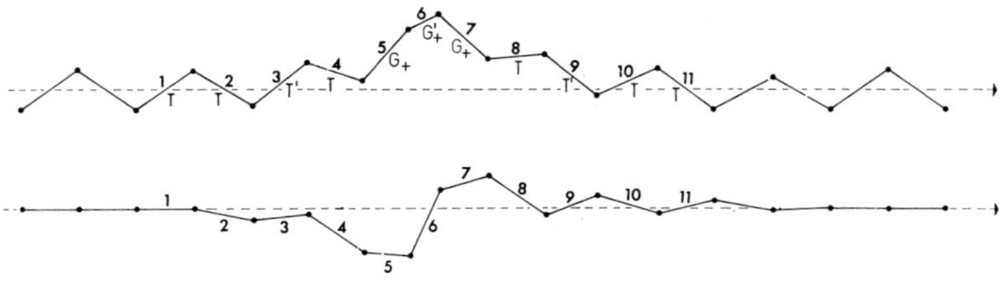

FIG. 9. Two perpendicular projections of one of the four equivalent conformations of a kink in the polyethylene structure, as recently proposed (Petraccone *et al.* 1972*a*).

the crystal environment and not arising from rigid bond-angle calculations, is shown in Fig. 9. The energy involved in such a point defect is no higher than 0.2–0.3 eV, but the defect seems able to move with a fairly low activation energy. In this case, bond angle deformations were necessarily considered.

We have made similar calculations (Corradini *et al.* 1971) for various kinds of chain folding in polyethylene (Fig. 10). In all cases so far studied, the conformational energy has always been notably lowered, when the flexibility of the bond angle was taken into consideration for crowded or constrained systems.

Finally, the study of the structure of compounds of low molecular weight (Benedetti *et al.* 1972) has recently provided evidence in favour of the presence of low energy jogs in nylon 6–6, arising from conformations of the adipic acid

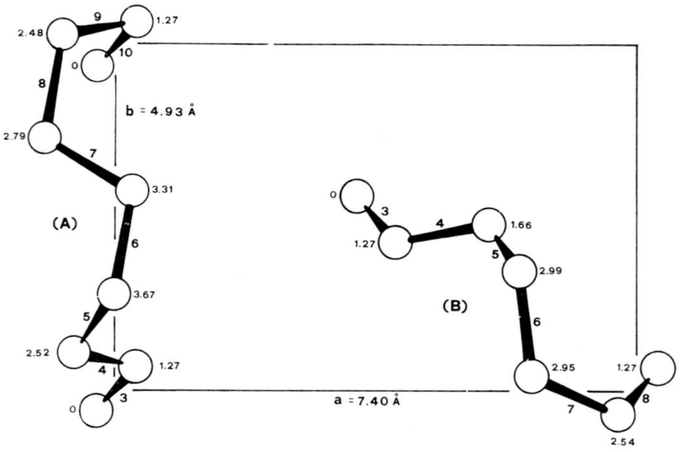

FIG. 10. Model of the conformation of folds A (along 200) and B (along 110) seen along the *c* axis of the unit cell of polyethylene. The numbers near the atoms are heights in ångströms.

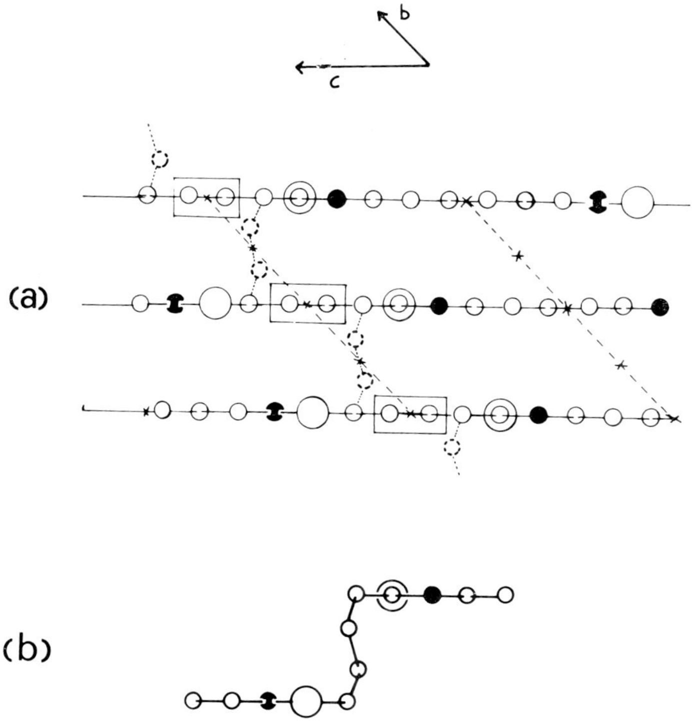

FIG. 11. Comparison between the conformation found in the crystalline state for N,N′-diethyladipamide (*b*) and a possible model of an array of jogs in nylon 6–6 (*a*).

moiety $S_+G_-TG_+S_-$ (Fig. 11). The presence of jogs may be postulated in the microstructure of nylon as well as in that of other polymers, and may be important in determining some physical properties (Dreyfus & Keller 1970).

The considerations put forward for crystalline or paracrystalline polymers may be extended to the amorphous state of polymers, which has been described according to two conflicting views. According to one view, polymer chains maintain a short-range parallel arrangement, due to the need of dense space occupancy, which lowers the energy of the system; according to the other, polymer chains take a completely coiled conformation, dictated by the need of maximum entropy gain without any inter-chain correlation.

The X-ray and electron diffraction data favour the first view, as was shown for instance by V. A. Kargin (1968, unpublished); also Pechhold (1968) has recently shown that, from the viewpoint of statistical thermodynamics, bundles of nearly parallel chains may be a model for the 'amorphous' state of polymers.

It is reasonable to say that for most polymers a large number of conformations are still available, as required for entropy reasons, while the chain remains locally straight, provided adequate consideration is given to the flexibility which arises from the cooperative coupling of bond angle deformations with changes in the internal rotation angles, as we have seen with the high temperature form of poly(1,4-*trans*-butadiene).

References

Abe, Y. & Flory, P. J. (1971) *Macromolecules* **4**, 219–237
Allegra, G. & Bassi, I. W. (1969) *Adv. Polym. Sci.* **6**, 549–574
Borisova, N. P. & Birshtein, J. M. (1963) *Polym. Sci. (U.S.S.R.)* **4**, 907–916
Benedetti, E., Ciajolo, M. R. & Corradini, P. (1972) *Eur. Polym. J.* in press
Corradini, P. (1960) *Rend. Accad. Naz. Lincei* **28** (8), 1–7
Corradini, P. (1968) in *The Stereochemistry of Macromolecules* (Ketley, A. D., ed.), Part III, pp. 1–60, Dekker, New York
Corradini, P. (1969) *Trans. N.Y. Acad. Sci.* **31**, 215–230
Corradini, P. & Avitabile, G. (1968) *Eur. Polym. J.* **4**, 385–393
Corradini, P., Frasci, A., & Martuscelli, E. (1969) *Chem. Commun.* 778–780
Corradini, P., Petraccone, V. & Allegra, G. (1971) *Macromolecules* **4**, 770–771
De Santis, P., Giglio, E., Liquori, A. M. & Ripamonti, A. (1963) *J. Polym. Sci.* **51**, 1383–1404
Dreyfus, P. & Keller, A. (1970) *Polym. Lett.* **8**, 253–258
Kitaigorodskij, A. I. (1965) *Acta Crystallogr.* **18**, 585–590
Lifson, S. & Warshel, A. (1968) *J. Chem. Phys.* **49**, 5116–5129
McMahon, P. E., McCullough, R. L. & Schlegel, A. A. (1967) *J. Appl. Phys.* **38**, 4123–4139
Natta, G. & Corradini, P. (1960) *Suppl. Nuovo Cimento* **15** (10), 9–39
Natta, G., Corradini, P., Sianesi, D. & Moreno, D. (1961) *J. Polym. Sci. 5-1*, 527–539
Nyburg, S. C. (1954) *Acta Crystallogr.* **7**, 385–392
Pechhold, W. (1968) *Kolloid Z. Polymer* **228**, 1–38
Petraccone, V., Allegra, G. & Corradini, P. (1972*a*) *J. Polym. Sci.* in press
Petraccone, V., Ganis, P., Corradini, P. & Montagnoli, G. (1972*b*) *Eur. Polym. J.* **8**, 99–105
Scott, R. A. & Scheraga, H. A. (1966) *J. Chem. Phys.* **45**, 2091–2101

Discussion

Flory: I must take issue with your conclusion about order in an amorphous polymer. Stress-temperature studies on elastic (amorphous) networks show that the temperature coefficient of the chain dimensions is unaffected by dilution. The temperature coefficient is in good agreement with the value obtained by measurements on dilute solutions of linear chains. The specific effect of dilution on the stress itself (apart from its temperature coefficient) is very small. If there were any appreciable change in the conformation of the chain it would surely be reflected in the elasticity measurements. Other evidence is furnished by the optical anisotropy, as observed, for example, in strain birefringence and in depolarized scattering. Such studies do indeed show evidence of inter-chain correlations, but on closer inspection the range of correlation is certainly much less than the range of correlation of bond directions along the chain. The optical anisotropy should be extremely sensitive to any considerable degree of correlation—in fact it is almost too sensitive for the purpose. Results will shortly be forthcoming on *n*-alkanes, where there is an effect, to be sure, but on analysis it will be seen to be much less than anything that should be expected if there were significant alterations in the conformation as required by bundle-like aggregates. The rubber elastic measurements clearly prove such structures to be absent.

Corradini: I agree there is much evidence in favour of an entirely random conformation of the chains in the amorphous state. However this must somehow be reconciled with other evidence. For instance diffraction studies on amorphous polyethylene have shown that some degree of parallelism in bundles of chains packed in a sort of pseudo-hexagonal array is also maintained in the melt. V. A. Kargin (1968, unpublished data) has shown that some correlation among chains is maintained up to radial distances of 1.5–2 nm. There must be some way of reconciling these data. As you mentioned this morning, when one tries to fit a chain into a given space it is very difficult to achieve dense packing with a completely random conformation. I believe it is very difficult to 'fill space' with entirely random chains in entirely random orientations. You would probably end up with a density in the amorphous polymer which is much too low, when compared with the experimental data.

Flory: The theory I discussed briefly (pp. 109–121) leads to exactly the opposite conclusion. That is, the difficulties of packing chains to high density are *not* relieved by *partial* ordering. One could as well assert that these difficulties produce a greater disorder by requiring increased flexing of the chains at high densities. Models adopted to treat such systems, their limitations notwithstanding, are emphatic in showing that the states of *complete order* (not to deny imperfections, however) and of *complete disorder* are states of minimum free energy. Intermediate states are of higher free energy. Thus, from a statistical thermodynamic point of view the ordered and disordered states are discontinuously related in the sense that they are separated by a phase transition. Both the theory of solutions of rod-like particles (Flory 1956*b*) and of semi-flexible chains (Flory 1956*a*) lead to the same conclusion, the physical basis for which is readily apparent. Theories and the models on which they rest should not be trusted too far, but they certainly do not support the case for order in liquid, or amorphous, polymers.

So-called diffraction studies on liquid polyethylene and other amorphous polymers do indeed show X-ray 'reflections', but these are closely akin to the haloes exhibited by liquids. They are diffuse, although somewhat less so than for low molecular liquids. They bespeak periodicity in the radial distribution function but little more than just that. Of course, the distance to the first coordination sphere of units is more sharply defined than for a liquid like benzene. But I submit that there is no real evidence from X-ray scattering for the kind of order mentioned.

Lifson: The surprisingly large entropy gained in *trans*-1,4-butadiene by solid-to-solid phase transition may have a quantum-mechanical interpretation. In this transition, the fibre shrinks along its axis and swells in the two perpendicular directions. This makes room for internal rotation of parts of the chain along the chain axis, because some of the C—C single bonds are directed along the fibre axis and can serve as axes of rotation. Such internal relations have high values of entropy. When you melt the crystal this internal rotation must be lost, and therefore the gain of entropy by melting is reduced by loss of internal rotation. I wonder if this interpretation can be checked quantitatively.

Corradini: I hope that somebody will do that.

Katchalski: You have implied, Dr Corradini, that in crystals possessing the faults or defects that you have described, segments of linear chains can rotate and change conformation without affecting the rest of the chain. Such a mechanism might then explain some of the conformational changes occurring in biopolymers and might be particularly pertinent in dealing with relatively small conformational changes like those observed, for example, in the haemoglobin molecule.

Corradini: If two points in the chain are fixed, there is plenty of freedom for the conformation between these fixed points to change. I am sure that in the synthetic polymers one can fix the positions of, say, two atoms 20 bonds apart and then achieve any kind of change in conformation in between. By analogy, changes within small segments in biopolymers must be possible.

Luzzati: Lipids may provide a biological analogue of the system that you described, Dr Corradini. The hydrocarbon chains of biological lipids are highly heterogeneous, typically 10 to 20 carbon atoms long. Usually at room temperature the chains take up a highly disordered conformation, similar to that of a liquid hydrocarbon. At lower temperature the chains often take up a more ordered structure, usually a two-dimensional hexagonal array of stiff chains, with rotational disorder.

I wish to add a word of caution about the interpretation of X-ray and electron diffraction experiments on liquid, since it is still not unusual to see broad bands interpreted in terms of Bragg's law, which is nonsense.

Elliott: There is some evidence from infrared spectroscopy on the conformation of chains of polyethylene. There is a band at 720 cm^{-1} (a rocking mode of the CH_2 groups) in which several CH_2 groups take part. This has no appreciable intensity unless a length of about four or five CH_2 groups is in the *trans* form. The band is quite strong in the crystalline polyethylene spectrum. If the polymer is melted at a sufficiently high temperature and then quenched, the band is very weak, so presumably there are then very few lengths of five successive CH_2 groups in the *trans* position. But if the temperature has been raised only a few degrees above the melting point the band has appreciable intensity and the polymer then has a considerable number of straight parts.

Chapman: As one raises the temperature of a phospholipid, e.g. a lecithin when in its gel-crystalline condition, increasing amounts of movement along the chain occur. This is shown by n.m.r. experiments (Veksli *et al.* 1969). When the transition point is reached one observes the sudden onset of a highly mobile condition corresponding to melting of the hydrocarbon chains. This is clearly a parallel to the behaviour observed with polyethylene.

References

FLORY, P. J. (1956*a*) *Proc. R. Soc. (Lond.) A* **234**, 60
FLORY, P. J. (1956*b*) *Proc. R. Soc. (Lond.) A* **234**, 73
VEKSLI, Z., SALSBURY, N. J. & CHAPMAN, D. (1969) *Biochim. Biophys. Acta* **183**, 434

Assembly in biological systems

E. KELLENBERGER

Abteilung für Mikrobiologie, Biozentrum der Universität Basel, Basel

Abstract The characteristic features of protein assemblies are discussed, that is: (*a*) the type and strength of interactions; (*b*) the regular crystalline arrangement in spatially folded and deformed two-dimensional crystal sheets; and (*c*) the characteristic determination of form (size and shape).

To obtain a better understanding of this determination of form two polymorphic systems are further discussed:

(1) The protein subunit of tobacco-mosaic virus is form-determining and needs no further protein for its assembly into rods. As Durham, Klug, and collaborators have shown, at least two physical variants of the rods occur—the helix and the stacked disc—depending on the ionic strength and pH. Both are relatively stable products which, at the end, are rather resistant towards dissociation.

(2) More complex systems (bacteriophages) provide examples of genetically determined polymorphisms. In phage T4, several aided assemblies are observed. The one on head-related particles is particularly interesting. According to what the active 'helping' genes are, tubular forms and tau particles (including their variants) are formed, made essentially of the 'uncut' gene product P23 (molecular weight 61 000). The capsids of isometric, normal (prolate) and giant heads are made of $P23^c$, the 'cut' form of the same gene product (mol. wt. 49 000).

A preliminary report is given of experiments by R. Bijlenga, D. L. Anderson and E. Lüscher which suggest that a species of tau particles is a precursor to phage heads and that the cutting of P23 to $P23^c$ occurs *in situ* on the particle in formation. Some involvement of phage DNA in this process is suspected and experimentally suggested.

Many enzymes, virus shells, bacterial flagella, bacterial pili, microtubules and other biological structures are assemblies of protein subunits. Each differs in the number of different protein species involved and in the number of identical subunits used.

Oligomeric enzymes are assemblies of from two to six identical protomers, while multifunctional enzymes are composed of as many non-identical subunits as there are functions, each subunit appearing mostly once, but sometimes twice or more. Larger structures may also be made up of a large number of identical protomers (bacterial flagella, pili, or some viruses such as tobacco mosaic virus [TMV]), or—in more complex systems—contain different numbers of various protein species. In such complex cases as bacteriophages T4 and lambda we can also distinguish morphologically distinct structural parts, most of which consist mainly of one merid, that is of an ordered assembly of a relatively large number of identical protomers.

GENERAL FEATURES OF PROTEIN ASSEMBLIES, PARTICULARLY MERIDS

In view of the previous contributions to this symposium, it may be interesting to outline some of the characteristic and unique features of protein assemblies. In contrast to what we have heard about biopolymers, protein assemblies are not covalently bonded, although the rare occurrence of disulphide bonds has not yet been ruled out. It is generally believed that weak interactions such as hydrophobic, electrostatic and hydrogen bonding are prevalent. These are the same types of interactions as occur in the tertiary structures of proteins.

Within or between merids the bonding of protomers is of variable strength. In some cases intramerid bonds are so strong that they can be opened only by denaturants at concentrations which also fully denature proteins. For example the following merids, enumerated in order of decreasing bond strength, are found in bacteriophage T4 (To *et al.* 1969): contracted tail sheath, tail fibre, head capsid, tail tube. The known intermerid bonds are much weaker and—in the above series—follow after the tail tubes. The order in the above series is independent of the chemical nature of the denaturant, in other words of organic solvent concentration (replacement of water: hydrophobic bonds), and of the nature and concentrations of ions (electrostatic interactions). The action of high ion concentrations and that of organic solvents are generally combined (acetic acid, guanidine hydrochloride, urea pH). Ionic strength and pH alone are frequently sufficient to disrupt bonds, as well as urea alone if at adequate concentration. Ionic detergents at room temperature were found to be active only in disrupting intermerid bonds of T4, inducing for example the sheath contraction. For rupture of the stronger interactions within merids, sodium dodecyl sulphate (SDS) has to be used in conjunction with high temperature—conditions in which proteins are likely to be fully denatured.

These and other observations strongly suggest that intramerid bonds are usually mixtures of several different types of weak interactions: the pattern of their arrangement within the interacting area would account for the very high specificity.

The high flexibility of most large protein assemblies, as deduced from electron microscopy, therefore cannot be accounted for by single bonds with a flexible angle. Since the nature and the strength of bonding within the interacting area are similar to those of intramolecular bonding, one must necessarily conclude that the deformation associated with this flexibility is not confined to the interaction area between protomers but extends into the tertiary structure of the protomers themselves. For the same reason protomers in quasi-equivalent positions in virus shells (Caspar & Klug 1962) must have slightly modified conformations. In the cases of polymorphism, one species of protomer can be assembled into several distinct morphological variants. In these variants, the bonding between protomers is frequently of different strengths and thus it can be predicted that these protomers will also have different conformations.

The following model is therefore postulated for such protein assemblies: on interaction, the protomers adjust their tertiary structure so as to form a continuous hydrophobic core which is not interrupted between protomers.

Many studies, based mainly on the work and influence of Klug, now make it certain that the merids are regular arrays which may be considered as two-dimensional 'biocrystals' folded and bent into three dimensions. I am deliberately introducing the term 'biocrystals' here because of the unique features which distinguish them from 'normal' crystals:

(1) These two-dimensional biocrystals have the property that in at least one dimension, and more frequently in both, they grow only to a precisely determined specific size.

(2) With very few exceptions (which in reality are abnormal non-physiological cases), finished biocrystals are not in dynamic equilibrium with the solution of their protomers. In a physiological environment the equilibrium shifts to a unidirectional reaction, apparently without the aid of a catalytic enzyme.

These two unique and specific features are certainly linked to high cooperativity within the biocrystals. Again, it is very difficult to explain such high cooperativity without introducing conformational changes in each of the component protomers.

In the discussion that follows, I shall select observations concerning polymorphism at different levels and introduce the concepts of sequential regulation through successively induced conformational changes.

Protomer assembly ranges from the simplest case of self-assembly, where all the information for the form of the final assembly product is contained in the geometry and bonding properties of the protomer itself, to the more complex cases, where the final form is attained only through the previous or concomitant help of other proteins (morphopoietic gene products: Kellenberger 1966). A protomer might be composed of several different proteins, but this case is obviously trivial and can be brought back to that of self-assembly. The situation becomes fundamentally different when proteins are involved that are *not* part of the protomer. Whether a protein found in an assembly is a helper protein or a constituent part of a protomer therefore has to be decided by experiments. If the numbers of molecules of two or more species of proteins involved in a complex structure are equal or integer multiples of one another, then at least one must suspect that together they might form a single protomer species. A real helper protein is certainly involved, for example, in all those cases where it can be demonstrated that it has only a transient function and is no longer present in the end-product. (For discussion of these cases see Kellenberger & Edgar 1971.)

ASSEMBLIES OF FULLY FORM-DETERMINING SUBUNITS

Among the representatives of this case, TMV has been studied most adequately. Purified virus protein can be reassembled into rods. This process has been investigated in thermodynamic terms and it was found that the bonding

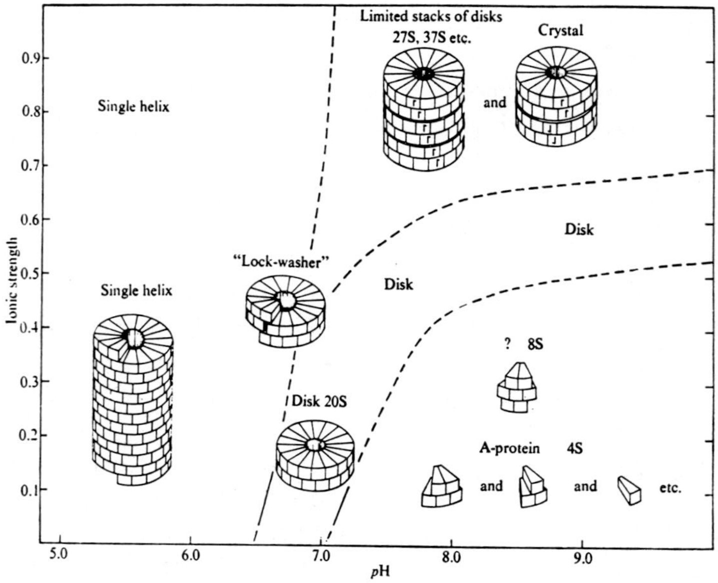

FIG. 1. Preponderant forms of assembly of subunits of TMV virus as a function of pH and ionic strength (Durham & Klug 1971)

is preponderantly of the hydrophobic type; the energies involved are related to organized water which, before aggregation, has to be 'melted' away from the interacting sites. The energy involved is indeed comparable to that of melting ice (Stevens & Lauffer 1965). Consequently, aggregation occurs at high temperature and dissociation at a lower temperature.

Reassembled rods are in chemical equilibrium with their subassemblies (Lauffer & Stevens 1968). A complete study of the equilibrium between different subassemblies and rods, as a function of pH and ionic strength, has been published recently (Durham & Klug 1971; Durham *et al.* 1971), and Dr Klug will discuss these results here (Klug 1972). The situation is summarized by the phase diagram of Fig. 1. Most important is the finding that discs are the subassembly through which either stacked discs or helical rods are assembled. The two-layer discs are formed by two rings of the same polarity, each consisting of 17 subunits, while in the helical form the rings are open and slightly changed, to contain only 16.34 subunits per turn. This *physical polymorphism* could well be related to the two carboxyl groups whose abnormal behaviour in titration Caspar (1963) has already pointed out. It is likely that these carboxyl groups interact differently, depending on the environmental concentration of protons and thus control the conformation of each protein subunit. Two pathways lead to structures of which the equilibrium becomes 'irreversible': in certain conditions, not yet clearly defined, the stacked discs assemble into long rods of stacked discs (Fig. 2) in which the protomers are now bound very strongly (A. Klug, personal communication).

The other pathway, which involves the viral RNA as well as discs, leads to the infectious viral particles. These particles suddenly become very resistant to dissociation and only strong denaturants such as 60% acetic acid are able to solubilize them again into subunits.

Among other interesting systems of self-assembly we might mention bacterial flagella (Asakura 1968) and the shells of some spherical viruses (reviewed by Hohn & Hohn 1970). The catalytic effect of nucleic acids or nucleic acid analogues on the assembly of the capsids of small spherical viruses is worth mentioning because here—unlike TMV—other nucleic acids, and even analogues, seem to be nearly as efficient as the one proper to the species (Hohn 1969).

AIDED ASSEMBLY OF MERIDS IN COMPLEX SYSTEMS

Aided assembly probably also occurs in relatively simple systems. Aid is however easier to recognize in those systems where genetics can be used as an experimental tool. Well-developed genetics is available for rather complex structures like bacteriophages T4 and λ. Fortunately, the complexity does not seem to be fundamental but is due rather to a high number of structural parts. Many of these structural parts considered individually seem to follow simple pathways of assembly. In these phages mutants are available for most genes, so that each gene action can be studied (see Epstein *et al.* 1963 for T4; and Kellenberger & Edgar 1971 for a review of λ and a summary of the experimental procedures).

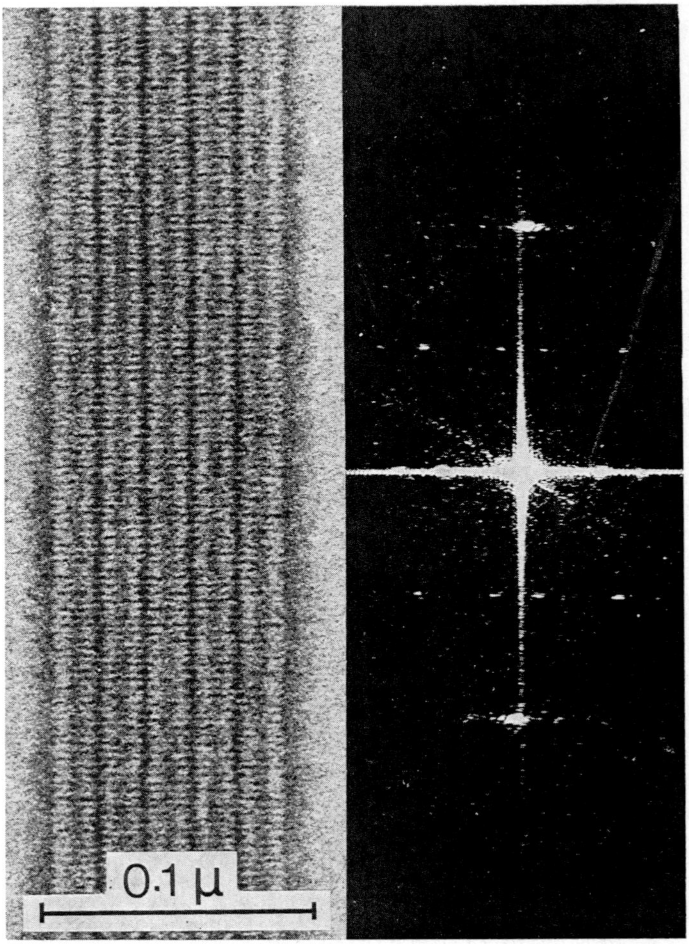

FIG. 2. A negatively stained preparation of stacked disc rods of TMV protein with the corresponding optical diffraction, showing the arrangement of rings in discs and the disc as repeat units. (Preparation kindly provided by A. C. H. Durham, Cambridge, micrograph and diffraction by J. Dubochet, Basel.)

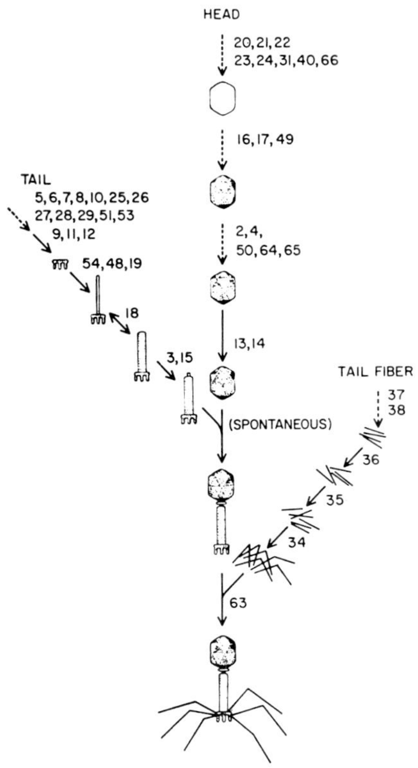

FIG. 3. Pathways of assembly of bacteriophage T4 according to Wood *et al.* (1968; reproduced by permission of *Federation Proceedings*). The numbers indicated for each step are those of the genes known to be involved at this step. A more recent path for head maturation is given in Fig. 9.

The pathways of assembly of phage T4 have been elucidated most elegantly by Edgar, Wood and collaborators (Wood *et al.* 1968); they are shown schematically in Fig. 3.

The assembly of these complex viruses occurs under most interesting conditions: the proteins involved all belong to the same class of the so-called 'late proteins'. In other words they all start being made simultaneously. Within a cell, the assembly of phage particles is far from being synchronized. The pathways in Fig. 3 are therefore not the result of a chronologically timed production of each gene product; the correct sequence of interactions is locally achieved through another mechanism. Such a mechanism has to account for the fact that the two partners of a specific protein-protein interaction do not react appreciably with each other when both are free in solution, but only when one of the partners is sitting in or on the maturing particle. There are three pos-

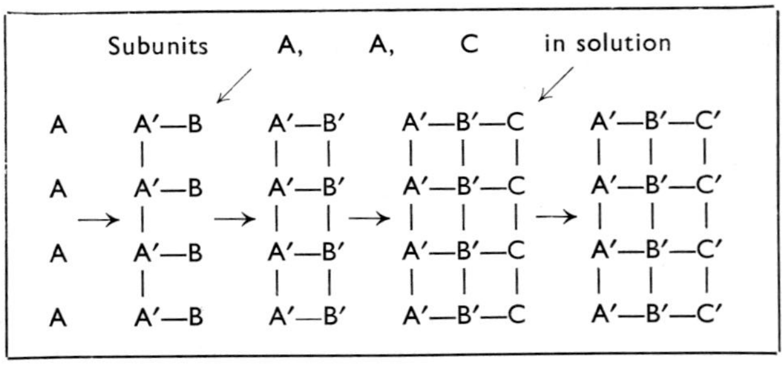

FIG. 4. Schematic mechanism of sequential regulation through induced conformational changes. Subunits A, B, C, etc. are free in solution (or on a membrane) but do not interact. B interacts with A′ only; A′ is induced through polymerization of A. The same is true for the pair B, C which free in solution does not interact except when B is modified into B′ after interaction with A′ (and possibly also polymerization).

sibilities which have to be considered, alone or in combination: (1) The maturing precursor particle may act as a nucleating centre and help to initiate the first steps of assembly of a merid. It is indeed known that the initial phases of assembly from subunits in solution are more difficult and more energy-consuming than later, when at least three subunits are bound together (Durham & Klug 1971). The immobilization of the first two subunits on a 'substrate' (precursor particle, cell membrane) would help to overcome this initial difficulty. (2) New specific sites of interaction may be made by the juxtaposition of proteins. (3) New specific sites may appear through induced conformational changes (Fig. 4).

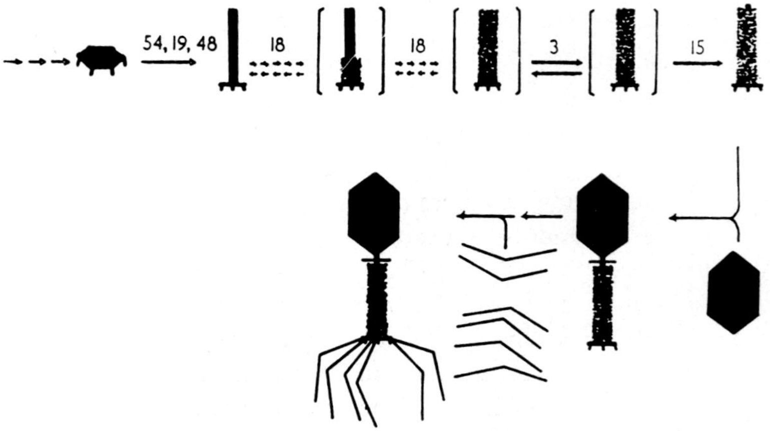

FIG. 5. Detailed scheme of the tail assembly according to King (1968; reproduced by permission of Academic Press). The action of genes 3 and 15 is required to stabilize the finished sheath. Until this takes place, sheaths can be dissolved again.

A good example for this sequentially induced specificity is given by the tail assembly, as shown in Fig. 5 (King 1968): according to recent studies P19 is the sole component of finished tubes (F. A. Eiserling, personal communication). Tail tubes without base-plates are never found *in vivo*. Hence the assembly of P19 into tail tubes is induced on the already assembled base-plates. When the tail tube is finished, or possibly even before, P18 is added to form the tail sheath in its extended form. This assembly is not stable and is still reversible. Only when further gene products are added does it become an irreversibly stable

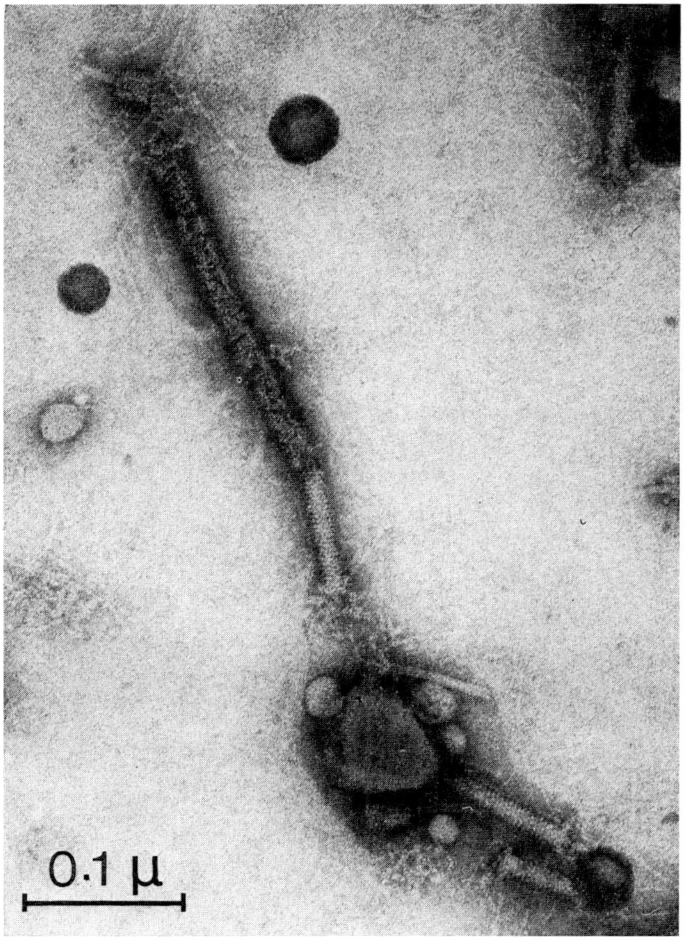

FIG. 6. A negatively stained preparation showing different forms of assembly of phage T4 tail sheath subunits: from upper left to lower right, a contracted sheath still sitting on the tail tube, polysheath showing 'crystal defects', a tail with extended sheath, a tail tube and a complete phage with artificially distorted head.

tail which can now, and only now, interact with a finished head. When no tail tubes are available P18 can assemble, after a certain interval, into polysheaths of undetermined length (Kellenberger & Boy de la Tour 1964). This polysheath has the same structure as a contracted sheath (Moody 1967). Contraction of the sheath occurs, for example, *in vivo* after absorption of phage to bacteria. The different variant forms of sheath are shown in Fig. 6.

This potentially very interesting system shows that the normal extended sheath has an aided assembly, through the tail tube acting as a morphopoietic core. The length of sheath is predetermined by the length of the tube. But how is the length of the tube determined? When measured in a fresh lysate, tail

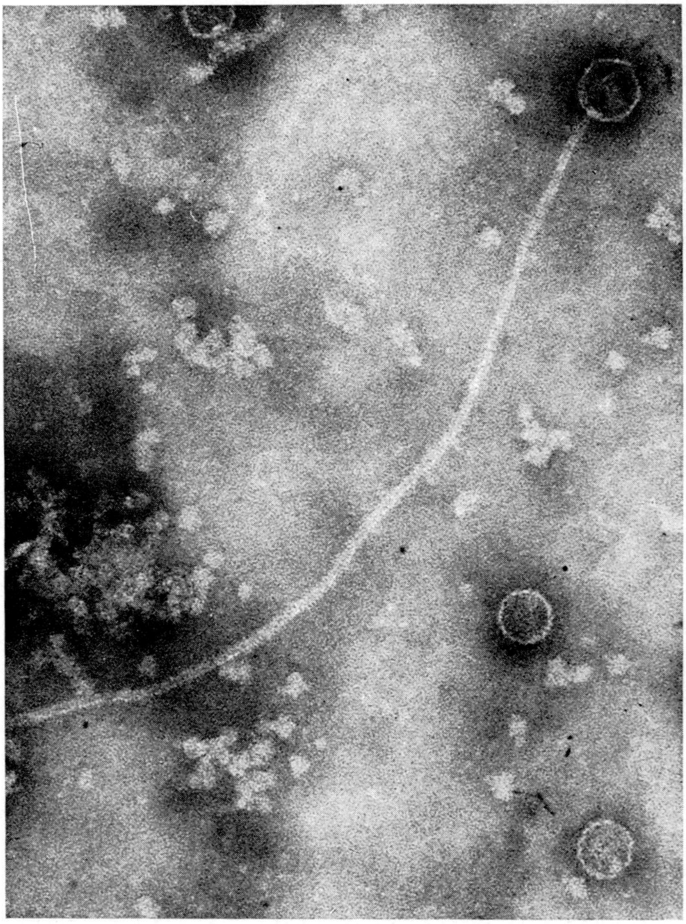

FIG. 7. A giant tail of bacteriophage lambda observed in a lysate of a mutant in gene *U* (Mount *et al.* 1968; reproduced by permission of the editors, *Virology*).

tubes produced *in vivo* show an extremely narrow length distribution, suggesting a constant number of rows of subunits (22 rows in tubes 95 nm in length, according to Eiserling [personal communication]). Only in old lysates (Epstein *et al.* 1963) and in old suspensions of separated tail tubes (To *et al.* 1969) is the distribution of lengths found to be much broader; these observations show that some conditions exist in which the length determination is not functioning. Unlike the situation in λ, no gene has yet been found that controls length in T4. Lambda tails are made of gene product *V*; gene *U*, however, determines the length without being part of the final tail (Murialdo & Siminovitch 1971). If gene *U* is not acting, tails grow to undetermined lengths (Mount *et al.* 1968) (Fig. 7).

The T4 tail tube may provide a good experimental system for testing the hypothesis that length is determined through cumulated strain (Kellenberger 1969; see also this volume, p. 296).

Another example of genetic polymorphism is given by the T4 capsid and capsid-related structures (= capsoids). The product of gene 23, in its native or modified form, is the major component of seven different head-related variants; their capsids or capsoids are shown schematically in Fig. 8, where they are

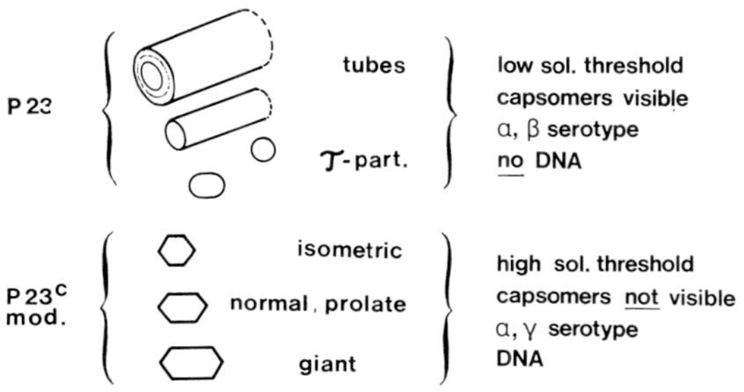

FIG. 8. T4 head polymorphism. The different variants are shown: multilayered tubular forms, single-layered tubular forms and τ particles (prolate and spherical) all made of P23, and phage heads made of P23^{C} in the three variant forms: isometric (Eiserling *et al.* 1970), normal prolate, and giant (A. H. Doermann & F. A. Eiserling, unpublished).

enumerated in the hierarchy of an increasing number of morphopoietic genes needed for contributing supplementary information (Laemmli *et al.* 1970). Through the study of lysates in SDS-acrylamide gels, Dickson *et al.* (1970), Hosoda & Cone (1970) and Laemmli (1970) have found that P23 undergoes a modification of electrophoretic mobility during phage development. A similar modification occurs also for some other gene products (Laemmli 1970). Simultaneously we showed that capsoids of the tubular form as well as lumps are made of the uncut native gene product (61 000 apparent mol. wt.) while the normal capsids are made of the cut (or cleaved) P23^c (49 000 apparent mol. wt.) (Kellenberger & Kellenberger 1970). Since then, A. H. Doermann and F. A. Eiserling in our laboratory (unpublished) have also shown that the capsids of both the small (short) and the giant (long) variants of T4 heads (containing DNA) are made of P23^c. According to whether they are constituted of P23 or P23^c, the seven variants therefore belong to one of two classes, as shown in Fig. 8. The protomers in the capsoids made of P23 are bonded much less strongly than in capsids made of P23^c.

The molecular weight differences were deduced from SDS–acrylamide gels. It is likely, but not yet proven, that the molecular weight modification is real and due to a reduction of about 20% in the length of the polypeptide chain. Such a cutting obviously provides a most efficient way to change the conformation of the protomer.

Recent experiments in our laboratory (to be published) suggest very strongly that the modification P23 to P23^c occurs *in situ*, on the maturing particles. The tentative pathway of head maturation that is emerging is summarized in Fig. 9:

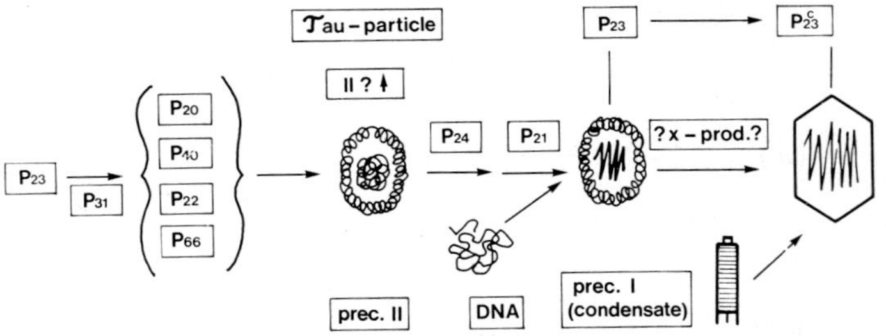

FIG. 9. Tentative pathway of head formation, as discussed in more detail in the text. The identity of precursor particle II with the τ particle observed in the electron microscope is not yet definitely established. Accordingly the aspect of this precursor II is assumed to be that of τ particles. The exact time of DNA packing has not been definitely established. (From unpublished data of R. Bijlenga, D. L. Anderson and E. Lüscher.)

a first precursor ('head precursor II') is made of P23, probably independently of the presence or absence of phage DNA. It is later filled with DNA, increases in size and becomes 'head precursor I', a very fragile structure still made with uncut P23. This precursor is likely to be identical with the so-called condensates (Kellenberger *et al.* 1959, 1968) and the particles produced by mutants in gene 50 (U. K. Laemmli, personal communication). Only during the last steps of head maturation would the transformation P23 to P23^c occur and lead to the stable strongly bonded capsid.

Most of these results were obtained by using a temperature-sensitive reversible mutant in gene 24 and by following the fate of radioactively labelled P23 (unpublished work with R. Bijlenga). P23 is pulse-labelled and chased before a shift is made from a non-permissive (40.5 °C) to a permissive (30 °C) temperature. After the shift, most P23 is transformed into $P23^c$ and is found in mature phage particles. Simultaneously, the τ particles produced at the non-permissive temperature decrease in number, as evidenced by electron microscopy counts. This suggests rather strongly that the P23 containing 'precursor II' is identical with the τ particles. We are now doing experiments to prove this identity.

According to an older concept, DNA condenses first and capsids form around these condensates (Kellenberger *et al.* 1959). At that time it had already been clearly demonstrated that protein ('condensing principle') is necessary for this condensation to occur. It was also postulated that one or several proteins involved in condensation might act independently of DNA and form particles (Kellenberger 1963, 1966, 1969). Later it was shown that the condensates possess a very fragile membrane (Kellenberger *et al.* 1968). Luftig *et al.* (1971) then demonstrated experimentally, by using a temperature-sensitive (*ts*) mutant in gene 49, that organized protein particles are made first and then filled with DNA. However, in the interpretation of their very elegant experiments they assumed that these protein particles are final capsids, an assumption which now seems to be only a very crude approximation. We think that most of the 'empty capsids' observed in the lysate of mutant 49 are breakdown products of initially DNA-containing but extremely fragile particles; this is suggested by our two observations: (1) in thin sections most of the particles indeed contain DNA in variable amounts (Granboulan *et al.* 1971), and (2) the capsids isolated from lysates are made of $P23^c$ (unpublished). The observed small proportion of pre-shift protein which is filled with post-shift DNA could be due to some precursor II which is also present. Fortunately the model of Fig. 9 reconciles most of the experimental data. Experimental proof for each point of the model is complicated by the great fragility of all the postulated precursors. They are indeed very difficult to observe in unaltered form by electron microscopy; DNA is particularly easily lost (Granboulan *et al.* 1971), τ particles cannot be purified without losing their supposedly proteinaceous core, and condensates lose their DNA on lysis and become so distorted that they can no longer be recognized on micrographs.

It was previously shown that the τ particles are already under the control of the form-determining genes, as for example those responsible for the short-headedness (Eiserling *et al.* 1970). All previous experiments demonstrated τ particles as abortive products (work done with U. K. Laemmli, unpublished; D. Scraba *et al.*, unpublished). With our new experimental data obtained on a reversible *ts*-mutant, these particles now become eligible candidates for head precursors. Hence, one has to search for the mechanisms of form determination at the level of τ particles.

References

ASAKURA, S. (1968) A kinetic study of in vitro polymerization of flagellin. *J. Mol. Biol.* **35**, 237–239

CASPAR, D. L. D. (1963) Assembly and stability of the TMV particle. *Adv. Protein Chem.* **18**, 37–118

CASPAR, D. L. D. & KLUG, A. (1962) Physical principles in the construction of regular viruses. *Cold Spring Harbor Symp. Quant. Biol.* **27**, 1–24

DICKSON, R. C., BARNES, S. L. & EISERLING, F. A. (1970) Structural proteins of bacteriophage T4. *J. Mol. Biol.* **53**, 461–474

DURHAM, A. C. H. & KLUG, A. (1971) Polymerization of tobacco mosaic virus protein and its control. *Nat. New Biol.* **229**, 42–46

DURHAM, A. C. H., FINCH, J. T. & KLUG, A. (1971) States of aggregation of tobacco mosaic virus protein. *Nat. New Biol.* **229**, 37–42

EISERLING, F. A., GEIDUSCHEK, E. P., EPSTEIN, R. H. & METTER, E. J. (1970) Capsid size and DNA length: the petit variant of bacteriophage T4. *J. Virol.* **6**, 865–876

EPSTEIN, R. H., BOLLE, A., STEINBERG, E. M., KELLENBERGER, E., BOY DE LA TOUR, E., CHEVALLEY, R., EDGAR, R. S., SUSMAN, M., DENHARDT, G. H. & LIELAUSIS, A. (1963) Physiological studies of conditional lethal mutants of bacteriophage T4D. *Cold Spring Harbor Symp. Quant. Biol.* **28**, 375–394

GRANBOULAN, P. H., SÉCHAUD, J. & KELLENBERGER, E. (1971) On the fragility of phage T4 related particles. *Virology* **46**, 407–425

HOHN, T. (1969) Role of RNA in the assembly process of bacteriophage fr. *J. Mol. Biol.* **43**, 191–200

HOHN, T. & HOHN, B. (1970) Structure and assembly of simple RNA bacteriophages. *Adv. Virus Res.* **16**, 43–98

HOSODA, T. & CONE, R. (1970) Analysis of T4 phage proteins. I. Conversion of precursor proteins into lower molecular weight peptides during normal capsid formation. *Proc. Natl Acad. Sci. U.S.A.* **66**, 1275–1281

KELLENBERGER, E. (1963) Morphogenesis of phage and its genetic determinants. In *New Perspectives in Biology* (SELA, M., ed.), pp. 234–245, Elsevier, Amsterdam

KELLENBERGER, E. (1966) Control mechanisms in bacteriophage morphopoiesis. In *Principles of Biomolecular Organization* (WOLSTENHOLME, G. E. W. & O'CONNOR, M., eds.), pp. 192–226, Churchill, London

KELLENBERGER, E. (1969) Studies on the morphopoiesis of the head of phage T4. VII. Polymorphic assemblies of the same major virus protein subunit. In *Symmetry and Function of Biological Systems at the Macromolecular Level* (ENGSTROM A. & STRANDBERG, B., eds.), pp. 349–366, Almquist & Wicksell, Stockholm

KELLENBERGER, E. & BOY DE LA TOUR, E. (1964) On the fine structure of normal and 'polymerized' tail sheath of phage T4. *J. Ultrastruct. Res.* **11**, 545–563

KELLENBERGER, E. & EDGAR, R. S. (1971) Structure and assembly of phage particles. In *The Bacteriophage Lambda* (HERSHEY, A. D., ed.), pp. 271–295, Cold Spring Harbor Laboratory, Cold Spring Harbor, N.Y.

KELLENBERGER, E. & KELLENBERGER-VAN DER KAMP, C. (1970) On a modification of the gene product P23 according to its use as subunit of either normal capsids of phage T4 or of polyheads. *FEBS (Fed. Eur. Biochem. Soc.) Lett.* **8**, 140–144

KELLENBERGER, E., SÉCHAUD, J. & RYTER, A. (1959) Electron microscopical studies of phage multiplication. VI. The establishment of the DNA-pool of vegetative phage and the maturation of phage particles. *Virology* **8**, 478–498

KELLENBERGER, E., EISERLING, F. A. & BOY DE LA TOUR, E. (1968) Studies on the morphopoiesis of the head of phage T-even. III. The cores of head-related structures. *J. Ultrastruct. Res.* **21**, 335–360

KING, J. (1968) Assembly of the tail of bacteriophage T4. *J. Mol. Biol.* **32**, 231–262

KLUG, A. (1972) The polymorphism of tobacco mosaic virus protein and its significance for the assembly of the virus. *This volume*, pp. 207–214.

LAEMMLI, U. K. (1970) Cleavage of structural proteins during the assembly of the head of bacteriophage T4. *Nat. New Biol.* **227**, 680–685

LAEMMLI, U. K., MÖLBERT, E., SHOWE, M. & KELLENBERGER, E. (1970) Form-determining function of the genes required for the assembly of the head of bacteriophage T4. *J. Mol. Biol.* **49**, 99–113

LAUFFER, M. A. & STEVENS, C. L. (1968) Structure of the tobacco mosaic virus particle; polymerization of tobacco mosaic virus protein. *Adv. Virus Res.* **13**, 1–63

LUFTIG, R. B., WOOD, W. B. & OKINAKA, R. (1971) Bacteriophage T4-head morphogenesis. On the nature of gene 49 defective heads and their role as intermediates. *J. Mol. Biol.* **57**, 555–573

MOODY, M. F. (1967) Structure of the sheath of bacteriophage T4. I. Structure of the contracted sheath and polysheath. *J. Mol. Biol.* **25**, 167–200

MOUNT, D. W. A., HARRIS, A. W., FÜRST, C. R. & SIMINOVITCH, L. (1968) Mutation in bacteriophage lambda affecting particle morphogenesis. *Virology* **35**, 134–149

MURIALDO, H. & SIMINOVITCH, L. (1971) The morphogenesis of bacteriophage lambda III. Identification of genes specifying morphogenetic proteins. In *The Bacteriophage Lambda* (HERSHEY, A. D., ed.), pp. 711–723, Cold Spring Harbor Laboratory, Cold Spring Harbor, N.Y.

STEVENS C. L. & LAUFFER, M. A. (1965) Polymerization—depolymerization of TMV protein. IV. The role of water. *Biochemistry* **4**, 31–37

TO, C. M., KELLENBERGER, E. & EISENSTARK, A. (1969) Disassembly of T-even bacteriophage into structural parts and subunits. *J. Mol. Biol.* **46**, 493–511

WOOD, W. B., EDGAR, R. S., KING, J., LIELAUSIS, A. & HENNINGER, M. (1968) Bacteriophage assembly. *Fed. Proc.* **27**, 1160–1166

Discussion

Levitt: Is there a convincing mechanism for length determination that does not involve a template? You mentioned the core of T4 tails. Here two gene products are involved; one could imagine that one gene made the subunit and the other a long α-helical molecule that determined the length of the core.

Kellenberger: The tail core or tube seems to be composed of one type of subunit only (product of gene 19). The two other gene products seem to be necessary to initiate the tube on the base plate. If they or other genes were involved in the length determination through an α-helix, for example, one should find amber mutants which influence the length. But even if the final tail tube is composed of only one type of subunit this does not exclude a transient action of one or several other proteins during assembly; after completion they would disappear from the structure.

Levitt: It is rather difficult to imagine length determination that does not involve the use of a template.

Kellenberger: We have proposed the 'cumulated strain' model (Kellenberger 1969).

Levitt: That mechanism would give a Gaussian-like distribution of lengths; the width of the Gaussian curve (i.e. the range of lengths) would depend on the increase in strain energy per added subunit. Cumulative strain may give a single length with a few subunits, but here you are dealing with hundreds of subunits.

Kellenberger: Some calculations are probably needed. It will certainly not work with hundreds of subunits! But the tail tube seems to have 22 rows of rings. It is very difficult to make predictions about the energies involved in the deformations. We need the help of protein chemists. Personally I do not think that templates are involved. Do you think of RNA or DNA as templates?

Levitt: No; in this case it would be a long α-helical protein. How long is the core of the phage tail?

Kellenberger: About 90 nm. It should be possible to investigate the length-determining mechanism experimentally on the T4 phage tail system. On base plates as initiators the *in vitro* assembly of gene product P19 to tail tubes should be possible, and also it would provide an excellent system for all sorts of studies.

Katchalski: Dr Kellenberger, can you give some more details concerning the proteolytic enzyme and the assembly mechanism that you have mentioned?

Kellenberger: Experiments are now being done in several laboratories. Nobody has yet been able to identify the proteolytic protein which is responsible for cutting the major head protein, P23. Mutants in about ten different genes result in not cutting the protein P23. The cutting is therefore an obligatory corollary to the right assembly process. Indeed several aberrant assemblies can prevent the cleavage from taking place. We have observed an interesting phenomenon with a temperature-sensitive (*ts*) mutation in the 23 gene itself. Under non-permissive conditions so-called 'crummy' heads are produced (Favre *et al.* 1965). We have now found that in these heads the protein subunit is not cleaved. This does not necessarily mean that the specific site for cleavage on P23 is altered; it is more likely that the wrong type of assembly is abortive and cannot proceed to those later steps of head maturation which involve the cleavage. In summary, we cannot yet say exactly when and how the cleavage occurs in the scheme shown in my Fig. 9 (p. 200). When we know more about precursors I and II and possible steps in between, we can then try to investigate the process of cleavage more thoroughly.

Klug: Although the details are not known, it seems clear from Laemmli's work (1970) that there is a large assembled structure as an intermediate. In other words assembly precedes cleavage, so that the cleavage is not simply a matter of preparing a protein for assembly.

Kellenberger: From recent work in our laboratory (unpublished, with D. L. Anderson and E. Lüscher) with a triple mutant of T4 phage in which the production of late proteins is no longer linked with DNA synthesis (Riva *et al.* 1970), it appears that cleavage needs the physical presence of DNA. It has to be definitively established, however, that the small proportion of cleaved $P23^c$ is dependent on the amount of DNA present in the cell. Organized capsid-like particles are found in very small numbers, barely accounting for the $P23^c$ found. Most of the P23 made is uncut and aggregated in (unorganized) 'lumps'. We have combined the triple mutant with a mutation in gene 21 which normally leads to τ particles. The quadruple mutant still produces τ particles but with only about 30% of the normal yield. This reduction corresponds roughly to the general reduction of phage proteins synthesized. The yield of τ particles in the quadruple mutant is, however, about 100 times higher than that of the capsid-like particles in the triple mutant. Several inconsistencies will have to be settled by experiment before a final model emerges.

Tanford: Do all your numbers from 1 to 60 represent distinct polypeptide chains, or are some of them mutations in different places on the same polypeptide chain?

Kellenberger: Each of these numbers represents one gene, and therefore—by definition—corresponds to one peptide chain. In many genes several mutations in different places are available. They have numbers and letters—e.g. amber-mutation N90 in gene 21, abbreviated as 21 (am N90).

Tanford: So this virus has something of the order of 60 structural pieces?

Kellenberger: More than 80 genes are known, but we estimate T4 to have 100 to 200 genes. About half the known genes are involved in the replication of phage DNA, the others in assembly.

Tanford: Are many of the proteins not incorporated in the virus, but just used to synthesize it and then thrown away?

Kellenberger: It is not yet really known which proteins go into the final structure or, if they are found on the final structure, whether they are structurally relevant there or not. We can easily imagine that a protein, after it has done its job, still stays adsorbed, but no longer has any function. One can test this by seeing whether the functions and activities of the virion are unimpaired by partial extraction of some proteins. Furthermore, the interpretation of electrophoretic lines follows two 'philosophies': in one, the more lines there are, the better it is; in the other, one tries to get a minimum number. Critical work has still to be done.

References

FAVRE, R., BOY DE LA TOUR, E., SEGRÉ, N. & KELLENBERGER, E. (1965) Studies on the morphopoiesis of the head of phage T-even. I. Morphological, immunological and genetic characterization of polyheads. *J. Ultrastruct. Res.* **13**, 318–342

KELLENBERGER, E. (1969) Studies on the morphopoiesis of the head of phage T4. VII. Polymorphic assemblies of the same major virus protein subunit. In *Symmetry and Function of Biological Systems at the Macromolecular Level* (ENGSTROM, A. & STRANDBERG, B., eds.), pp. 349–366, Almquist & Wicksell, Stockholm

LAEMMLI, U. K. (1970) Cleavage of structural proteins during the assembly of the head of bacteriophage T4. *Nat. New Biol.* **227**, 680–685

RIVA, S., CASCINO, A. & GEIDUSCHEK, E. P. (1970) Uncoupling of late transcription from DNA replication in bacteriophage T4 development. *J. Mol. Biol.* **54**, 103–119

The polymorphism of tobacco mosaic virus protein and its significance for the assembly of the virus

A. KLUG

MRC Laboratory of Molecular Biology, Cambridge

Abstract At acid pH, the free protein of TMV aggregates into a helix similar to that found in the intact virus, but at neutral pH the predominant aggregate is the disc which consists of two rings of subunits, both facing the same way, as do successive turns of the helix. The control between these two modes of aggregation resides in an unfavourable juxtaposition, in the helical form, of two carboxyl pairs per protein subunit. This 'negative switch' can be overcome either by protonation at acid pH, or by the energy supplied by interaction between the protein and the RNA, in either case converting the disc form into the helix. This control mechanism ensures that the primary aggregation of the protein is into the disc. This acts as a subassembly for the initiation of virus growth, by recognizing a special sequence of bases at the 5′-end of the TMV-RNA, and also for the subsequent extension of the nucleoprotein rod.

In his paper Professor Kellenberger (1972) mentioned the extensive polymorphism exhibited by the protein of tobacco mosaic virus (TMV) and reproduced the phase diagram my colleagues and I obtained a year or two ago (Durham *et al.* 1971). I should like to add a few remarks on the significance of the polymorphism and the mechanism of its control. Most of the results have now been published or are in the press. I shall therefore deal only with the salient points which may well turn out to have parallels in other systems.

We now know that this polymorphism, far from being an adventitious and bothersome feature associated with a 'sticky' protein, is the basis for the rapid and specific assembly of the virus from its protein and RNA (Butler & Klug 1971). We believe that all the different aggregates are the result of the same specific interactions as are found in the virus but are slightly modified in the different polymorphic forms (cf. Caspar 1963). In the assembly of the virus the interaction with the RNA switches the mode of bonding of the protein from that of one polymorphic form, the disc, to the other important form, the helix.

MODES OF AGGREGATION

The aggregation of TMV protein can occur in one of two modes. First to be discovered and understood was the helix, in which the arrangement and conformation of the protein units is identical to that in the virus (Fig. 1*a*). The aggregation of the protein in this mode can be described as quasi-crystallization and the resulting rods are of indeterminate length, involving many thousands of molecules.

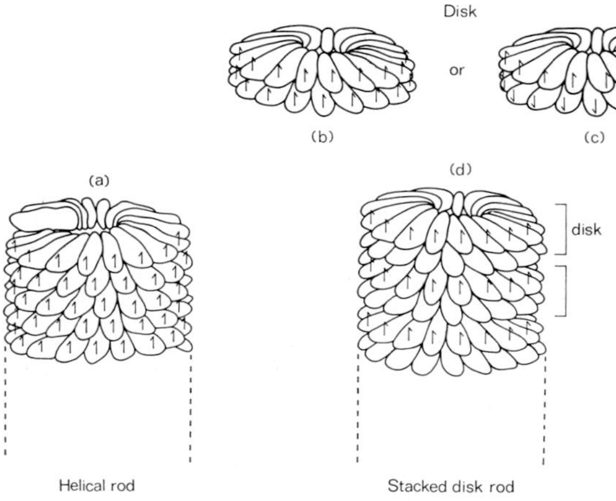

FIG. 1. The relationship of the polarity of the disc form of TMV protein to that of the helix.

The second mode can be described as two-layer polymerization and the principal structure exemplifying this mode is the 'disc', which consists of two rings or layers, each of 17 protein subunits (Fig. 1*b* or *c*). The existence of a stable structure of this kind poses a structural problem. The lateral bonding between subunits in a disc is clearly similar to that in the virus helix because the number of subunits per ring, 17, is close to the number of subunits in a turn of the virus helix, 16.34. If, however, the disc structure were a simple variant of the helical structure, in which successive rings of the disc face in the same direction as do successive turns of the helix (Fig. 1*b*), then it is difficult to see how a polar stacking of this kind would stop at the two-ring stage and not continue indefinitely. A far simpler explanation of the pairing between rings would be that they face in opposite directions and are thus related by twofold axes perpendicular to the axes of the rings (Fig. 1*c*). This would provide a natural explanation for the aggregation being limited to two rings, but the bonding between the rings would then be polar and unrelated to that between turns of the virus helix. This question has been resolved by the analysis of electron micrographs of long rods in which the discs stack end-to-end (Fig. 1*d*). The three-dimensional image reconstruction shows that the structure is a polar one in which the two rings of the disc face the same way (Finch & Klug 1971). The outer parts of the subunits in the two rings comprising a disc move towards each other in an axial direction by distances of the order of 5 Å and it is this pairing which is presumably responsible for the two-layer structure of isolated discs. We therefore describe the second mode of aggregation of TMV protein as two-layer polymerization, aggregates of this type being stabilized by an additional 'pairing' interaction, which is not present in the helix, between subunits vertically above each other (Durham & Klug 1971).

THE CONTROL OF AGGREGATION

The driving force for aggregation of TMV protein is largely hydrophobic bonding, since increases in both temperature and ionic strength favour the formation of large aggregates (Lauffer & Stevens 1968). Neither, however, has a major effect in controlling the mode of aggregation, and the variable which appears to distinguish the mode is the pH. The exact pH at which the transition between the modes occurs depends slightly on the temperature and ionic strength, but it is always about pH 6 to 7, with the helical aggregation predominating below the transition and the two-layer aggregates above. This transition pH is found to correspond to the p*K* values of two carboxyl groups, which titrate abnormally in the virus (Caspar 1963) but not in the stacked disc rods (Durham & Klug 1971), and in the titration of the protein helix an abnormally sharp release or uptake of protons correlates with the disaggregation or formation of the helix (Klug & Durham 1971; Butler 1971).

Because the helical structure is known to occur at acid pH and the disc structure has basically the same kind of bonding as in the helix, it ought to be possible to convert protein directly from disc to helix by lowering the pH, without first dissociating it into separate units. Experiments in which the pH was suddenly lowered and the state of the protein followed with the electron microscope showed that short imperfect helical rods were formed within 5 seconds and then annealed over a period of about a day to give long perfect helices. Close examination of the imperfect helices shows that they are composed of two-turn segments of helix stacked on top of each other, with their ends not in register, each segment appearing to be formed from a single disc which has dislocated under the influence of the high proton concentration (Durham *et al.* 1971).

Since the subunits in any single disc must change their mode of aggregation together, any such transition is inherently cooperative and can be considered to be allosteric, with the protons as the allosteric effectors. The effectors act by protonating two pairs of carboxyl groups to give two carboxyl-carboxylate pairs. This abolishes the repulsion between the charged carboxyl groups and concomitantly the pairing interaction of the two-layer aggregate, enabling the subunits to take up the strictly equivalent positions of the helix (see Fig. 2*a*). Were it not for the carboxyl-carboxyl pairs the bonding in the helix would presumably be more favourable than that in the disc, but the unfavourable interaction of these groups acts as a negative switch to prevent the helix being the most favoured aggregate in all conditions.

THE FUNCTION OF THE DISC

In physiological conditions of pH and ionic strength the dominant form in the association equilibrium of TMV protein is the disc. This key position of the disc in the phase diagram suggested that it had an important biological function. One possible role for the disc is in the initiation or nucleation of assembly of the virus particle. Once the nucleoprotein helix has been started, the continuing growth is readily comprehended: subunits could add to the growing end of the rod, just as in growth at a screw dislocation in a crystal. The problem lies in the nucleation of helix growth: the formation of a single ring of protein subunits, each of which is bonded to only one neighbour on each side, is clearly a much less favourable process. The disc made of two layers of subunits represents a more stable bonding arrangement because the structure is a cross-strutted one and it could provide a pre-formed surface to which the RNA could bind; growth could then proceed from this first turn. It has already been seen that a disc is capable of being transformed into a small stretch of helix, when protons are added to overcome the negative switch provided by the carboxyl-carboxyl pairs. In the relevant biological process this could provide a mechanism for helix initiation, but brought about by interaction with the viral RNA rather than by protonation (Fig. 3).

The experiments begun with the aim of testing the above hypothesis that the protein disc is the nucleus for the formation of helix during reconstitution with RNA showed this hypothesis to be true, but unexpectedly they further showed that the disc is needed not only for initiation, but also for growth of the nucleoprotein helix (Butler & Klug 1971). When discs are present, growth of the virus does not proceed by the addition of single protein molecules, but rather by the incorporation of a whole disc at once. Although addition of one ring of subunits to a growing nucleoprotein helix presents no particular

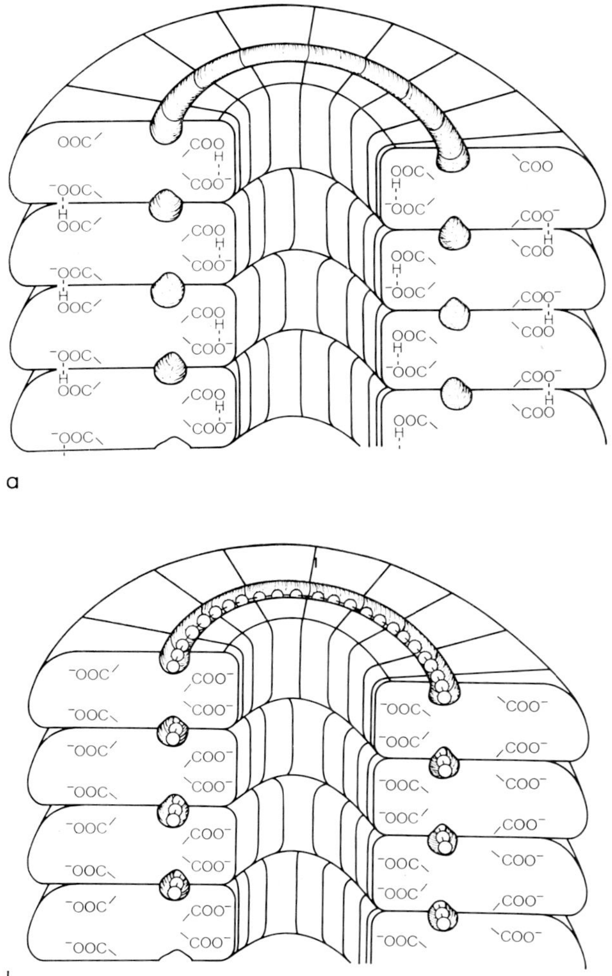

FIG. 2. (*a*) Two pairs of carboxylic acid groups, which have normal p*K* values in the small protein aggregates and the discs, are constrained to come close together in the helical structure, resulting in an abnormal p*K* of 7.1. The protein alone cannot, therefore, form the helix except at acid pH when protons are supplied which stabilize the structure by forming carboxyl-carboxylate hydrogen bonds.

The anomalously titrating groups are at distances of 24 Å and 84 Å from the helix axes (Caspar 1963).

(*b*) The helical structure can be formed at neutral and alkaline pH when the RNA is supplied. The interaction of the nucleotide sites (at a radius of 40 Å) provides enough energy to stabilize the helix against the repulsion of the pairs of juxtaposed carboxyl groups.

The diagram represents the state of the virus at a pH of, say, 8. At pH 7 the carboxylic groups will be half protonated, so partly relieving the unfavourable interaction between them.

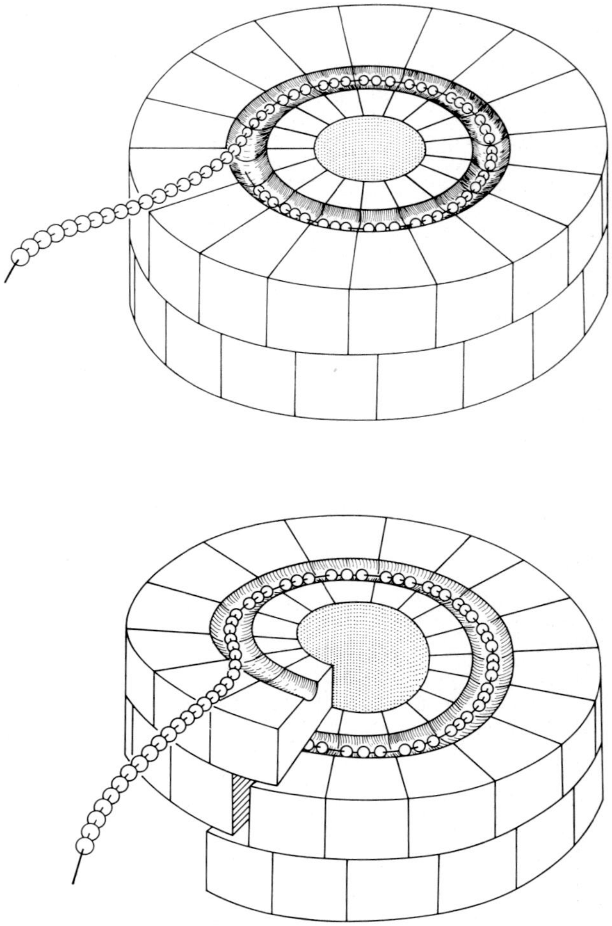

FIG. 3. The specific recognition of a special terminal sequence of TMV-RNA initiates conversion of the disc form of the protein into two turns of the helix.

conceptual difficulty, it is rather difficult to envisage how two rings can be added in such a way as to incorporate the RNA between the first and second turns. However, it is probably not too difficult for the RNA to intercalate between the first and second turns from the outside, because the three-dimensional density distribution of rod forms of the protein shows holes of considerable radial extent between subunits that are adjacent in the axial direction. The subunits can therefore presumably deform or move to help the RNA on its way in.

The nucleation of the helix assembly by the disc furnishes a mechanism for recognition by the protein of its homologous RNA. A particularly favourable interaction occurs between the disc and a special sequence of about 50 nucleotides at the 5′-hydroxyl end of TMV-RNA (Butler & Klug 1971). The initiation of nucleoprotein helix growth with other RNA's or synthetic polynucleotides, or indeed TMV-RNA itself when its 5′- end has been digested enzymically, is much slower than with TMV-RNA and so accounts for the selectivity of the TMV protein for its own RNA. If this interpretation is correct then it is possible to study the initiation of assembly by using one of these RNA's and so slowing down initiation until it, rather than growth, becomes the rate-limiting step. These experiments were done by Butler (1971), with poly(A) as the RNA, and his results show a striking difference from the kinetics of the assembly when TMV-RNA is used. He found that the rate of reaction depends approximately on the second power of the disc concentration and therefore suggests that two discs are involved in the initiation process.

The possible role of the discs in nucleation of TMV assembly was foreseen; indeed it provided the motive for our first experiments with discs and the RNA. However the continued growth of rods with discs as the protein source was completely unexpected since it was always assumed that the virus would grow by adding single subunits at the end of the growing helix, rather like crystal growth along a screw dislocation. With hindsight, the growth of the nucleoprotein rod by the incorporation of whole discs has several advantages. For example, it allows short unfavourable nucleotide sequences along the RNA to be coated by the protein, since the concerted transition of the disc to a helical segment produced by favourable interaction with nearby stretches of RNA will overcome any local unfavourable effect.

CONCLUSION

The major control in the assembly of the virus protein lies in the pairing of the carboxyl-carboxylate groups that was discussed earlier. By preventing the aggregation of the protein alone into the helical form, when RNA is not present, this ensures the availability of a supply of discs that can act as a subassembly for the complete assembly of the TMV particle. These discs are not, however, incorporated into the growing particle as such, but are transformed by the interaction with the RNA, which provides enough energy to overcome the negative switch and allows the protein subunits to take up their final mode of aggregation, the nucleoprotein helix (Fig. 2*b*).

In the language of allosteric theory, the protein is in the tense state in the virus helix because of the strained environment for the carboxyl groups. It is the disc and small aggregates of the protein which correspond to the relaxed state, presumably because movements of the subunits are allowed, which obviate the necessity for the carboxyl groups to approach close to one another. In this sense it can be said that the natural state of aggregation of TMV protein is the two-layer disc. By binding of an appropriate ligand, the disc undergoes a transition to a form with a different but related structure, the helix. The system of TMV protein association thus shows strong similarities with other biological systems, such as haemoglobin, in which there are ligand-controlled transitions between ordered forms with different quaternary structures.

The polymorphism of TMV protein is now seen to be the basis of significant control mechanisms, both at the initiation step of assembly and during elongation of the nucleoprotein helix. One might say that TMV is not only a self-assembling system, but also a self-nucleating one, and that the mechanism by which this is achieved is now understood in outline.

References

BUTLER, P. J. G. (1971) *Cold Spring Harbor Symp. Quant. Biol.* **36**, 461–468
BUTLER, P. J. G. & KLUG, A. (1971) *Nat. New Biol.* **229**, 47–50
CASPAR, D. L. D. (1963) *Adv. Protein Chem.* **18**, 37–121
DURHAM, A. C. H. & KLUG, A. (1971) *Nat. New Biol.* **229**, 42–46
DURHAM, A. C. H., FINCH, J. T. & KLUG, A. (1971) *Nat. New Biol.* **229**, 37–42
FINCH, J. T. & KLUG, A. (1971) *Philos. Trans. R. Soc. Lond. Ser. B*, **261**, 211–219
KELLENBERGER, E. (1972) Assembly in biological systems. *This volume*, pp. 189–202
KLUG, A. & DURHAM, A. C. H. (1971) *Cold Spring Harbor Symp. Quant. Biol.* **36**, 449–460
LAUFFER, M. A. & STEVENS, C. L. (1968) *Adv. Virus Res.* **13**, 1–63

Discussion

Lowey: At low pH, in the absence of RNA, how long do the TMV particles grow?

Klug: In the electron microscope John Finch has observed rods up to 2 μm long.

Lowey: There is then no regulation of the length by the protein alone.

Klug: I don't think the final length is regulated; the rods presumably just break.

Revel: How do you visualize the interaction between TMV-RNA and protein, specifically in the first disc? Is one site on the RNA recognized by the proteins and then the other discs added without RNA recognition?

Klug: I think the RNA recognition takes place only at the beginning. However it does not seem to involve merely a few bases but a stretch about 40 bases long, presumably with a special sequence which interacts particularly strongly with the sites on the protein which house the nucleotides. In the complete virus each protein unit has three nucleotides associated with it, so that each protein unit carries three sites on it. Over the whole length of the RNA, the sequence of bases is random, so any base must be able to go into any of the three sites, but presumably with different energies. That, incidentally, is probably why discs are used in rod growth rather than single subunits. If a single subunit were added, it would not add if there were a run of three unfavourable bases. But, by supplying a whole disc, a short unfavourable stretch can be overcome.

Returning to the question of initiation, one might have imagined *a priori* that there is a special triplet at the end of the RNA which triggers the whole thing off. But the evidence is that about 40 nucleotides have to be digested away from the 5′ end before the efficiency of TMV-RNA for reconstitution is reduced to the level found for, say, poly(A). This suggests that the initiation sequence on the RNA is recognized by the front face of the disc, which has $17 \times 3 = 51$ nucleotide sites on it. Because of the 17-fold rotational symmetry of the disc, this sequence may have repeating triplets in it.

We do not yet know what the sequence is. We do know that poly(A) will form rods with the discs (in contrast with poly(C) and poly(U), which do not), but the reconstitution is very slow compared with TMV-RNA. The efficiency is somewhat greater at lower pH, as Fraenkel-Conrat & Singer (1964) showed, presumably because this helps the protein towards the helical state: in a manner of speaking the poly(A) is dragged along.

Eigen: Could you cleave off the end and add it to poly(A) or to some other polynucleotide by using a ligase?

Klug: This might be feasible, but the special nucleation sequence is at the 5′ end and not easy to get. P. J. G. Butler is trying to extract it from partially reconstituted rods to find out what it is.

Gilbert: Is the initiation point at the 5′ end of TMV?

Klug: It definitely is the 5′ end. Digesting away the 3′ end has no effect on reconstitution. Even when the RNA chains are cleaved by ribonucleases, the initial stages of the reconstitution go on as fast as before because the number of 5′ ends is substantially the same.

Reference

FRAENKEL-CONRAT, H. & SINGER, B. (1964) *Virology* **23**, 354

Protein interactions in the myofibril

SUSAN LOWEY

Children's Cancer Research Foundation, and Department of Biological Chemistry, Harvard Medical School, Boston, Massachusetts

Abstract The myofibril, the contractile unit of muscle, is a complex assembly of a variety of proteins. Some of these proteins are globular in conformation, such as actin or troponin; some are highly α-helical, such as tropomyosin; while the most abundant protein, myosin, combines both globular and fibrous features in one large molecule. With the exception of actin, all of these proteins consist of more than one polypeptide chain. Myosin has the most unusual subunit structure, being composed of two major chains each with a molecular weight of 200 000 daltons and four small subunits in the range of 20 000 daltons. All these proteins exist as part of highly organized structures consisting of several hundred molecules; none function independently as monomers. In the thick filament, the predominant molecular species is myosin; evidence for other proteins is inconclusive. The thin filament, however, has been shown to contain the regulatory proteins troponin and tropomyosin in addition to actin. Other proteins identified in the myofibril are α-actinin in the Z band, and M protein(s) in the M line of the sarcomere. This paper will summarize the properties of the individual proteins, describe their assembly into filamentous structures, and discuss the kinds of interactions within filaments and between filaments which may lead to a contractile force.

The contractile unit of a muscle fibre, the myofibril, may be thought of as a highly concentrated suspension of proteins (about 20%) enclosed by a sleeve-like membranous network known as the sarcoplasmic reticulum. This membrane system is not connected directly to the fibre surface, but lies in close proximity to a second membrane system, the T system or transverse tubules, which is continuous with the plasma membrane (Fig. 1). When an action potential passes over the surface of a fast muscle fibre, a wave of depolarization spreads into the interior of the cell by way of the T system. These tubules enwrap each fibril either at the level of the Z line (in frog sartorius muscle) or at the level of the A- and I-band junctions (in mammalian and some fish muscles). Adjacent to the transverse tubules are morphologically distinct regions

FIG. 1. Longitudinal section through a frog muscle. Triads consisting of two terminal cisternae (tc) and a centrally-located transverse tubule (tt) are seen adjacent to the Z lines (Z). (From Peachy 1965.) ic: intermediate cisternae; fc, fc′: fenestrated collars; lt, lt′: longitudinal tubules; gly: glycogen granules; gr: small granules in I bands (I); A: A bands; H, L: H and L zones; M: M lines.

of the sarcoplasmic reticulum known as the 'terminal cisternae'. They release stored calcium into the myofibril as a result of the depolarization wave, thereby signalling the onset of contraction (see review by Peachy 1968). Relaxation is restored when calcium is transported back into the reticular vesicles. Thus contraction is controlled by a biochemical process within the muscle cell which regulates the concentration of ionized calcium (Weber 1966). How calcium transmits its signal to the contractile proteins is a fascinating example of the interplay of different proteins within a polymeric structure.

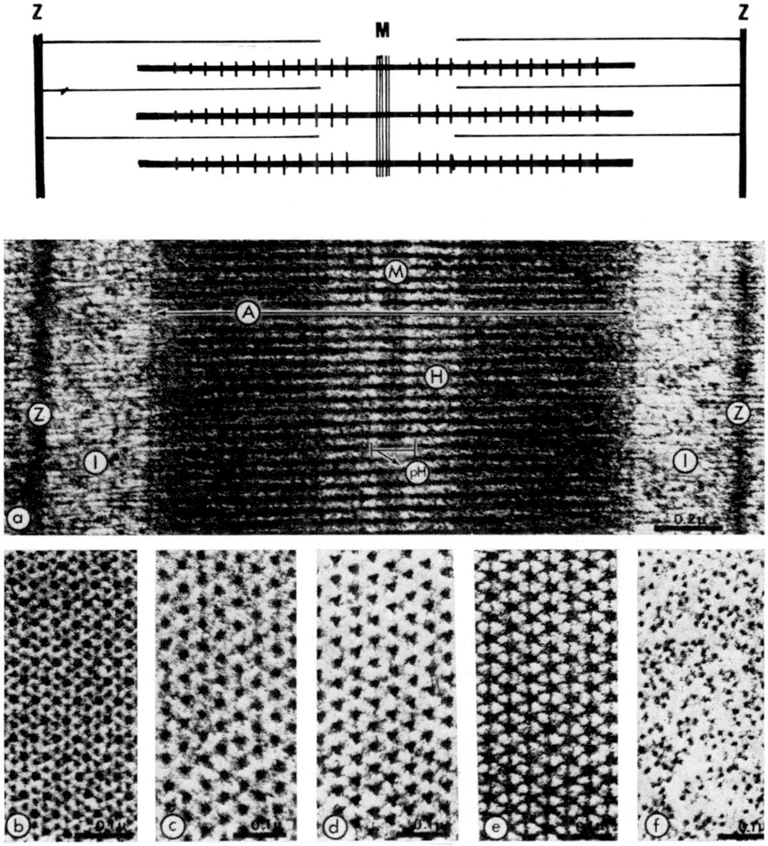

FIG. 2. Top: Schematic representation of the repeating unit (sarcomere) of a myofibril.
Below: From Pepe (1971, p. 323), by courtesy of Marcel Dekker, Inc.: (*a*) Longitudinal section through the sarcomere of the lateral muscles of the freshwater killifish.
Cross-section through the A band in the region of (*b*) overlap; (*c*) the H zone; (*d*) the pseudo-H(pH) zone; and (*e*) the M line. (*f*) Cross-section through the I band.

STRUCTURE OF THE MYOFIBRIL

Before discussing the properties of the individual proteins and their interactions, I shall summarize briefly the organization of the filaments within each sarcomere, the repeating unit of a myofibril. Electron micrographs of muscle reveal two sets of filaments arranged in hexagonal arrays (Hanson & Huxley 1953) (Fig. 2). The thick filament is approximately 1.6 µm long and consists of several hundred myosin molecules. Cross-bridges project along the entire length of the filament except for a bare region (pseudo-H zone) of about 0.2 µm at the centre of the filament. As we shall see, the unusual bipolar shape of the myosin filament clearly reflects the distinctive polar shape of the myosin molecule (Huxley 1963). Attached to the bare region of the thick filament is the M protein(s), which in cross-sections of certain muscles is seen to form bridges between the myosin filaments (Fig. 2e) (Pepe 1971).

The thin filament is about 1 µm long and composed chiefly of actin molecules and the less abundant regulatory proteins, tropomyosin and troponin (Ebashi & Endo 1968). The thin filaments appear to be anchored in the sarcomere by the Z line, a structure shown to possess a considerable degree of two-dimensional order (Huxley 1963). The protein composition of the Z line is still uncertain, but α-actinin has been identified as one of the constituents (Masaki *et al.* 1967).

A necessary condition for contraction is overlap between the thin and thick filaments (the A-band region). When the fibre is stimulated, the thin and thick filaments interact with one another and generate the necessary force to move the actin filaments, relative to the myosin filaments, towards the centre of the sarcomere (Huxley & Hanson 1954). One of the roles of the more minor Z- and M-line structures may be to preserve the integrity of these highly ordered arrays of filaments during the disruptive force of the contractile event.

PROTEINS OF THE MYOFIBRIL

Ever since Kühne (1868) discovered that a viscous protein could be extracted from muscle by strong salt solutions, people have been finding increasing numbers of proteins in muscle tissue. Kühne's original 'myosin' preparation was later shown to contain actin in addition to myosin and was renamed actomyosin (Straub 1942; Banga & Szent-Györgyi 1942). But the description of actomyosin did not end there; some twenty years later, Ebashi & Ebashi (1964) discovered that actomyosin, as extracted from muscle, contained two additional proteins, tropomyosin and troponin. And so the list of isolated myofibrillar proteins has grown to include at least half a dozen proteins whose function and location in the sarcomere are reasonably well-defined.

A general outline of procedures for the fractionation of the muscle proteins

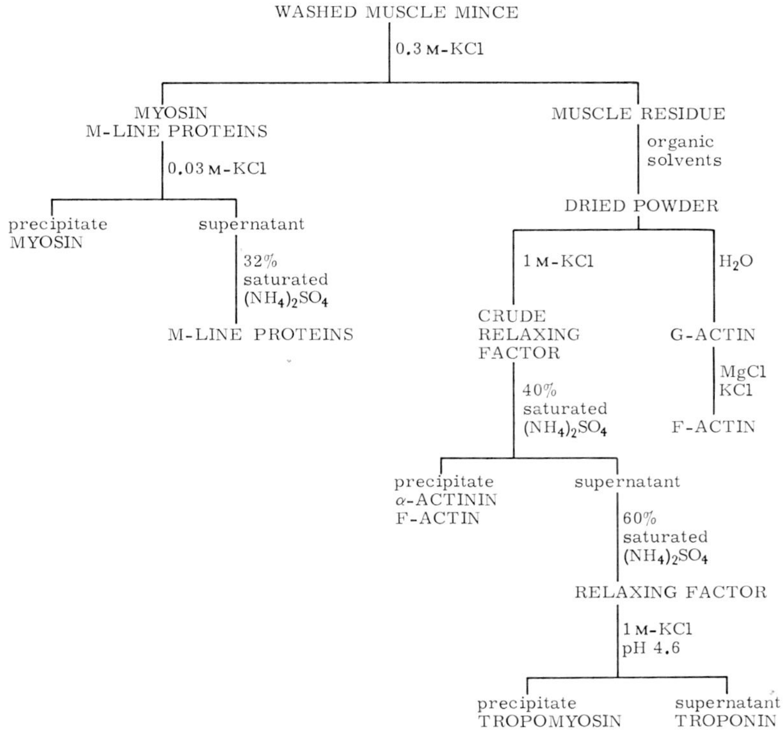

FIG. 3. Outline of general procedures for the fractionation of myofibrillar proteins. Final purification steps are omitted.

is presented in Fig. 3. The underlying rationale for the separation is based on two characteristics of the fibrous proteins: (1) fibrous proteins, with the important exception of myosin, are remarkably resistant to denaturation by organic solvents such as acetone or ether, and (2) the fibrous proteins cover a wide range of solubilities, with myosin the least soluble and G-actin the most soluble member of the group. The intermediate proteins can be separated by an appropriate choice of ammonium sulphate concentration and pH. (Instead of denaturing the residual myosin in the muscle residue with organic solvents, one could extract the residue with slightly alkaline water. However, in our experience, the yield of soluble proteins is much lower by this procedure.)

One of the most sensitive and informative techniques for determining the size and subunit composition of a protein is electrophoresis on acrylamide gels in the presence of sodium dodecyl sulphate (SDS) and a reducing agent. Fig. 4 gives a comparison of the electrophoretic profile of the total myofibrillar extract with that of the individual proteins. An important conclusion to be

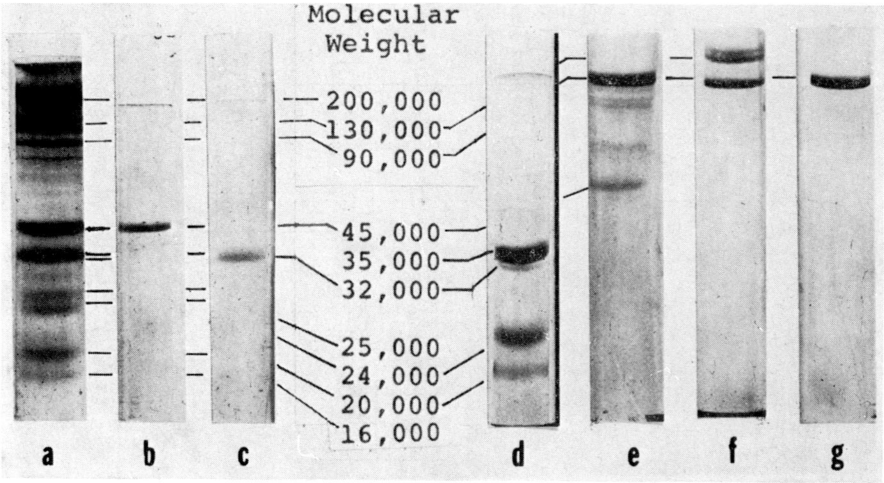

FIG. 4. Characterization of the myofibrillar extract and individual isolated proteins by electrophoresis on SDS polyacrylamide gels. (*a*) Total extract from chicken muscle; (*b*) rabbit actin; (*c*) chicken tropomyosin; (*d*) rabbit troponin; (*e*) chicken α-actinin; (*f*) chicken M-line proteins isolated by ammonium sulphate fractionation; (*g*) M-line protein isolated by ion-exchange chromatography of material in (*f*).

drawn from this study is that most of the proteins in the myofibril have probably been identified. The number of 'new' proteins to be discovered must be very limited. Another interesting feature of these gels is the diversity of subunit molecular weights: the sizes range from 16 000 to 200 000 daltons.

Table 1 lists the molecular weights of the fibrous proteins in solvents chosen to minimize aggregation and yet not denature the proteins. Several interesting points emerge from this table: with the exception of actin, all the fibrous proteins consist of more than one polypeptide chain (Young 1969). In the structural proteins tropomyosin (Woods 1967), α-actinin (Goll *et al.* 1971; Margossian 1972) and one of the M-line proteins (S. Lowey & R. G. Harrison, unpublished data), the stable functional unit consists of two chains of identical mass. Troponin is a more complex molecule; it appears to have three subunits in equimolar amounts (Fig. 4*d*) (Greaser & Gergely 1971), although claims for two subunits have also been reported (Ebashi *et al.* 1971). The complexity of the troponin molecule reflects its multiple functions. The 20 000 dalton subunit is the major site for calcium binding in the myofibril, while the higher molecular weight subunits bind to tropomyosin and inhibit the ATPase activity of actomyosin. The 20 000 dalton subunit, when saturated with calcium, acts to release the inhibition by the 24 000 component, thereby allowing hydrolysis of ATP to occur (Hartshorne *et al.* 1969). The myosin molecule is another example

TABLE 1

Properties of contractile proteins†

Localization in myofibril	*% Total protein*	*Protein*	*Sedimentation coefficient*	*Viscosity (ml/g)*	*Molecular weight*	*Subunit mol. wt.*	*% α-helix*
Thick filament	55	Myosin	6.4	210	470 000	200 000 20 000	57
Thin filament	25	G-actin	3.3	4	46 000	46 000	26
Thin filament	5	Tropomyosin	2.6	34	64 000	32 000	90
Thin filament	5	Troponin	4.0	4	80 000	35 000 24 000 21 000	35
Z line	tr	α-actinin	6.2	9	180 000	90 000	60
M line*	tr	M-protein(s)	9.5 (15.0) 6.0	— — —	180 000 (360 000) —	90 000 (90 000) 140 000	40 (40) <10

† Compiled from references in text.
* Assigned to M line on the basis of preliminary fluorescence antibody staining experiments.
tr: traces.

of a fibrous protein with subunits of different sizes. Here the differences are particularly striking: one type of chain has a molecular weight of about 200 000 daltons, and the other chains are about 20 000 daltons. In view of myosin's complexity in structure and function, a separate section will be devoted to a discussion of its properties.

Before leaving the subject of the protein composition of the myofibril, I want to give an example of how these proteins have been identified with a particular region of the sarcomere. If antibodies are prepared against purified α-actinin (or against the proteins presumed to be derived from the M line), a fluorescent dye can be covalently attached to the immune globulin and the fluorescein-labelled antibodies used to stain the myofibrils (Fig. 5) (Lowey & Steiner 1972). A comparison of the myofibrils by fluorescence and phase microscopy clearly shows that anti-α-actinin stains the Z line and that anti-M proteins stain the M-line region of the sarcomere (S. Lowey, R. G. Harrison & S. S. Margossian, unpublished data).

a a′

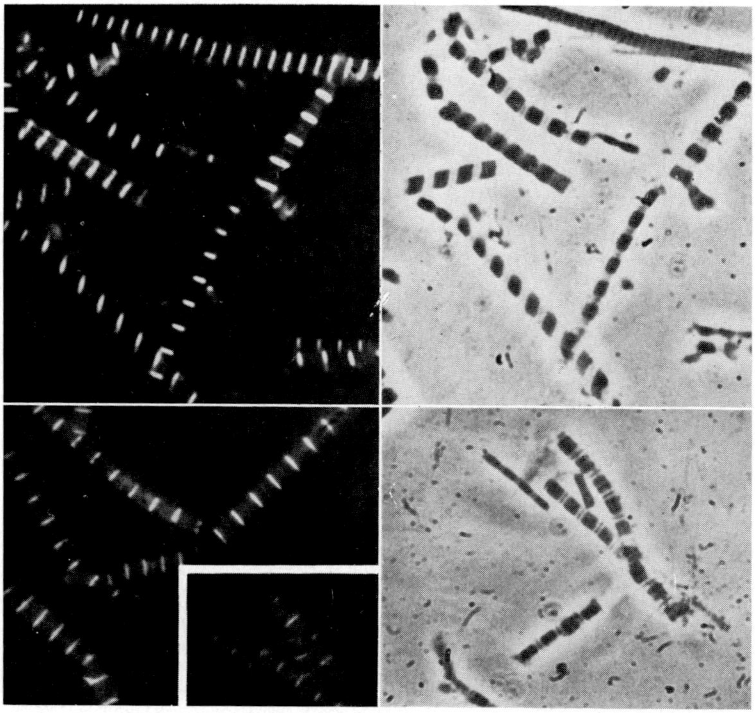

b b′

FIG. 5. Fluorescent (*a*, *b*) and phase-contrast (*a*′, *b*′) images of fields of myofibrils stained with antibodies prepared against unfractionated M-line proteins (*a*) and against α-actinin (*b*).

THE MYOSIN MOLECULE

The bulk of the myosin molecule is composed of two large polypeptide chains, each about 200 000 daltons (Gershman *et al.* 1969; Gazith *et al.* 1970; Lowey *et al.* 1969); the remaining 10 to 15% of myosin seems to consist of either two or three classes of low molecular weight subunits, the number depending on the type of muscle (Fig. 6). Myosin from fast skeletal muscles (Fig. 6*a*, *b*, *c*) contains approximately 4 moles of light chains: 2 mol of an 18 000 dalton subunit and a total of 2 mol of the 25 000 and 16 000 dalton subunits (Lowey & Risby 1971; Weeds & Lowey 1971). Myosin from slow muscles (Fig. 6*d*, *e*) seems to have equimolar amounts of a 27 000 and 20 000 dalton subunit (Lowey & Risby 1971). The relationship between the light chain pattern and the speed of the muscle adds support to the suggestion that the light chains may be involved in regulating the hydrolytic activity of myosin (Stracher 1969; Dreizen & Gershman 1970). However, one class of light chains, the 18 000

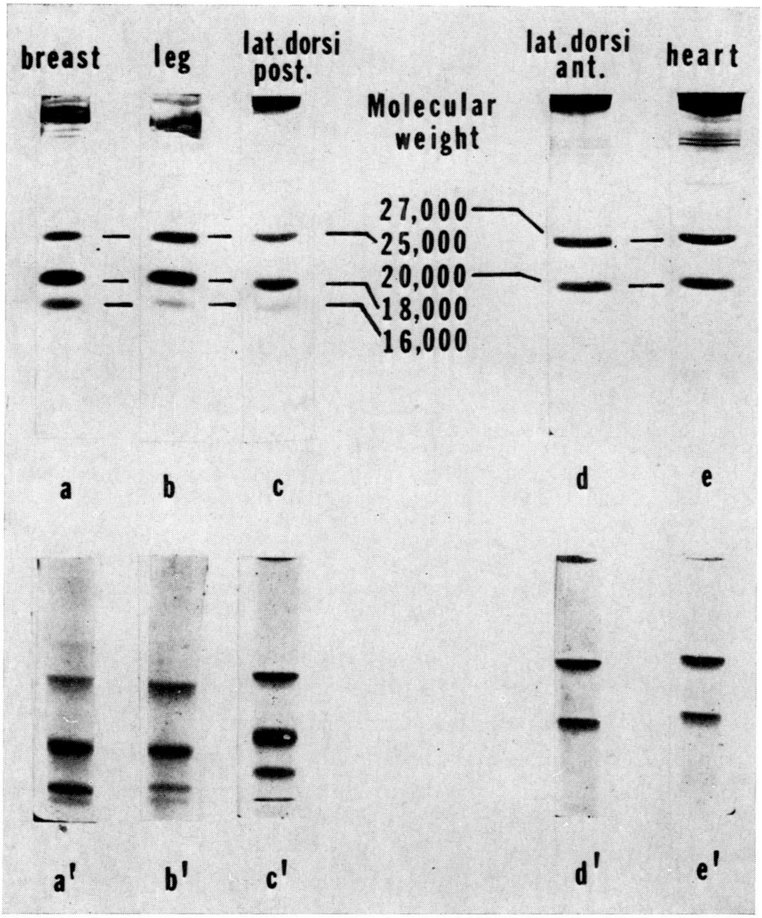

FIG. 6. Light chains from chicken muscles. (*a*) to (*e*) are 10% SDS polyacrylamide gels; (*a'*) to (*e'*) are standard gels of light chains isolated from the myosins depicted in (*a*) to (*e*). (From Lowey & Risby 1971, by permission of *Nature*.)

molecular weight species in fast muscle myosins, does not seem to be essential for ATPase activity. If myosin is reacted with 5,5′-dithiobis-(2-nitrobenzoic acid) (DTNB), an appreciable portion of the 18 000 dalton subunits are dissociated from myosin (Fig. 7*d*) without significantly altering the enzymic activity of the myosin (Fig. 7*b*) (Gazith *et al.* 1970). The remaining light chains can only be removed by denaturing the myosin (Fig. 7*c*). The function of the DTNB-light chains remains unknown. It is important to note that the composition of the DTNB-light chains is chemically quite distinct from that

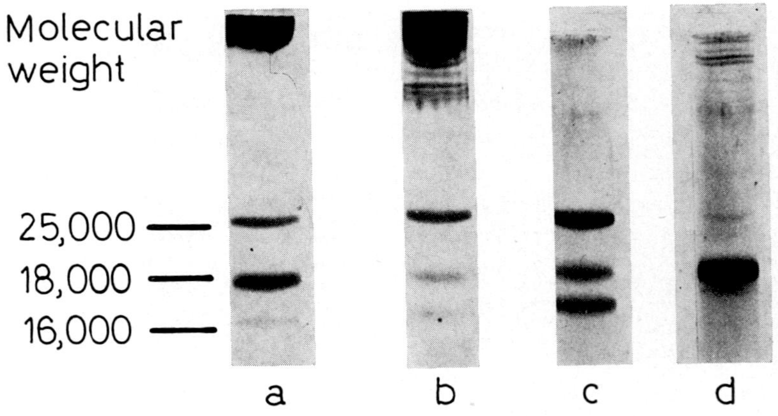

FIG. 7. (*a*) Electrophoresis of native myosin on SDS polyacrylamide gels; (*b*) myosin after treatment with DTNB; (*c*) light chains removed from DTNB-treated myosin with denaturing solvents; and (*d*) the light chain liberated from myosin by the DTNB reaction.

of the 16 000 and 25 000 dalton subunits which are chemically related in sequence (Weeds & Lowey 1971).

A schematic model of myosin based on the studies described above and on the many studies of native myosin and its proteolytic subfragments is shown in Fig. 8. Myosin is depicted as a polar molecule consisting of two similar halves; each half contains one heavy chain and a minimum of one light chain—probably two in myosin from fast muscles. The conformation and tertiary structure of myosin has been largely deduced from studies of its proteolytic fragments by

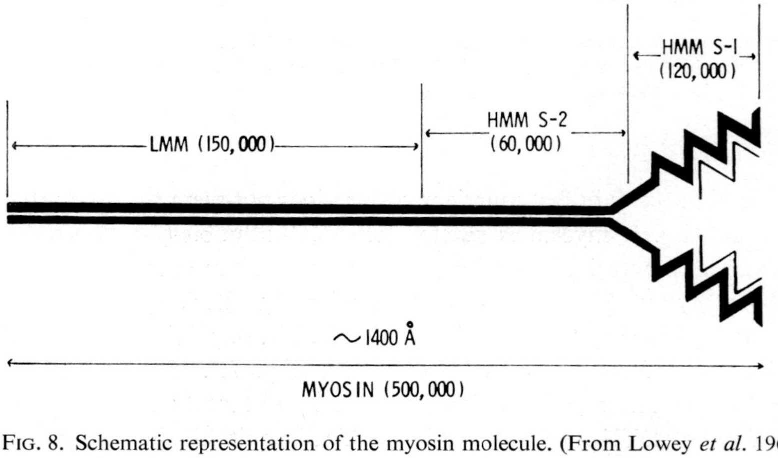

FIG. 8. Schematic representation of the myosin molecule. (From Lowey *et al.* 1969.)

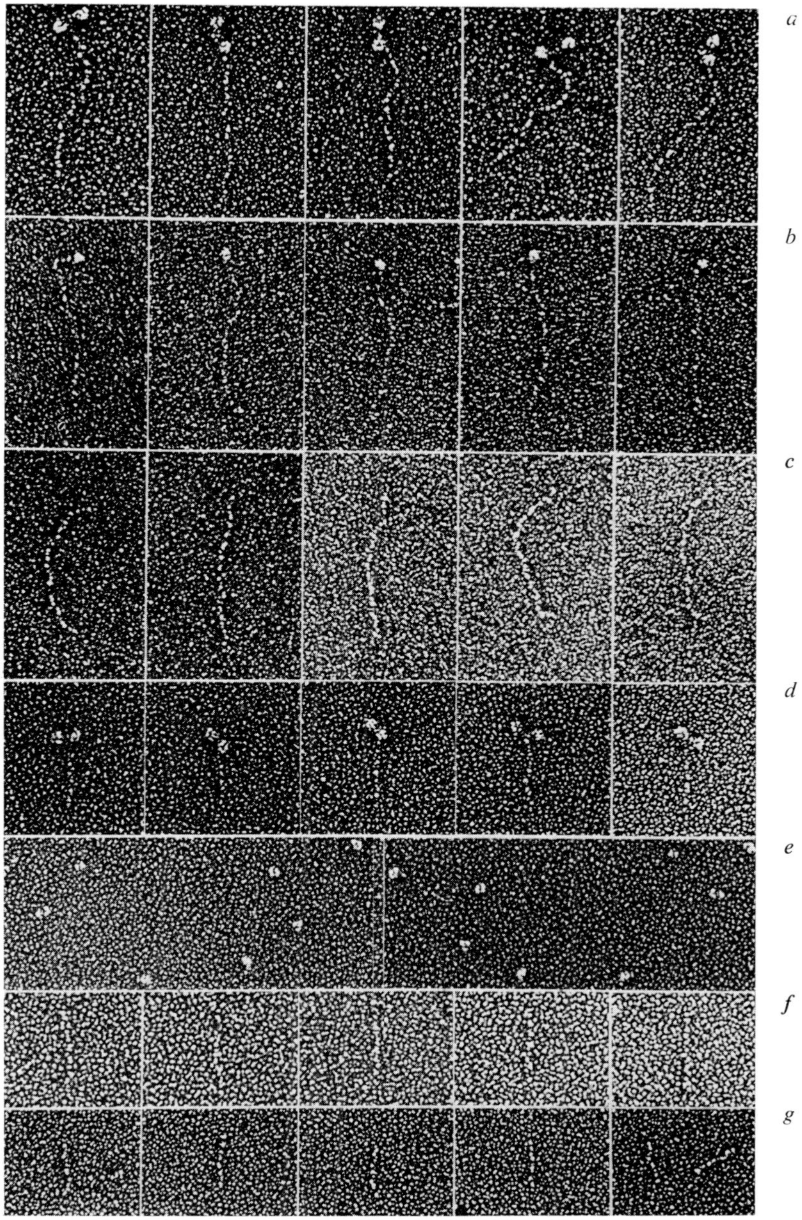

FIG. 9. A composite of selected myosin molecules and subfragments of myosin. From the top to the bottom are: (*a*) myosin, (*b*) single-headed myosin, (*c*) rods, (*d*) HMM, (*e*) HMM S-1, (*f*) LMM and (*g*) HMM S-2. (From Lowey *et al.* 1969.)

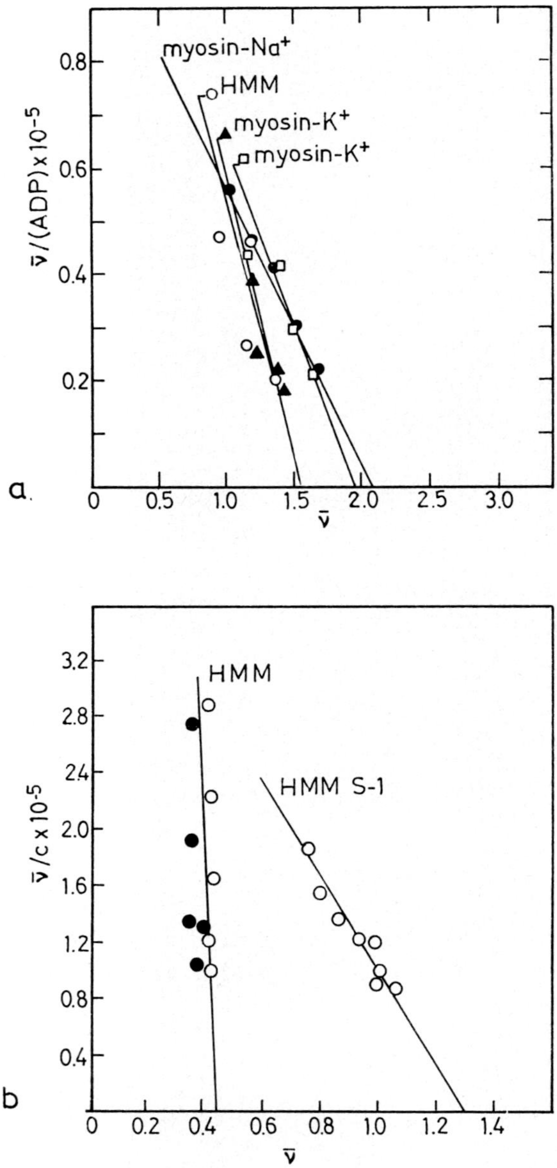

FIG. 10. (*a*) Scatchard plot for the equilibrium binding of ADP to myosin and HMM where $\bar{\nu}$ is the average number of moles ADP bound per mole protein (from Lowey & Luck 1969); (*b*) Scatchard plot for the binding of HMM and HMM S-1 to F-actin, where $\bar{\nu}$ is the average number of moles of HMM (or HMM S-1) bound per mole of actin monomer (S. S. Margossian & S. Lowey, unpublished data).

hydrodynamic methods and direct visualization by electron microscopy (Fig. 9) (Slayter & Lowey 1967; Lowey *et al.* 1969). By means of the enzyme papain, it is possible to prepare many different types of subfragments from myosin, including a single-headed myosin, isolated rods, single globular 'heads' (HMM S-1), the water-soluble portion of myosin containing the two heads attached to a 50 nm tail (HMM), the water-soluble portion of the rod (HMM S-2) and the water-insoluble region of the rod (LMM) (Fig. 9).

Optical rotatory dispersion measurements of the isolated rods give an unusually high α-helix content (> 90 %), consistent with an absence of proline residues in the amino acid composition of the rod. The HMM S-1 subfragments have a helical content of about 30 %, a fairly typical value for globular proteins. Thus, each of the two large polypeptide chains depicted in Fig. 8 is α-helical over a distance of about 140 nm and terminates in a globular conformation about 10 nm in diameter (Lowey *et al.* 1969). The α-helical segments stabilize each other by strong side-chain interactions along the length of the rod (Cohen & Holmes 1963), while the globular regions appear to have considerable flexibility. Each 'head' contains at least one site for binding nucleotides (Schliselfeld & Bárány 1968) and one site for actin binding (Young 1967). Fig. 10*a* shows the binding of 2 mol of Mg-ADP to myosin or HMM by equilibrium dialysis (Lowey & Luck 1969). Measurement of the binding of F-actin to HMM by centrifuging the complex and determining the concentration of free HMM in the supernatant gives a value of 2 mol actin bound/mol HMM (Fig. 10*b*) (Lowey 1970; S. S. Margossian & S. Lowey, unpublished data).

THE THICK FILAMENT

As the ionic strength of a myosin solution is gradually lowered from about 0.5 M to less than 0.1 M, small aggregates are seen to form at intermediate salt concentrations which have a central bare region of about 200 nm with globular appendages on either side. It is apparent from the shape of the myosin molecule that filament assembly is initiated by the antiparallel aggregation of the rod portion of myosin. Growth proceeds at either end by the addition of molecules pointing in the same direction (Huxley 1963). These synthetic myosin filaments look very similar to the native thick filaments (Fig. 11*d*) except that their length is more variable. For the cross-bridges to appear in equivalent positions on the surface of the filament, the rods must be tilted relative to the filament axis, as shown by the schematic drawing in Fig. 11*d*.

It has been suggested that a second protein component, whose role would be to regulate the length, may be present along the length of the native filament.

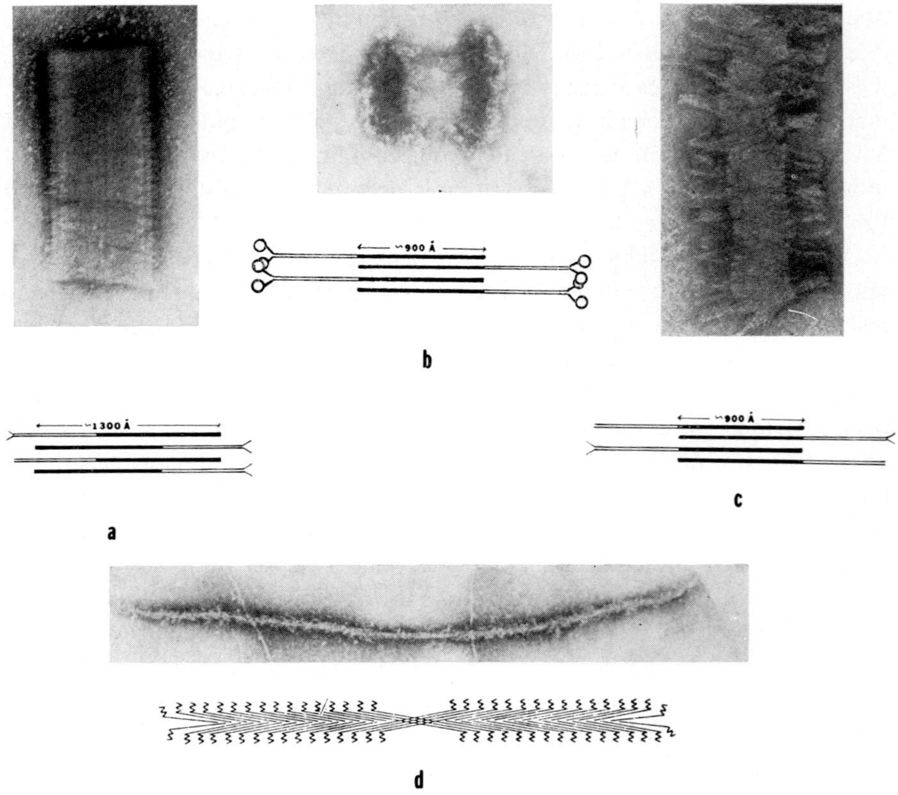

FIG. 11. (*a*) and (*c*) are two types of segments from myosin rods; (*b*) is a segment from myosin. Schematic representation of the molecular overlaps are shown below the electron micrographs. (From Harrison *et al.* 1971.) (*d*) Native thick filament from muscle, and schematic diagram of molecular packing. (From Huxley 1965.)

Conclusive evidence for such a protein has not been forthcoming, however, and until it is obtained, one prefers to think of the formation of the thick filament as a process of self-assembly under the control of specific ionic conditions (Harrington & Josephs 1968).

The myosin molecules are arranged with helical symmetry in the polar portion of the thick filament (Fig. 12*a*). The 'cross-bridges' (consisting primarily of HMM S-1) project from the filament in pairs related by a twofold axis; each pair must be rotated 120° and translated 14.3 nm to generate the helix (Huxley & Brown 1967). Although electron microscopy and X-ray diffraction can tell us about the arrangement of molecules in the polar region of the filament, they do not reveal the molecular packing in the bare region of the filament.

One approach to this problem was provided by the work of Cohen who showed that the simple, rod-shaped muscle proteins such as tropomyosin, paramyosin and the rod fragment of myosin form ordered aggregates *in vitro* which can be related to the native structures (Cohen & Longley 1966; Cohen *et al.* 1970, 1971*b*). Two distinct modes of aggregation have been observed with the rod subfragment of myosin: one with an overlap of 130 nm and another with an overlap of about 90 nm (Fig. 11*a, c*). Myosin itself can only form an aggregate with a 90 nm overlap (Fig. 11*b*). Both types of segments are consistent with a value of 145 nm for the length of the rod. By arranging molecules of this length on the surface lattice indicated by the X-ray results, one can generate the interactions found in the bipolar segments (Fig. 12*b*). Although a segment with a 43.5 nm overlap has not been found in myosin from striated muscles, this overlap is observed in rod segments from a smooth muscle myosin (Kendrick-Jones *et al.* 1971). These results suggest that the surface lattice which defines the packing of myosin molecules in the polar region of the thick filament may be directly related to the antiparallel packing of molecules in the *in vitro* aggregates (Harrison *et al.* 1971).

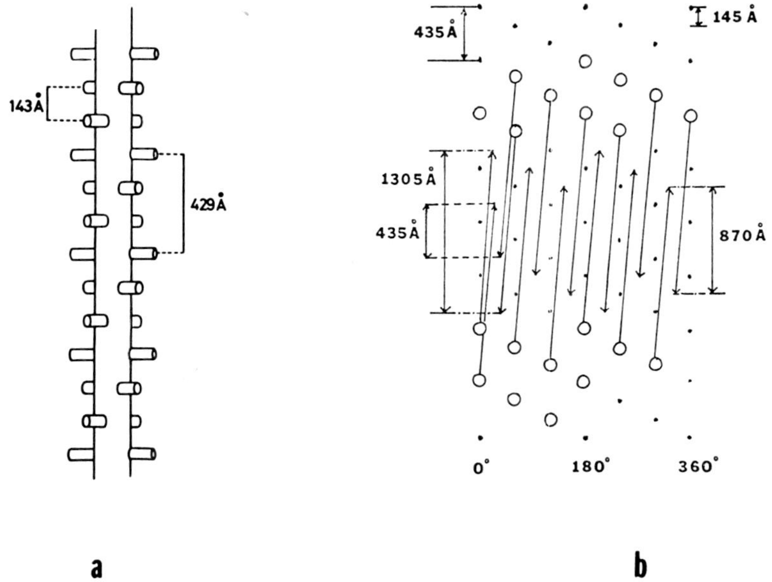

FIG. 12. (*a*) Schematic diagram of a myosin filament showing arrangement of cross-bridges on a 6/2 helix. (From Huxley & Brown 1967.) (*b*) Modified two-dimensional lattice of the cylindrical surface in (*a*). The length of the rod is taken as 145 nm, and the molecular axes are tilted relative to the fibre axis (see Fig. 11*d*). (From Harrison *et al.* 1971.)

THE THIN FILAMENT

In a salt-free solvent, actin exists as a monomeric unit with a molecular weight of 45 000 daltons. When salt is added, G-actin-ATP is polymerized to F-actin-ADP and inorganic phosphate is released as a consequence of the dephosphorylation. Negatively stained actin polymers appear indistinguishable from the isolated native thin filaments except for their length: the native filaments are uniformly 1 μm long, whereas the *in vitro* filaments are of indefinite length (Fig. 13*a*). Electron microscopy (Hanson & Lowy 1963) and X-ray diffraction (Huxley & Brown 1967) have shown the thin filaments to consist of actin monomers, each about 5.5 nm in diameter, arranged in a helix whose pitch is about 2 × 36 nm (Fig. 14). The specific orientation of the actin globules in the chains is revealed by the attachment of the HMM S-1 subfragments to actin. The distinctive 'arrowhead' appearance of the 'decorated' actin filaments emphasizes the underlying structural polarity of the thin filaments; this means that each actin monomer must be related to its four neighbours in such a manner that all the actin units are oriented in the same direction (Fig. 13*b*) (Moore *et al.* 1970).

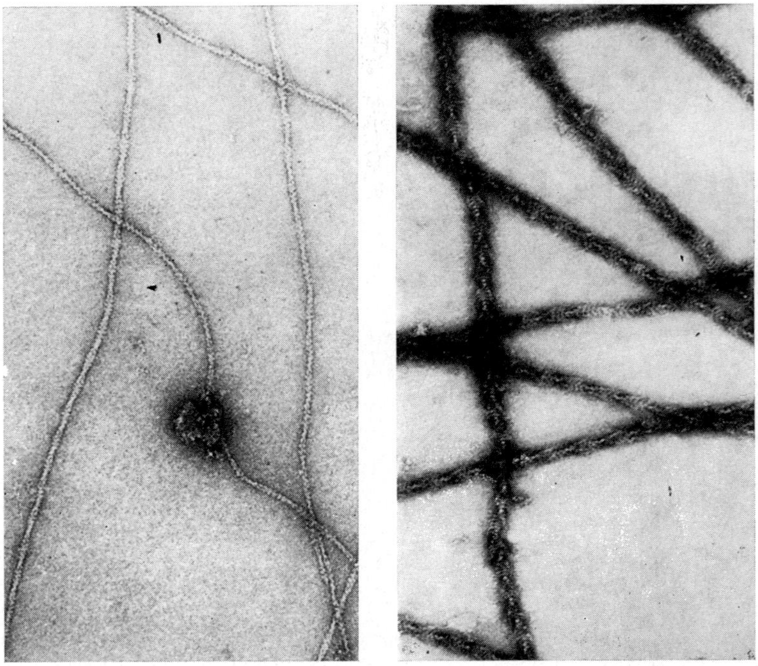

FIG. 13. (*a*) Purified F-actin filaments. (*b*) Actin filaments 'decorated' with HMM S-1. (From Moore *et al.* 1970.)

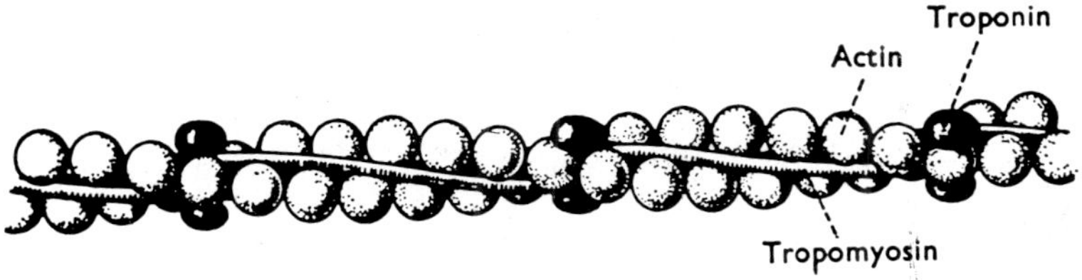

FIG. 14. A model for the thin filament. (From Ebashi *et al.* 1969, by permission of the authors and Cambridge University Press.)

Tropomyosin appears to lie in the two grooves of the actin helix, where it positions a troponin molecule about every 40 nm along the thin filament (Fig. 14). Evidence in support of this model comes from a periodicity of 40 nm observed in I-bands stained with antibodies prepared against troponin (Ohtsuki *et al.* 1967). This periodicity is also evident in the I-band of untreated muscles (Huxley & Brown 1967), and has been identified with the end-to-end bonding of tropomyosin in its *in vitro* aggregates (Cohen & Longley 1966). The tropomyosin filaments must undergo a periodic bending in order to fit into the helical grooves of the actin filament. This variable bending or 'supercoiling' of the tropomyosin filament is consistent with the data obtained from crystals of tropomyosin by electron microscopy and X-ray diffraction (Caspar *et al.* 1969; Cohen *et al.* 1971*a*). The 'kite-shaped' appearance of the nets in negatively stained preparations of tropomyosin crystals is a direct consequence of the bending imposed by the pattern of cross-connections between tropomyosin filaments (Fig. 15*a*) (Caspar *et al.* 1969). When troponin is co-crystallized with tropomyosin, the troponin molecule(s) can be seen attached to the middle of each long arm in the tropomyosin net (Higashi & Ooi 1968) (Fig. 15*b*). The presence of troponin has also been observed to produce changes in the unit-cell dimensions in the tropomyosin crystal. An implication of these variations is that troponin may be able to influence the conformation and interactions of the tropomyosin filament (Cohen *et al.* 1971*a*).

The function of troponin in muscle is to regulate the interactions of actin with myosin: by binding calcium, a conformational change may be induced in each troponin molecule which is somehow transmitted to the seven to eight actin units under its control (Ebashi & Endo 1968). Since tropomyosin interacts strongly with both troponin and actin, it may well be the mediator through which the 'on-off' messages are communicated. One can envisage a mechanism whereby conformational changes are initiated in tropomyosin by troponin; these changes may in turn affect the interactions between tropomyosin and actin and ultimately lead to a modification in the reactivity of the actin molecules.

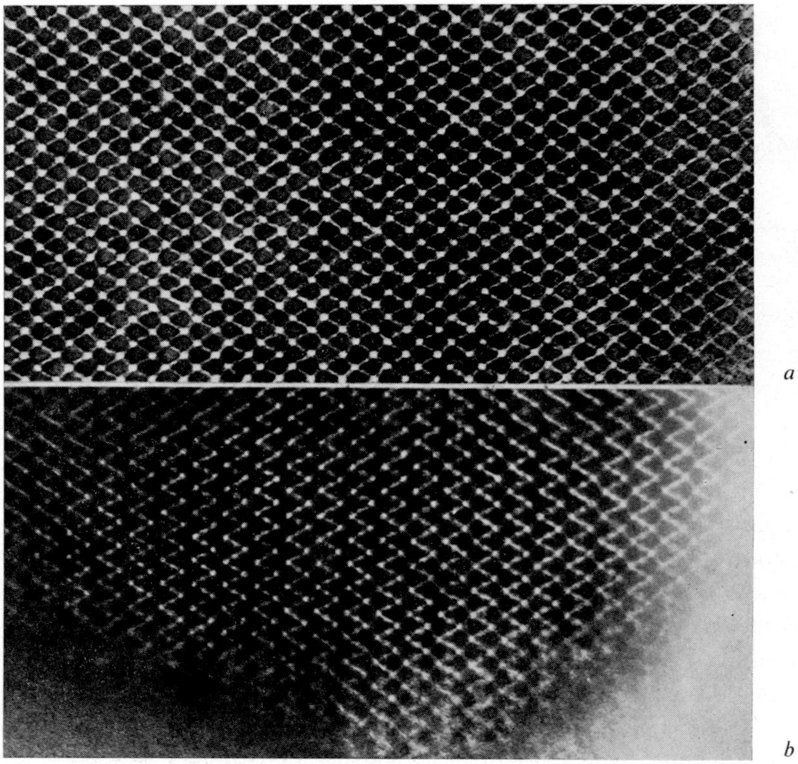

FIG. 15. Electron micrographs of crystals of (*a*) tropomyosin; (*b*) tropomyosin-troponin complex. (From Cohen *et al.* 1971*a*, by permission of the authors and Cold Spring Harbor Laboratory.)

INTERACTIONS BETWEEN FILAMENTS

It is reasonable to assume that during relaxation the helical portion of myosin interacts with the backbone of the thick filament and the HMM S-1 globules are arrayed on its surface (Fig. 16*a*, *b*). On stimulation of the muscle, a 'bridge' is formed between the thin and thick filaments (Fig. 16*c*). The attachment of HMM S-1 to actin must generate a force sufficient to move the actin filament relative to the myosin filament. One of the puzzling features of this interaction has been the problem of how a link can be maintained despite the increasing interfilament distance caused by contraction.* A plausible solution to this problem can be found by considering the solubility properties of the

* It is well-known that muscle contracts at 'constant volume', i.e. when the fibre shortens the filaments move further apart, and when the muscle is stretched the filaments move closer together (Huxley 1969).

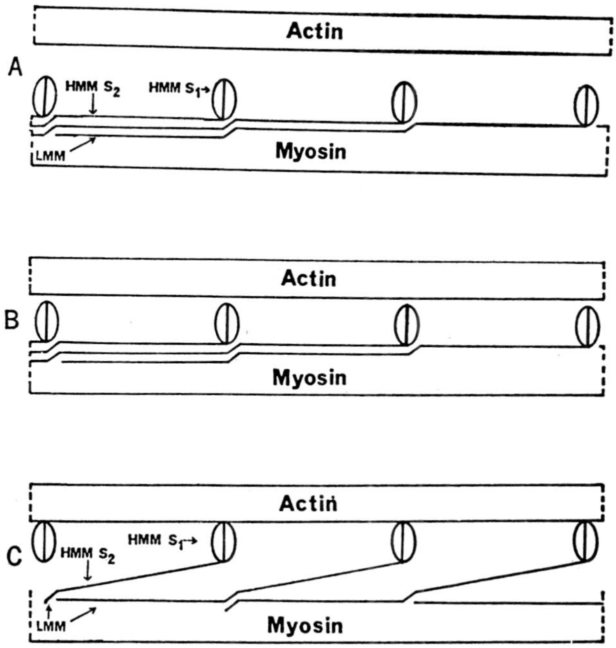

FIG. 16. Schematic representation of positions of filaments and cross-bridges at different interfilament distances. (*a*) 25 nm and (*b*) 20 nm corresponding to sarcomere lengths of about 2.0 and 3.1 µm. (*c*) During contraction the cross-bridge attaches to the actin filament by bending away from the axis of the thick filament. (After Huxley 1969, by permission; copyright 1969 Am. Ass. Adv. Sci.)

myosin molecule: LMM is the least soluble region of myosin, and hence is primarily responsible for anchoring myosin in the core of the thick filament. HMM S-2 is much more soluble than LMM and, correspondingly, has weaker interactions with the backbone of the filament. HMM S-1 is highly water-soluble and its interactions are probably limited to the actin filament. Thus it appears highly likely that the HMM S-2 moiety can bend reversibly away from the axis of the thick filament; the degree of bending will depend on the interfilament distance HMM S-1 must bridge to link with actin (Huxley 1969; Lowey *et al.* 1967, 1969; Pepe 1966). X-ray diffraction studies on whole muscle have provided strong evidence for the existence of such a mechanism (Huxley & Brown 1967; Huxley 1968).

Although the precise details of the interaction between myosin and actin are still largely unknown, a general picture of the sequence of events occurring during the cyclical attachment of myosin and actin is beginning to emerge. The clarification of the chemical events accompanying the structural changes has been provided largely through the kinetic studies of E. W. Taylor and his co-workers. For many years it was thought that actin activates the Mg-ATPase activity of myosin by forming an active complex. It has now been shown by rapid stopped-flow measurements that the activation of myosin ATPase is due to the displacement of the products of the hydrolysis reaction by actin (Lymn & Taylor 1971). The cycle of events can be summarized as follows: ATP rapidly dissociates the actomyosin complex by binding to the ATPase sites of myosin; free myosin then splits ATP and forms a relatively stable myosin–products complex; actin recombines with this complex and dissociates the products, thereby re-forming the actin–myosin bridge—which is presumably the step that generates force. Although numerous conformational changes may well accompany this cycle, present evidence for a structural change in myosin comes largely from electron microscope studies of insect flight muscle (Reedy *et al.* 1965), X-ray diffraction of living muscle (Huxley & Brown 1967) and three-dimensional image reconstructions of electron micrographs showing actin filaments decorated with HMM S-1 (Moore *et al.* 1970) (see Fig. 13*b*). All these studies suggest that HMM S-1 is in a tilted configuration when attached to actin, and that the configuration of HMM S-1 in the dissociated or relaxed state is probably quite different. Whether any cooperative interactions occur between the two heads of myosin during the contractile event is still largely a matter of speculation; present data argue for relatively independent activity by each head (S. S. Margossian & S. Lowey, unpublished data).

Finally, I want to comment briefly on the regulation of the actin-myosin interaction. As discussed previously, existing evidence implies that control of the interaction resides largely in the thin filament through the action of the calcium-binding troponin. This mechanism implies a relatively passive role for myosin and the thick filament as far as regulation is concerned. Although data favouring a more active role for myosin in vertebrate muscles are scant, it has been shown that molluscan muscles regulate contraction-relaxation primarily through the thick filament (Kendrick-Jones *et al.* 1970). Molluscan myosin preparations contain a light chain which binds calcium, and have a calcium-dependent ATPase activity in the presence of pure actin (Szent-Györgyi *et al.* 1972). These invertebrate muscles do not have a troponin regulatory system, even though the thin filaments do contain tropomyosin. The finding of a regulatory subunit of molecular weight about 20 000 daltons in molluscan myosin raises the question of whether the light chains of rabbit myosin, in particular the DTNB-light chains, might also have some hitherto undiscovered regulatory function in the myosin-actin interaction.

Another indication of myosin's participation in the control mechanism comes from the finding that, at low concentrations of ATP, myosin can 'turn on' actin even in the absence of calcium by forming rigor links (links formed when no or insufficient ATP is available: Bremel & Weber 1972). Furthermore, these actomyosin rigor links can affect the affinity of troponin for calcium, when the latter is present in the system. Thus even in vertebrate muscles, myosin can influence the flow of information in the thin filament. Clearly much remains to be discovered about the interplay of the various proteins in the myofibril before we can fully understand the intricate series of events culminating in a contractile force.

[*Note added in proof:* In a rapidly moving field such as muscle research, a review article unfortunately tends to be outdated even before its publication; for the latest information readers should refer to *Cold Spring Harbor Symp. Quant. Biol.* 37, on The Mechanism of Muscle Contraction.]

ACKNOWLEDGEMENTS

This work was supported by Public Health Service Grant AM-04762, Public Health Service Research Career Program Award K3-AM-10630, National Science Foundation Grant GB-8616, and a grant from the Muscular Dystrophy Associations of America.

References

BANGA, I. & SZENT-GYÖRGYI, A. G. (1941–1942) *Stud. Inst. Med. Chem. Univ. Szeged* **1**, 5
BREMEL, R. D. & WEBER, A. (1972) *Nature (Lond.)* **238**, 97
CASPAR, D. L. D., COHEN, C. & LONGLEY, W. (1969) *J. Mol. Biol.* **41**, 87
COHEN, C. & HOLMES, K. C. (1963) *J. Mol. Biol.* **6**, 423
COHEN, C. & LONGLEY, W. (1966) *Science (Wash. D.C.)* **152**, 794
COHEN, C., LOWEY, S., HARRISON, R. G., KENDRICK-JONES, J. & SZENT-GYÖRGYI, A. G. (1970) *J. Mol. Biol.* **47**, 605
COHEN, C., CASPAR, D. L. D., PARRY, D. A. D. & LUCAS, R. M. (1971*a*) *Cold Spring Harbor Symp. Quant. Biol.* **36**, 205
COHEN, C., SZENT-GYÖRGYI, A. G. & KENDRICK-JONES, J. (1971*b*) *J. Mol. Biol.* **56**, 223
DREIZEN, P. & GERSHMAN, L. C. (1970) *Biochemistry* **9**, 1688
EBASHI, S. & EBASHI, F. (1964) *J. Biochem. (Tokyo)* **55**, 604
EBASHI, S. & ENDO, M. (1968) *Prog. Biophys. Mol. Biol.* **18**, 123
EBASHI, S., ENDO, M. & OHTSUKI, I. (1969) *Q. Rev. Biophys.* **2**, 351
EBASHI, S., WAKABAYASHI, T. & EBASHI, F. (1971) *J. Biochem. (Tokyo)* **69**, 441
GAZITH, J., HIMMELFARB, S. & HARRINGTON, W. F. (1970) *J. Biol. Chem.* **245**, 15
GERSHMAN, L. C., STRACHER, A. & DREIZEN, P. (1969) *J. Biol. Chem.* **244**, 2726
GOLL, D. E., SUZUKI, A. & SINGH, I. (1971) *Biophys. Soc. Annu.Meet. Abstr.*, 107a
GREASER, M. L. & GERGELY, J. (1971) *J. Biol. Chem.* **246**, 4224
HANSON, J. & HUXLEY, H. E. (1953) *Nature (Lond.)* **173**, 530
HANSON, J. & LOWY, J. (1963) *J. Mol. Biol.* **6**, 46

Harrington, W. F. & Josephs, R. (1968) *Dev. Biol. Suppl.* **2**, 21
Harrison, R. G., Lowey, S. & Cohen, C. (1971) *J. Mol. Biol.* **59**, 531
Hartshorne, D. J., Theiner, M. & Mueller, H. (1969) *Biochim. Biophys. Acta*, **175**, 320
Higashi, S. & Ooi, T. (1968) *J. Mol. Biol.* **34**, 699
Huxley, H. E. (1963) *J. Mol. Biol.* **7**, 281
Huxley, H. E. (1965) *Sci. Am.* **213**, 18
Huxley, H. E. (1968) *J. Mol. Biol.* **37**, 507
Huxley, H. E. (1969) *Science* **164**, 1356
Huxley, H. E. & Brown, W. (1967) *J. Mol. Biol.* **30**, 383
Huxley, H. E. & Hanson, J. (1954) *Nature (Lond.)* **173**, 973
Kendrick-Jones, J., Lehman, W. & Szent-Györgyi, A. G. (1970) *J. Mol. Biol.* **54**, 313
Kendrick-Jones, J., Szent-Györgyi, A. G. & Cohen, C. (1971) *J. Mol. Biol.* **59**, 527
Kühne, W. (1868) *Physiol. Chem. Leipzig*
Lowey, S. (1970) *8th Int. Congr. Biochem. Abstr.* 28
Lowey, S. & Luck, S. (1969) *Biochemistry* **8**, 3195
Lowey, S. & Risby, D. (1971) *Nature (Lond.)* **234**, 81
Lowey, S. & Steiner, L. (1972) *J. Mol. Biol.* **65**, 111
Lowey, S., Goldstein, L., Cohen, C. & Luck, S. M. (1967) *J. Mol. Biol.* **23**, 287
Lowey, S., Slayter, H. S., Weeds, A. G. & Baker, H. (1969) *J. Mol. Biol.* **42**, 1
Lymn, R. W. & Taylor, E. W. (1971) *Biochemistry* **10**, 4617
Margossian, S. S. (1972) *Fed. Proc.* **31**, 866 (Abstr.)
Masaki, T., Endo, M. & Ebashi, S. (1967) *J. Biochem. (Tokyo)* **62**, 630
Moore, P. B., Huxley, H. E. & DeRosier, D. J. (1970) *J. Mol. Biol.* **50**, 279
Ohtsuki, I., Masaki, T., Nonomura, Y. & Ebashi, S. (1967) *J. Biochem. (Tokyo)* **61**, 817
Peachy, L. D. (1965) *J. Cell Biol.* **25**, 209
Peachy, L. D. (1968) *Annu. Rev. Physiol.* **30**, 401
Pepe, F. A. (1966) *J. Cell Biol.* **28**, 505
Pepe, F. A. (1971) in *Subunits in Biological Systems* (Timasheff, S. N. & Fasman, G. D., eds.) p. 323, Dekker, New York
Reedy, M. K., Holmes, K. C. & Tregear, R. (1965) *Nature (Lond.)* **207**, 1276
Schliselfeld, L. H. & Bárány, M. (1968) *Biochemistry* **9**, 3206
Slayter, H. S. & Lowey, S. (1967) *Proc. Natl Acad. Sci. U.S.A.* **58**, 1611
Stracher, A. (1969) *Biochem. Biophys. Res. Commun.* **35**, 519
Straub, F. B. (1942) *Stud. Inst. Med. Chem. Univ. Szeged* **2**, 3
Szent-Györgyi, A. G., Szentkiralyi, E. M. & Kendrick-Jones, J. (1972) *J. Mol. Biol.*, in press
Weber, A. (1966) in *Current Topics in Bioenergetics* (Sanadi, D. R. ed.) p. 203, Academic Press, New York
Weeds, A. G. & Lowey, S. (1971) *J. Mol. Biol.* **61**, 701
Woods, E. F. (1967) *J. Biol. Chem.* **242**, 2859
Young, M. (1967) *Proc. Natl Acad. Sci. U.S.A.* **58**, 2393
Young, M. (1969) *Annu. Rev. Biochem.* **38**, 913

Discussion

Huxley: There may not be any positive chemical evidence that there is an extra component in the thick filament but there is quite strong evidence from X-ray diffraction that a structure with a periodicity different from myosin is present. It has a repeating periodicity of about 44 nm and there appear to be at least ten such repeats on either side of the M line, so that it extends quite a long way along the thick filament.

Lowey: But myosin alone is able to regulate its own growth to a large extent. Josephs & Harrington (1968) obtained fairly narrow length distributions of myosin filaments simply by adjusting the pH and ionic strength. Myosin is unique in this respect and different from TMV protein. There may well be another protein component in the thick filament, but myosin by itself can certainly aggregate into a filament resembling the native form.

Tanford: Are any of these polypeptide chains glycopolypeptides?

Lowey: No, not as far as we know.

Chapman: Is any membrane interaction involved in calcium stimulation of the troponin?

Lowey: Only in so far as the calcium seems to be stored in the membrane vesicles and gets released from them during stimulation.

Schwyzer: Is the arrowhead formation calcium-dependent?

Huxley: The arrowheads only form when the myosin has no ATP on it. Myosin without ATP will attach to actin whether calcium or troponin are there or not, and the arrowheads don't look any different in each case. It is the attachment of the myosin, presumably in a more perpendicular configuration, when the myosin is loaded with ATP or its split products, which is likely to be blocked by the troponin/tropomyosin system. This would correspond to its attachment at the beginning of the working stroke, with the configurational change following so that it finishes up in a tilted configuration. But that tilted configuration may occur in the absence of ATP whether the calcium or the two regulatory proteins are there or not.

Changeux: How large is the conformational change of troponin caused by calcium binding?

Lowey: I don't think anyone has really seen any conformational change in the troponin itself on calcium binding. People have reported different migrations on gels but that may not be too relevant. Cohen *et al.* (1971) have found that the dimensions of tropomyosin crystals change when troponin is introduced, so troponin appears to affect the tropomyosin interactions in some manner.

Changeux: You mentioned that there were different chains of myosin in 'fast' and 'slow' muscle. It is known that the nerve is apparently responsible for the slow or fast character of the muscle. Do you have any idea about the regulatory mechanism involved?

Lowey: People studying the properties of myosin from cross-innervated fast and slow muscles are planning to look at the light chain pattern, but no results have been published yet.

Huxley: You can change the speed of the muscle, and you can change the myosin ATPase. What this does to the light chains may be known in about six months.

Elliott: Since you can see the two heads of the myosin molecule in shadowed preparations, but not ordinarily in negatively stained ones, might you see more of the structure of the muscle filament on a shadowed preparation?

Lowey: I think there is too much disorder to see more.

Katchalski: Would you like to give us a general picture of the mechanism of muscle contraction as you see it now, Dr Huxley?

Huxley: The most likely scheme at the moment is that a site on the actin filament is blocked by the regulatory system in the absence of calcium, so that the head of the myosin is not able to attach. This is providing that the head is carrying either ATP or—from the work of Taylor and Lymn and their colleagues (e.g. Lymn & Taylor 1971)—the split products of ATP, that is ADP and phosphate. In the relaxed muscle the first stage of the ATP splitting has probably already occurred, but the cycling of ATP on the myosin alone is extremely slow, because the products are released extremely slowly from the myosin in the ionic conditions present in muscle. A very crude steric unblocking mechanism may act if the tropomyosin molecule is pulled out of the way by some structural change in the troponin, and thereby allow the myosin heads to attach to the actin. This complex formation first of all results in the release of the product and is then followed by some structural change in the myosin head–actin complex. We have no evidence for this, but it seems reasonable that the myosin would first attach in an almost perpendicular configuration and then change over to a new configuration, tilted by about 45°. There is evidence for this type of tilted configuration from observations on insect flight muscle and from three-dimensional analysis of the decorated actin filaments. When it is in the tilted configuration, the far end of the myosin head would have moved by about 7 nm. Given that there is a system of forces in the muscle which keeps the filaments approximately the same distance apart as long as the overlap of the filaments does not change very much, then if you tilt the head over you necessarily translate the two filaments past each other. Given that the myosin molecules are arranged in the thick filaments with approximate polarity, this must pull the I filaments in towards the centre of the sarcomere. Given that there is ATP in solution, the cross-bridge can be dissociated again, return and attach to another actin monomer, and go through the process again and again as long as the muscle is active.

The main gaps in the evidence are that we have no way of looking at this initial attachment to see whether it is perpendicular, and we do not know how this site is unblocked. Of course what we would like best would be a crystalline preparation of S1 so that we could do a proper structural analysis of the head. Although the model is moderately plausible, a lot of evidence is still needed for it.

Katchalski: Is there any information about how the calcium initiates the process?

Huxley: There is a lot of evidence that the concentration of calcium needed is something like 10^{-6} M and that three or four calcium molecules per troponin are needed. It looks as though there is only one troponin to seven actins. Yet it also looks as though all these actins can be switched on and off by the calcium going onto that one troponin. It seems likely that that is mediated in some way by the tropomyosin, which is a rod-shaped molecule in close proximity to all the actins. But there is very little evidence at all about what change takes place in the troponin. One can imagine something very crude like one piece of troponin being separated from the others, so that instead of the tropomyosin being held in a particular position it is much freer to be pushed out of the way by the incoming myosin.

Eigen: The muscle receives the stimulus but how does it propagate down the muscle?

Huxley: The big problem used to be how it propagated across the cross-section of the muscle. The diffusion time from the outside was much too long for anything to diffuse from the outside membrane. Then people discovered that the outside membrane had invaginations which extended all the way into the middle and joined up with other membranes. In fact virtually every myofibril, the unit which is about 1 μm in size, has its own piece of membrane around it, and it is from this membrane that the calcium is apparently released. It only has to diffuse a distance of 1 μm at the most and there is now no problem about the time it takes to switch the system on.

Chapman: Is there no direct connection between the membrane and the first protein in the sequence?

Huxley: No. The diameter of the fibrils is something of the order of 1 μm and the repeating distance is about 2.5 μm. Between virtually each fibril and the next is a thin sheet of membrane. The calcium is either in or on this membrane in the relaxed muscle. There is a junction, known as a triad, which in frog muscle is at the level of the Z line; so that, between the reticulum in one sarcomere and the next, there is a central tubule, which in fact seems to be an invagination of the outside membrane. It looks as though the action potential is conducted inwards along these tubules, though it is not quite clear whether the transverse tubules are excitable or not. Whatever it is that happens at the junction between transverse tubule and longitudinal reticulum, it seems to lead ultimately to the release of calcium by these structures. The muscle is then switched on, but if one stops sending signals in, the calcium is bound again by these structures and the muscle switches itself off again.

Katchalski: Can you check the binding of calcium by troponin, in an *in vitro* experiment? Are there any conformational changes in the troponin as a result of the binding?

Huxley: One can certainly show that troponin binds calcium. Some preliminary experiments on troponin in the ultracentrifuge have not shown anything too decisive yet about configurational changes. We are trying to look at purified systems containing actin and tropomyosin and troponin in the electron microscope, and by X-ray diffraction, to see whether there are any structural changes in the actin–tropomyosin system when calcium is present. We can see changes in whole live muscle by X-ray diffraction which probably arise from structural changes in the thin filaments when the muscle becomes active. Mostly all the reflections in a muscle are lost after stimulation; the myosin layer line reflections are lost. But in very long contractions, one can see new reflections coming up which are not there in the resting muscle and which could very likely be due to a change in the position adopted by the tropomyosin, moving more deeply into the groove of the actin helix.

Revel: Is the ATP hydrolysis related to the dissociation between myosin and actin?

Huxley: It has not been shown that it is necessary to split ATP in order to get dissociation. It is perfectly plausible that the ATP dissociates the actin and the myosin very rapidly. The hydrolytic step then occurs but the products remain bound and the myosin enzyme gets stuck at this point unless actin can combine and release the product.

Katchalski: Is heavy meromyosin really pure? Can you carry out reactions with this fragment and actin?

Lowey: It depends on what you mean by 'pure'. In the isolated head region of myosin (HMM S-1), 'pure' chromatographically means that on ion-exchange columns one gets a single peak with HMM S-1. In the analytical centrifuge one would get a single sedimenting peak. HMM S-1 binds very strongly to actin, and its actin-activated ATPase activity is very high. If HMM S-1 is put on SDS acrylamide gels some proteolytic cleavage is evident. Since the native HMM S-1 head has a molecular weight of the order of 110 000, one might expect to see one band of about 70 000 or 80 000 daltons and three light chain bands on an acrylamide gel. But actually one usually sees a split band in the 80 000 dalton region, meaning there is additional cleavage in the heavy chain. One also sees another low molecular weight band. I think there is evidence now that this band is not a light chain, but comes from the cleavage of the heavy chain. Thus, the so-called 'pure' HMM S-1 does have some additional cleavages, but they are in a sense minimal when one considers that a non-specific protease such as papain is being used to prepare the fragment.

Engel: Does the ATPase activity of F-actin in its pure form, without tropomyosin, have biological relevance?

Lowey: I would think it is unrelated to the contractile mechanism.

References

COHEN, C., CASPAR, D. L. D., PARRY, D. A. D. & LUCAS, R. M. (1971) Tropomyosin crystal dynamics. *Cold Spring Harbor Symp. Quant. Biol.* **36**, 205–216

JOSEPHS, R. & HARRINGTON, W. F. (1968) On the stability of myosin filaments. *Biochemistry* **7**, 2834–2847

LYMN, R. W. & TAYLOR, E. W. (1971) Mechanism of adenosine triphosphate hydrolysis by actomyosin. *Biochemistry* **10**, 4617–4624

The *lac* repressor and the *lac* operator

WALTER GILBERT

Department of Biochemistry and Molecular Biology, Harvard University, Cambridge, Massachusetts

Abstract The *lac* repressor is a tetrameric protein, 150 000 in molecular weight. It interacts specifically with a unique region of the *E. coli* chromosome, selecting with high affinity and efficiency one sequence out of the four million base pairs available. The specific region with which the repressor interacts, the operator, can be isolated as a double-stranded DNA fragment by using the ability of the repressor to protect this region against digestion by DNase. This isolation yields a DNA fragment which still interacts with the repressor with the original specificity. Estimates of the length of this double-stranded fragment suggest that it is about 20 base pairs long.

The *lac* repressor carries out its function by interacting with a unique point on a DNA molecule. This protein product of the control gene of the lactose operon in *E. coli*, the *i*-gene, binds to a single site on the DNA of *E. coli*, the operator, located immediately between the promoter site and the beginning of the first structural gene of the operon. When it binds to this site on a DNA molecule the repressor prevents the RNA polymerase from transcribing the structural genes of the operon.

The *lac* repressor is a tetramer, 150 000 in molecular weight, containing four identical subunits. Each subunit has a binding site for the low molecular weight inducer of the lactose operon. The inducer is a sugar analogous to the sugar lactose, the substrate of the enzymes of this operon. Lactose itself is not the inducer (Burstein *et al.* 1965) and the true inducer has recently been identified as allolactose (Jobe & Bourgeois 1972), a rearrangement of the lactose made as a by-product by β-galactosidase. A variety of non-metabolizable analogues of lactose can serve as inducers, the most potent of which is isopropylthiogalactoside (IPTG).

The repressor is an acidic protein; however the molecule binds to a phosphocellulose column at neutral pH. This property presumably reflects the existence of a basic region on the molecule which is capable of interacting with the phosphates on DNA. The molecule has at least one binding site which can recognize the operator sequence on a DNA molecule with high specificity and high affinity. The repressor binds to the operator sequence with a dissociation constant varying from 10^{-13} M in low salt buffers to 10^{-11} M in more physiological conditions (Riggs *et al.* 1970*a*). The repressor binds as strongly as this to only one site in the entire *E. coli* genome: the *lac* operator itself. Mutations in that operator locus, o^c (operator constitutive) mutations, damage the degree of repression observable *in vivo* and hinder the binding of repressor to DNA *in vitro*. This fact shows directly that the *lac* repressor binds to a single locus on a DNA molecule. The interaction of the *lac* repressor with DNA can be studied *in vitro* either by following the radioactive repressor bound to unlabelled DNA as the DNA sediments down a glycerol gradient (Gilbert & Müller-Hill 1967) or more easily by using the filter binding assay developed by Suzanne Bourgeois and her co-workers (Riggs *et al.* 1968, 1970*a*). This assay exploits the fact that the lactose repressor protein can bind to a cellulose nitrate filter: if the protein molecule binds a small molecule, such as radioactive IPTG, that radioactive ligand is trapped on the filter; if the protein molecule is bound to a radioactive DNA molecule, that radioactive DNA molecule is retained on the filter. The inducer, which has a dissociation constant of the order of 10^{-6} M, weakens the binding of the repressor to DNA sufficiently to make that binding undetectable. This can be most directly observed with the filter binding assay: once DNA molecules have been trapped by the repressor on the filter, rinsing the filter with a solution of IPTG causes the DNA molecules to be released from the filter.

The direct action of the repressor in blocking RNA synthesis from the *lac* operon *in vitro* has been studied by deCrombrugghe *et al.* (1971) and Chen *et al.* (1971), and by Eron & Block (1971), who showed that the repressor directly blocks RNA synthesis. However the control of *lac* genes in the wild-type cell is more complicated than this simple picture of a control protein binding to DNA, preventing the RNA polymerase from functioning, but falling off the DNA when the protein is bound to the small molecule inducer and permitting the RNA polymerase to function. The reason for this is that there is a pleiotropic positive control mechanism which *E. coli* uses to restrain the synthesis of a wide variety of enzymes which can catabolize other metabolites than glucose.

This pleiotropic system, once called 'catabolite repression', involves the catabolite activator protein (CAP), a 50 000 molecular weight protein which senses the intracellular level of cyclic AMP (Zubay *et al.* 1970; Emmer *et al.* 1970). The intracellular cyclic AMP level monitors the availability of glucose, rising sharply when the cells pass into glucose starvation (Makman & Sutherland 1965). CAP has been shown to have a cyclic AMP-dependent binding to DNA (Riggs *et al.* 1971) but a specific binding of CAP to unique sites on DNA has not yet been established. One might hypothesize, however, that CAP binds to the promoter of the lactose operon, because a deletion of that part of the promoter most distal to the structural genes abolishes the requirement for CAP *in vivo*; the residual low level functioning of this mutant promoter then seems insensitive to the cyclic AMP control (Silverstone *et al.* 1969). Thus one could picture the pleiotropic positive control, in the absence of glucose, as being through the binding of the CAP-cyclic AMP complex to the promoters of the various operons, the binding of this protein then permitting the RNA polymerase to interact with and initiate upon these promoters with high efficiency. However for each operon there is a specific further control element, either negative as with the lactose repressor, or positive as for the arabinose C-gene protein (Greenblatt & Schleif 1971); this specific control element senses the need for the synthesis of the products of the operon. Mutant strains are available which have by-passed the pleiotropic CAP control; there are mutations in the *lac* promoter, such as UV5 (Silverstone *et al.* 1970), which permit the *lac* genes to be expressed *in vivo* in the absence of CAP, and which permit *lac* DNA to be read *in vitro* by the RNA polymerase alone and controlled solely by the *lac* repressor (Eron & Block 1971). For these mutant strains the simplest picture for the control is true.

How does the *lac* repressor recognize the operator? In order to find a unique site on the DNA of *E. coli* the *lac* repressor must be able to recognize, at the very minimum, a sequence of 12 base pairs in order to pick out one such sequence among four million base pairs. The ability to recognize such a large piece goes naturally with the high affinity that the *lac* repressor shows for the operator. The binding energy is of the order of 15 to 18 kcal, easily accounted for by 12 to 15 contacts. There has been no evidence that the *lac* repressor unwinds the DNA strands. *In vitro* the repressor shows no affinity for denatured DNA. One would presume that the protein recognizes the outside of the bases in the large groove of the DNA molecule; a variety of hydrogen bonds can be made there to read a DNA sequence. Bonds could be made to the amino groups, to the keto groups, or to the N-7 position on the purines.

To study further the interaction of repressor with DNA I have sought to isolate, as a fragment of a DNA molecule, the region that binds to the *lac* repressor, by using the *lac* repressor to protect the operator against the action of pancreatic DNase. The repressor protein itself is now easily obtainable. The low level in the wild-type cell, about ten copies per gene, has been raised 50-fold by mutations of the promoter for the *i*-gene, and a further 20-fold by placing this mutant gene on a phage which could multiply within the cell (Müller-Hill *et al.* 1968; J. Miller, unpublished). This yields 2–3% of the cell's protein as *lac* repressor, easily purified (Müller-Hill *et al.* 1971). The operator region, however, represents such a small fraction of the DNA that it cannot be purified in a single step. I begin, not with *E. coli* DNA, but with a copy of the operator region carried on the DNA of phage lambda: lambda p*lac*5, a non-defective phage which carries as a replacement of an inessential region of the phage genome the *lac* promoter, operator and *z*-gene. This is almost a 100-fold enrichment of the operator sequence, since the phage carries one such sequence for every 3×10^7 daltons, one for each 50 000 base pairs. In order to isolate fragments of DNA that carry the operator, I make use of the properties of the filter binding assay: fragments can be bound to the filter by the repressor and eluted with IPTG. This provides an extremely convenient one-step purification for any fragment that carries the operator sequence. Sonicating the phage DNA breaks it up into pieces of about a half million in molecular weight, of the order of 1000 base pairs long. Binding a mixture of these fragments to the filter with the repressor and subsequent elution with IPTG results in the purification of those fragments that bear the *lac* operator (Bourgeois & Riggs 1970). These sonic pieces, which are obtained in a 1% yield from the original phage DNA, provide starting material for the final isolation of an operator fragment.

Although the repressor binds to DNA with a very high affinity, its reaction with DNA is non-covalent, in equilibrium, and thus the presence of repressor on a DNA sequence cannot protect that sequence indefinitely against the action of nuclease. The very tight binding of the repressor to DNA has been analysed into a fast forward rate of association but nonetheless a slow decay of the repressor-DNA complex (Riggs *et al.* 1970*b*). In the conditions used here, the repressor-DNA complex decays with a half-time of 15 minutes, illustrated in Fig. 4. This decay is easily measured by mixing the repressor with labelled fragments containing the operator region, adding a great excess of unlabelled DNA containing the operator region and filtering samples at various times. As the repressor molecule leaves a piece of radioactive DNA and becomes bound to one of the non-radioactive molecules, the number of counts that can be retained on the filter falls. This slow decay, however, can be exploited. If the repressor-DNA complex is treated with DNase for a few minutes, one might expect to digest all the excess DNA, or at least to convert it into a form which could not bind to the cellulose nitrate filter, in a time short compared to the time the repressor remains on the operator. If the digestion is done with more and more DNase, for a fixed time, the amount of radioactive DNA bound by the repressor to the filter should decrease and level off, becoming independent of the amount of DNase. Fig. 1 shows such an experiment.

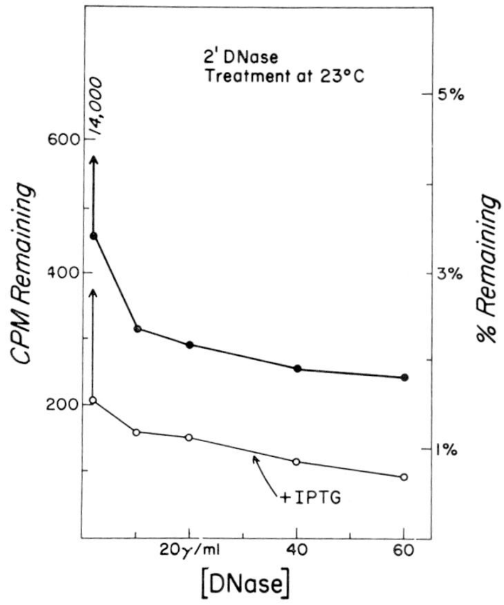

FIG. 1. The *lac* repressor protects a fraction of DNA against DNase digestion. Radioactive sonicated DNA fragments were made by growing a temperature-inducible lysogen of $\lambda C_I857plac_5S_7$ at 34 °C in a glucose-tris medium in 3×10^{-3} M-PO_4, heating to 42 °C for 15 min at a cell density of 4×10^8/ml; then, after a further 15 min growth at 34 °C, washing and resuspending the cells at a density of 8×10^8 in the same medium with ^{32}P. Twenty to 100 mCi of ^{32}P were used in 10 ml cultures containing either limiting phosphate (10^{-6} to 10^{-5} M) or 10^{-4} M-phosphate, in which case incorporation continued for a two-hour period. The cells were washed, suspended in 2 ml of TE buffer (0.01 M-tris—0.001 M-EDTA buffer, pH 7.5), sonicated for six 15-second bursts and extracted with phenol; the aqueous phase was shaken with ether and the ether removed with N_2. The mixture of radioactive DNA and RNA was then made up to about 6 ml with BB (binding buffer containing 0.01 M-Mg^{++}, 0.01 M-KCl, 0.01 M-tris pH 7.5, 10^{-4} M-EDTA, 10^{-4} M-dithiothreitol, 5% dimethylsulphoxide and 50 μg/ml bovine serum albumin). Twenty μg of *lac* repressor was added and aliquots of this mixture filtered on six to eight 25 mm Schleicher and Shuell B6 filters. The filters were presoaked with BB without BSA; the filtration was moderately fast. Each filter was rinsed with 2×0.3 ml of BB, then transferred to a different apparatus and eluted with 2×0.25 ml washes with 10^{-3} M-IPTG in BB. The eluate was phenol-extracted and dialysed against TE. This one-step process yields sonic fragments containing the *lac* operator; the material will bind again to filters with about 50–70% efficiency.

For the DNase treatment: to [^{32}P]DNA in 0.5 ml of BB at room temperature, 0.05 to 0.2 μg of repressor were added; after a few seconds, varying amounts of DNase were added. After 2 min, the 0.5 ml was filtered at full suction and the filter rinsed with a further 0.5 ml of BB. IPTG was used at 10^{-3} M.

Radioactive sonicated fragments are mixed with a moderate excess of the repressor and the mixture is treated with DNase at room temperature and then filtered on a cellulose nitrate filter. As a control, an exactly parallel reaction is carried out in the presence of IPTG which would prevent the *lac* repressor from binding to and protecting the operator. Fig. 1 shows that there is such an effect; the protected amount levels off at about 1 % of the input DNA. One per cent of the input DNA, of course, would represent a fragment ten base pairs in length, but the efficiency of trapping the DNA fragment by the repressor is not 100 %; it is closer to 50–70 %. Furthermore, as I shall show later, the complex decays appreciably during the digestion.

For this protected material to be characterized it has to be rescued from the filter before being destroyed by the DNase. When the filter is shaken in buffers containing 0.1 % SDS (sodium dodecyl sulphate) the DNase is inactivated, the

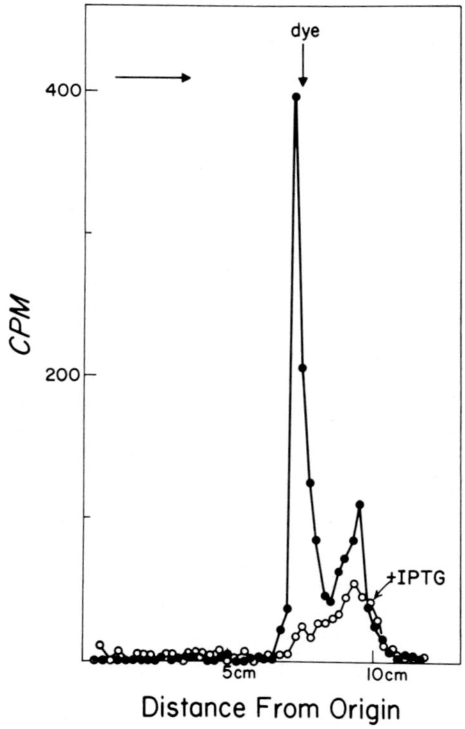

FIG. 2. Gel electrophoresis of the protected DNA. After a 2-min treatment of [^{32}P]DNA and repressor with 40 μg/ml pancreatic DNase, filtration and washing as described for Fig. 1, the filter was vortexed with 0.2 ml of 0.1 % SDS in 0.04 M-tris-acetate, pH 8.3. This sample was heated for 5 min at 100 °C and applied to a 15 cm 12 % polyacrylamide gel made up in 0.04 M-tris-acetate, pH 8.3, with 0.1 % SDS. At the end of the run, the gel was cut into slices and each slice counted without scintillator, using the Cerenkov radiation from the ^{32}P electrons. The dye is bromophenol blue. The control sample was from a DNase digestion in the presence of 10^{-3} M-IPTG.

repressor is denatured and the label comes off in solution. (Only about half of the background label from the 'IPTG' filters comes off in this treatment.) To characterize this material, the labelled fragments released from the filter were denatured and electrophoresed on a 12% polyacrylamide gel which will resolve small nucleic acids according to size. Fig. 2 shows the pattern on such a gel of the material released from the filter with SDS. Running immediately behind the dye in the sample protected by the repressor is a major peak of label, but this is absent in the control sample in which IPTG was present during the digestion. On these gels transfer RNA migrates at 0.75 of the mobility of the dye, material in the range 20 bases long migrates with the dye and the forward peak of material appearing on the gels consists of di- and trinucleotides.

This experiment shows that the protected fraction has a coherent short length. One can examine the protected fragment more closely by recovering it from the SDS wash. Phenol extracts the SDS, ether removes the phenol, and nitrogen bubbled through the solution blows off the ether. This yields a solution of ^{32}P-labelled material in a usable form. The most striking property of this material is that a fraction of the labelled material will bind again to a repressor molecule, as assayed by filter binding. About 20% of the labelled material

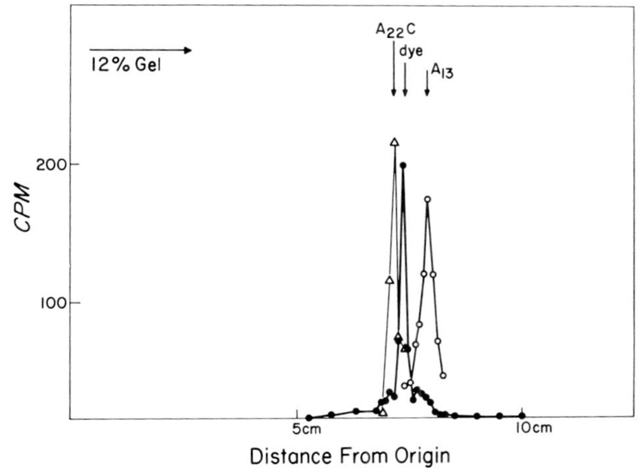

FIG. 3. The size of the operator fragment, purified by rebinding to the *lac* repressor. After filtration and rinsing of a digestion mixture, as described for Fig. 2, the ^{32}P was released from the filter by vortexing it with 0.1% SDS in SSC (0.15 M-NaCl, 0.015 M-Na citrate). After phenol extraction and ether extraction, nitrogen was bubbled through the solution and the material was diluted 1 to 10 into BB, 0.2 μg of repressor was added, and 0.5 ml samples were filtered and washed with 0.5 ml of BB. The filter was then vortexed with 0.2 ml of BB containing 10^{-3} M-IPTG and part of the eluted material was adjusted to 0.1% with SDS; it was then denatured and electrophoresed as described for Fig. 2. The markers were run on parallel gels. Their positions are all determined relative to the dye mobility.

binds directly to the filter as a background; another 20% can be bound to the filter by the repressor and can be eluted from the repressor with IPTG. Such eluted material is displayed on a gel in Fig. 3. This material, which has gone through the DNase digestion and been bound to and eluted from the repressor, runs as a single peak on these gels slightly behind the dye position. On this gel are superimposed patterns of markers run on parallel gels; these were single-stranded RNA markers 13 and 23 bases long (kindly provided by Paul Doty's laboratory). The denatured operator fragment runs between these markers at an estimated length of 20 bases. Of course we do not know that RNA fragments and DNA fragments run equivalently on these gels, but it is unlikely that we are making a gross error in the assignment of length.

We can obtain operator fragments, protected by the repressor, which have a length of approximately 20 bases. These DNA molecules are double-stranded and capable of binding again to the repressor. Their double-stranded nature is further demonstrated by the observation that on denaturation they no longer bind to the repressor and have a melting temperature, a T_m, of 67 °C in SSC. The operator molecule also sediments differently in a native or denatured state and moves differently on a Sephadex G-100 column before or after heat treatment. On the polyacrylamide gels the molecule appears always to be denatured. I believe that the low ionic strength and filtering activity of the gel denature this DNA fragment.

The best evidence that this fragment represents the operator sequence is that this double-stranded piece of DNA binds again to the repressor and is released by IPTG—characteristics solely of the *lac* operator. The affinity of this fragment for the repressor is still comparable to the original affinity of the *lac* operator; this is shown indirectly by the fact that binding is detectable on filters when 10^{-10} and 10^{-11} M solutions of the repressor are used. However the interaction of the repressor with this small fragment is not precisely the same as its interaction with a region of a longer DNA chain. The lifetime of the repressor-operator fragment complex is appreciably shorter than the lifetime of the repressor-large DNA complex. Fig. 4 shows directly that the decay of the repressor-operator complex is an order of magnitude faster than the decay of the repressor from the operator embedded in a large piece of DNA. A 2-minute half-time characterizes the decay of the repressor from the operator fragment; the same time is observed if one measures the decay of protection against DNase digestion. This more rapid decay from the small fragment must be reflected in a weaker binding of the repressor to the small fragment than to normal DNA.

The faster decay might be interpreted in several ways. The operator fragment may differ slightly from the normal operator; a slight unwinding of the structure or the presence of free phosphates at the ends of the molecule close

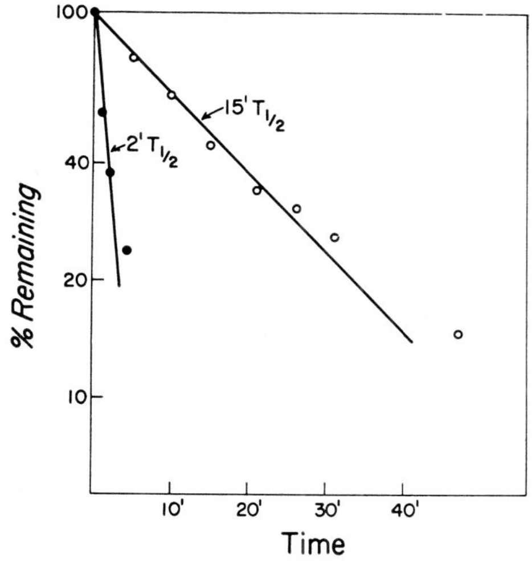

FIG. 4. Decay of repressor-operator complexes. The decay was measured by mixing radioactive DNA, either sonic fragments (which behave the same as whole phage DNA molecules) or repurified operator DNA, in BB with 0.01 μg/ml repressor. After filtration of a first 0.5 ml sample (0.5 ml wash), 20 μg/ml unlabelled λ*plac*$_5$ DNA was added and further 0.5 ml aliquots were filtered at different times. The open circles are sonicated DNA, approximately 1000 base pairs in length. The closed circles show the decay from operator DNA fragments, approximately 20 base pairs in length.

to the protein might account for the slight change in affinity. Alternatively, the rapid binding of repressor to the operator site has been interpreted as involving a search along the DNA molecule for that site; the presence of arms of DNA around the operator might slow up the decay by permitting the repressor to wander off the operator site and then be recaptured. Yet again, the binding of the repressor to the operator might be thought of as having an overall electrostatic component which would not be as great for this small molecule as for a longer region of DNA. The affinity of the repressor for this operator fragment is not down by the large factor that one would expect if, for example, I had isolated only half of the correct binding site, picturing the repressor as interacting with two sites on the DNA and the DNase being able to cut between them. In such a case one might expect the affinity of the repressor for each site separately to be described by half of the binding energy and thus that the dissociation constant would be of the order of 10^{-6} M: such weak binding would be undetectable in these assays.

This operator fragment is approximately 7 nm long. This is comparable in size to the repressor protein which, if spherical, should have a diameter of the order of 7 nm. The fragment is larger than the minimal estimate of 12 base pairs and so could reasonably contain all the points at which the protein must make contact with the DNA. A similar isolation of the operators of phage lambda has been carried out by V. Pirrotta and is being pursued by T. Maniatus, working in Mark Ptashne's laboratory at Harvard. The striking difference between the two systems is that the lambda operators are much longer DNA pieces than this *lac* operator, with the protected region ranging up to 80–100 base pairs in length; possibly protection is by a polymerization of several repressor molecules around the DNA. The *lac* operator behaves as though it is protected by a single repressor molecule.

What is the sequence of this *lac* operator fragment? The most promising approach now is to copy this fragment into RNA using the RNA polymerase. In order to do this I produce operator on a large scale, beginning with 100 mg amounts of phage DNA, and scaling up the binding and DNase digestion to the maximal capacity of the filters, which is 20 μg repressor per square centimetre of surface. When denatured strands of DNA are copied with the RNA polymerase holoenzyme, using commercially available alpha-labelled triphosphates as precursors, I get a 40-fold amplification of the DNA strands. The input into a single reaction is about 3 picomoles of DNA strands and we recover about 100–150 pmol of RNA strands. Twenty to fifty per cent of the input of [α-^{32}P]triphosphates is incorporated when a single labelled triphosphate is used in limiting amounts, with an excess of the other three. The RNA product is approximately the correct size: peculiarly, the enzyme does not seem to make small fragments, but behaves as though it picks up the DNA strand at one end and goes to the other. The enzyme initiates with guanine and an alkaline phosphatase treatment of a guanine labelled RNA product yields one end for every 6.8 guanines or a crude overall estimate of 27 base pairs for an average length. Allan Maxam and I are currently analysing T1 and pancreatic fingerprints of these RNA products. We have identified almost all the pancreatic fragments and many of the T1 fragments, but we have not yet assembled a sequence. Several difficult problems will remain to be overcome even after a sequence for the RNA products has been established. Does the enzyme make a faithful copy of the DNA strand or does it make errors? My present hope is that we shall be able to predict enough of the properties of the DNA sequence, to be verified by studying the pyrimidine tracts from operator fragments, to enable us to prove the sequence. In principle, we can use mutant operators, isolated from very weak o^c mutants, to identify those base pairs that interact with the protein. There will be severe problems, however, because the mutant operators will bind to the repressor less well and the entire isolation will be more difficult.

What does the binding site of the repressor look like? The operator sequence may have no symmetry, because the DNA might interact with a single subunit of the repressor (there would then be four binding sites on the tetrameric repressor molecule), or because the four tetramers of the repressor fit together to form a binding site that does not reflect any symmetry in the protein. A second possibility has been that the repressor would bind on a dyad axis between two of the subunits. The DNA sequence then should have a dyad symmetry; this has been argued on genetic grounds by Sadler & Smith (1971). Such a symmetry would permit the formation of two looped-out structures on the DNA molecule as was suggested first by Gierer (1966). Alternatively, there might be some reflection of the fourfold structure of the repressor. This might involve the formation of ears with a four-fold symmetry, each capable of recognizing a monomer of the tetramer (Gierer 1966). Another possibility is that the operator might be made up of a primary sequence, fourfold repeated, with each repeat pointing in the same direction. The thought here is that the repressor subunits might be in the form of a hindered helix, with the four subunits arranging themselves in a helical pattern around the DNA and each subunit being in contact with the same sequence a little further along the DNA molecule. This theory has been espoused most recently by Benno Müller-Hill and his co-workers (Adler *et al.* 1972) who propose that a protuberance at the amino terminal end of the repressor monomer makes contact with DNA, a similar contact being made by each of the four subunits to four sequences arranged in tandem along the DNA molecule. We do not yet have a complete sequence and thus lack sufficient evidence permanently to exclude any of these possibilities. The current evidence on the sequence, however, does not suggest that there is any symmetry or any repeat. We recover more pancreatic fragments than we would expect if there were either fourfold or twofold symmetry. The numbers of fragments and their sequences are such as to suggest that the final sequence will not have any symmetry, but will represent an apparently arbitrary contact between the protein and DNA.

We shall pursue our studies of this interaction, but complete understanding will probably have to wait until X-ray analysis has unravelled the structure of the protein and a direct analysis can be made of crystals of the protein with the operator sequence.

ACKNOWLEDGEMENTS

I thank Terry Platt for providing *lac* repressor, Charlotte Hering for RNA polymerase, Mary Archer and Jeanne Winsten for technical assistance, Allan Maxam for the RNA fingerprinting, and the NIHGMS for their support of this work under grant GM No. 09541.

References

ADLER, K., BEYREUTHER, K., FANNING, E., GEISLER, N., GRONENBORN, B., KLEMM, A., MILLER, J., MÜLLER-HILL, B., PFAHL, M. & SCHMITZ, A. (1972) How *lac* repressor binds to DNA. *Nat. New Biol.* **237**, 322–327

BOURGEOIS, S. & RIGGS, A. D. (1970) The *lac* repressor-operator interaction. IV: assay and purification of operator DNA. *Biochem. Biophys. Res. Commun.* **38**, 348–354

BURSTEIN, C., COHN, M., KEPES, A. & MONOD, J. (1965) Rôle du lactose et de ses produits métaboliques dans l'induction de l'operon lactose chez *Escherichia coli. Biochim. Biophys. Acta* **95**, 634–639

CHEN, B., DECROMBRUGGHE, B., ANDERSON, W. B., GOTTESMAN, M. E., PASTAN, I. & PERLMAN, R. L. (1971) On the mechanism of action of *lac* repressor. *Nat. New Biol.* **233**, 67–70

DECROMBRUGGHE, B., CHEN, B., ANDERSON, W. B., NISILEY, P., GOTTESMAN, M. E., PASTAN, I. & PERLMAN, R. L. (1971) *Lac* DNA, RNA polymerase and cyclic AMP receptor protein, cyclic AMP, *lac* repressor and inducer are the essential elements for controlled *lac* transcription. *Nat. New Biol.* **231**, 139–142

EMMER, M., DECROMBRUGGHE, B., PASTAN, I. & PERLMAN, R. L. (1970) Cyclic AMP receptor protein of *E. coli*: its role in the synthesis of inducible enzymes. *Proc. Natl Acad. Sci. U.S.A.* **66**, 480–487

ERON, L. & BLOCK, R. (1971) Mechanism of initiation and repression of *in vitro* transcription of the *lac* operon of *Escherichia coli. Proc. Natl Acad. Sci. U.S.A.* **68**, 1828–1832

GIERER, A. (1966) Model for DNA and protein interactions and the function of the operator. *Nature (Lond.)* **212**, 1480–1481

GILBERT, W. & MÜLLER-HILL, B. (1967) The *lac* operator is DNA. *Proc. Natl Acad. Sci. U.S.A.* **58**, 2415–2421

GREENBLATT, J. & SCHLEIF, R. (1971) Arabinose C protein: regulation of the arabinose operon *in vitro. Nat. New Biol.* **233**, 166–170

JOBE, A. & BOURGEOIS, S. (1972) The natural inducer of the *lac* operon. *J. Mol. Biol.*, **69**, 397-408.

MAKMAN, R. S. & SUTHERLAND, E. W. (1965) Adenosine 3′, 5′-phosphate in *Escherichia coli. J. Biol. Chem.* **240**, 1309–1314

MÜLLER-HILL, B., CRAPO, L. & GILBERT, W. (1968) Mutants that make more *lac* repressor. *Proc. Natl Acad. Sci. U.S.A.* **59**, 1259–1264

MÜLLER-HILL, B., BEYREUTHER, K. & GILBERT, W. (1971) *Lac* repressor from *Escherichia coli.* In *Methods in Enzymology* (Grossman, L. & Moldave, K., eds.), pp. 483–490, Academic Press, New York

RIGGS, A. D., BOURGEOIS, S., NEWBY, R. F. & COHN, M. (1968) DNA binding of the *lac* repressor. *J. Mol. Biol.* **34**, 365–368

RIGGS, A. D., SUZUKI, H. & BOURGEOIS, S. (1970*a*) *Lac* repressor-operator interaction. I: equilibrium studies. *J. Mol. Biol.* **48**, 67–83

RIGGS, A. D., BOURGEOIS, S. & COHN, M. (1970*b*) The *lac* repressor-operator interaction. III: kinetic studies. *J. Mol. Biol.* **53**, 401–417

RIGGS, A. D., REINESS, G. & ZUBAY, G. (1971) Purification and DNA-binding properties of the catabolite gene activator protein. *Proc. Natl Acad. Sci. U.S.A.* **68**, 1222–1225

SADLER, J. R. & SMITH, T. F. (1971) Mapping of the lactose operator. *J. Mol. Biol.* **62**, 139–169

SILVERSTONE, A. F., MAGASANIK, B., REZNIKOFF, W. S., MILLER, J. H. & BECKWITH, J. R. (1969) Catabolite-sensitive site of the *lac* operon. *Nature (Lond.)* **221**, 1012–1014

SILVERSTONE, A. E., ARDITTI, R. R. & MAGASANIK, B. (1970) Catabolite-insensitive revertants of *lac* promoter mutants. *Proc. Natl Acad. Sci. U.S.A.* **66**, 773–779

ZUBAY, G., SCHWARTZ, D. & BECKWITH, J. (1970) Mechanism of activation of catabolite-sensitive genes: a positive control system. *Proc. Natl Acad. Sci. U.S.A.* **66**, 104–110

Discussion

Changeux: Does only one subunit interact with DNA or are four of them involved? Are there monomeric mutants of the repressor protein where a single subunit is known to bind to the operator?

Gilbert: No one has reported a monomeric subunit which can interact with DNA. The conformational transition that separates the DNA binding form from the DNA non-binding form is very easy to trigger *in vitro*. So with any process that generates monomers one can always argue that a monomer is left in the form with low affinity for DNA and not in the form with high affinity. On the other hand one could say that the fact that no one has observed monomers that bind to DNA is the crude expectation from the tetramer model.

Changeux: Did you try complementation between different mutants?

Gilbert: The complementation that happens *in vivo* works the other way. One can observe an interaction between 'good' (i^+) and 'bad' (i^-) molecules generating 'bad' molecules. This again leaves you with an argument which you could use in either form. Either the structure is terribly delicate in its balance between binding and non-binding or the structure is evidence for a direct effect. If the binding is to a tetrameric form and if one does put a bad subunit in, of course that destroys the binding to the DNA because that subunit contributes a reasonable fraction of the binding site.

Silman: Have you tried cross-linking the bound repressor to the DNA to see how many subunits remain afterwards?

Gilbert: That has not been tried. It all depends on what the cross-linking agent is. It would be difficult to discover how many subunits are left because the subunits are identical. Can one identify which parts of the protein are in contact with the DNA? A recent model (Adler *et al.* 1972) asserts that the interesting part of the molecule, the part most sensitive to changes that interfere with DNA binding, is the amino terminal end of the molecule. (This is supported genetically. There are a large number of genetic changes in the repressor (i^{-d}) which abolish DNA binding while leaving the rest of the structure of the protein moderately intact. These map in the gene very close to the amino terminal end.) Adler *et al.* (1972) suggest that the amino terminal end of each monomer makes direct contact with the DNA; four monomers line up along the DNA. However, there is also the possibility that the amino terminal end is involved in the structure of the tetramer so that if one damages the amino terminal end the molecule cannot take up a DNA binding conformation. T. Platt & K. Weber (to be published) have localized a number of these i^{-d}

mutants in the protein sequence. We now know a number of groups on the protein in which a change prevents DNA binding.

Katchalski: Can you estimate the maximum number of bases that a subunit of molecular weight 40 000 could accommodate, on geometrical considerations only?

Gilbert: I tried this by making a model with a lump of clay with a protein of molecular weight 40 000 and squeezing it to see how much DNA I could reasonably cover. By squeezing considerably I can cover 20 bases along the DNA. There is a lot of material in a protein of 40 000 molecular weight.

Changeux: In the model with four identical subunits the subunits are not equivalent in the manner they bind to DNA. This is quite a problem. Could something other than the nature of the base be recognized in the DNA structure?

Gilbert: There has always been a question of whether the bases are modified in some way. But what new enzyme is to recognize the sequence to be modified with high specificity?

Changeux: What part of the DNA structure is recognized?

Gilbert: There is no problem with recognizing the outside of the DNA structure. In the big groove, there are hydrogen bond acceptors and donors enough to establish the sequence.

Levitt: Can the repressor recognize mispaired bases—for example, G·U or A·C—that would be unique to the sequence of the recognition site on the DNA?

Gilbert: Not in a DNA structure—it is against the rules. In the RNA structure there are mispaired bases at corners. In the DNA structure we have not seen any strange bases. We have not exhausted the possibility of modified bases in operators. Modified bases take one back to exactly the same problem of how to recognize a region with high specificity and therefore a regression. There are groups easily visible in the large groove and a similar set of groups in the small groove. If one builds models of, say, α-helices and asks whether the α-helices can fit in a reasonable way against a DNA molecule and make hydrogen bonds into the big groove, the answer is yes.

Luzzati: I do not quite understand why you give yourself the rule that the repressor has to recognize a double-stranded helix. This point was discussed by Crick (1971), who gave cogent arguments in favour of the notion that DNA–protein recognition is likely to involve unfolded portions of DNA penetrating the protein.

Gilbert: The *lac* repressor and the λ repressor do not behave as one would expect if they were reacting with a single-stranded region of DNA. These proteins do not recognize a free single-stranded piece of DNA.

Tanford: When you say that the individual subunit doesn't bind, have you only looked for binding with an affinity of 10^{-13} M or have you also looked for binding with an affinity of 10^{-6} M? If, for example, the binding free energy is shared between two subunits, one subunit would have half the free energy and the binding constant would be the square root of the binding constant of the dimeric protein.

Gilbert: You are perfectly right. In fact the techniques we use for looking at the binding of protein and DNA are all techniques which exploit affinities or dissociation constants of 10^{-9} M and below. To see the interaction of protein with DNA at a concentration of 10^{-6} M is extremely difficult.

Changeux: What is the proof that the bound form of the repressor protein is a tetrameric one?

Gilbert: There is no real proof. There is no cooperativity in the binding of the protein to DNA for the *lac* repressor, which is a tetrameric solution. For the λ repressor, which is not a tetrameric solution but is a monomer in solution, cooperativity is visible in the binding curve. But there is no direct evidence that only four subunits are binding and not more.

Changeux: Is the stoichiometry between repressor subunits and *lac* operator actually known?

Gilbert: It has never been done properly.

Schwyzer: The interaction between cyclic AMP (cAMP) and its 'binding proteins' has a mean dissociation constant of about 2×10^{-9} M. For CAP it probably should be much larger, because there seem to be only very few molecules around. Do you know approximately the order of magnitude of the binding constant of cAMP to CAP?

Gilbert: About 10^{-7} M. It is tighter than the usual enzyme substrate. The binding constant of the *lac* repressor for its small molecule is 10^{-6} M.

Katchalski: Are there any enzymes in which the site of binding and orientation of the attached oligonucleotides have been determined?

Gilbert: The only thing that has been studied is ribonuclease. The enzyme recognizes pyrimidines rather than purines but in fact it recognizes the Watson-Crick side of the base rather than anything else.

References

ADLER, K., BEYREUTHER, K., FANNING, E., GEISLER, N., GRONENBORN, B., KLEMM, A., MILLER, J., MÜLLER-HILL, B., PFAHL, M. & SCHMITZ, A. (1972) *Nat. New Biol.* **237**, 322–327

CRICK, F. H. C. (1971) *Nature (Lond.)* **234**, 25

Protein-lipid interactions

D. CHAPMAN

Department of Chemistry, Sheffield University

Abstract The way in which proteins and lipids interact is of considerable importance for a number of biological situations, and particularly the organization of cell membranes.

The lipids of membranes have received extensive study and a variety of classes are known to occur; much less is known about the proteins. In this paper the interaction forces between the lipids and proteins are discussed, model systems are described and recent chemical, enzymic and physical studies of real cell membranes are summarized to provide an overall view of protein-lipid interactions in membrane structure. The heterogeneity of lipid organization in the bilayer of some membranes and the question of where the protein is situated in these membranes are discussed.

Membrane biogenesis is also discussed, including the question of whether newly synthesized protein associates preferentially with newly synthesized lipid.

The interaction between protein and lipid is a fundamental process of biological systems which is beginning to demand increasing attention. The biological systems in which protein–lipid interactions are considered to be of importance include blood coagulation processes, certain enzyme and receptor systems, serum lipoproteins, and cell membrane structures.

I shall concentrate here mainly on the question of protein–lipid interactions in cell membranes, summarizing the information available on (*a*) the nature and type of the lipids and proteins concerned in cell membrane structure, (*b*) the interaction forces involved, (*c*) model systems which simulate the ways in which lipids and proteins may interact, (*d*) the chemical and physical studies of real membranes, and (*e*) the biosynthetic process of cell membrane construction.

Since many enzymes (Rothfield & Romeo 1971) and receptor lipoproteins (Changeux *et al.* 1967) are thought to be organized in the membrane structures of cells, these will be encompassed naturally within the discussion on cell membrane structure.

THE LIPIDS AND PROTEINS OF CELL MEMBRANES

Let us consider the numbers and different types of lipids and proteins involved in cell membranes.

Lipids

The term 'lipid' covers a wide range of molecular structures. It includes glycerides (tri, di and mono), cholesterol esters, the phosphoglycerides (e.g. lecithin, phosphatidylethanolamines), the sphingolipids (e.g. the cerebrosides and sphingomyelins) and the gangliosides and other glycolipids (Chapman 1969*a*).

Even if we exclude the cholesterol esters and the tri, mono and diglycerides which are present in the serum lipoprotein and limit ourselves to those lipids

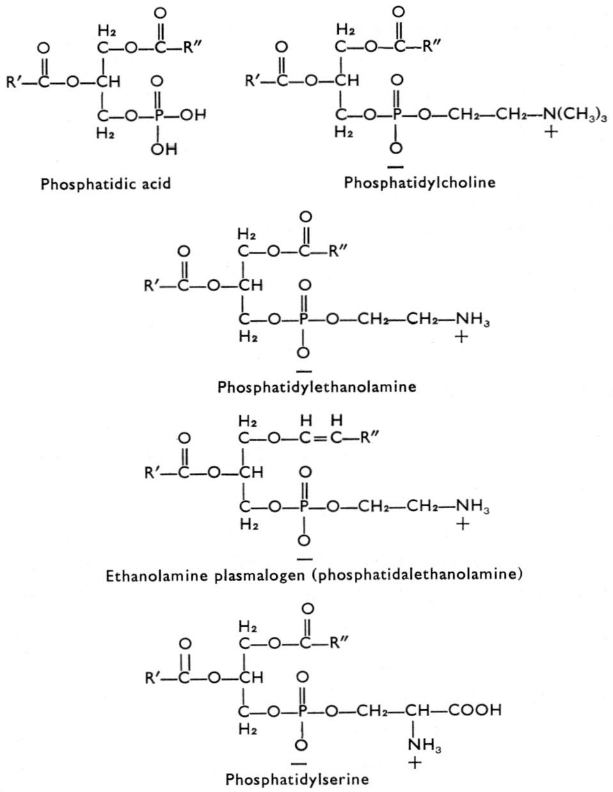

FIG. 1. Phosphoglyceride structures.

which form the basic organization of cell membranes, we find a tremendous variety of structures present. The structures of some of these lipids are shown in Figs. 1 and 2. They contain various groups, such as OH, NH_2, $N^+(CH_3)_3$, COOH, PO_4, and various sugar groupings. These functional groups can combine in several ways and, as a result, lipids may contain zwitterion type structures or negatively charged structures. Metal ions are also known to be intimately involved with some lipid structures but not with others.

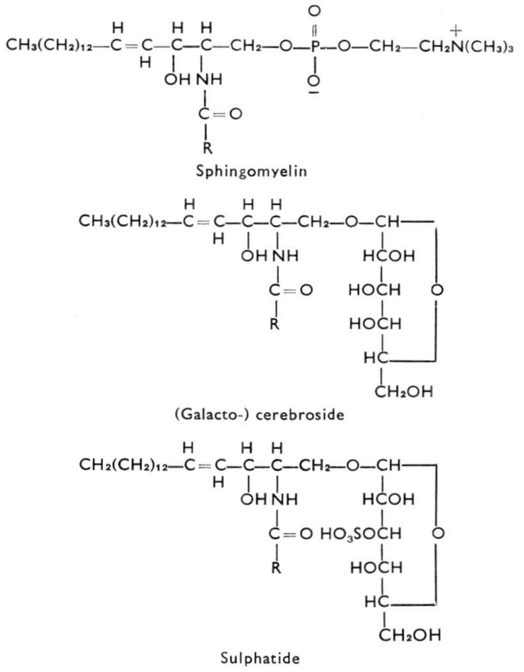

FIG. 2. Sphingolipid structures.

Associated with each of these different lipid classes is a range of fatty acids of various chain lengths and amounts of unsaturation. In the myelin membranes there are long fatty acids 24 carbon atoms in length and of limited unsaturation, whilst in mitochondrial membranes there are large amounts of highly unsaturated C_{18} fatty acids (O'Brien 1965).

The relative ease with which lipid components can be separated has thus allowed the lipid content of membrane systems to be carefully evaluated. A similar situation does not yet apply to membrane proteins. A difficulty in the study of membrane proteins arises from the ability of the proteins to bind lipid tenaciously.

Proteins

Some of the techniques that have been used for the isolation of membrane proteins (Chavin 1971) are indicated in Table 1.

TABLE 1

Some techniques used for isolation of membrane proteins (after Chavin 1971)

Technique or reagent	*Protein and source*	*Comments*
Osmotic shock	High affinity binding proteins, Gram-negative bacteria	Proteins released are probably pericytoplasmic or loosely bound to membrane rather than integral components of membrane
Distilled water	'Structural' protein from ghosts of human red cells	Membrane completely solubilized only if ghosts have been thoroughly deionized
Low ionic strength	ATPase, *Streptococcus faecalis*	Enzyme solubilized by repeated washes with 1 mM-tris HCl buffer in absence of Mg^{2+}
High ionic strength	Protein from ghosts of human red cells	0.8 M-NaCl solubilizes about 40% of membrane protein
Chelating agents	'Spectrin', ghosts of human red cells	5 mM-EDTA, low ionic strength buffer, plus β-mercaptoethanol
Triton X-100	Acetylcholinesterase, electroplax electric organ	0.5% w/w detergent solution for 48 h at 4 °C. Sonication before detergent treatment. Enzyme partially inactivated
Deoxycholate	Acetylcholine receptor, electroplax electric organ	1% w/w detergent solution. Acetylcholinesterase also solubilized

One approach to the solubilization of membrane proteins consists of the application of techniques used for aqueous systems. A variable, and sometimes large, amount of protein can be solubilized by fairly mild conditions, such as high ionic strength (Rosenberg & Guidotti 1969), chelating agents (Marchesi *et al.* 1970), and even distilled water (Mazia & Ruby 1968). Large amounts of protein can be solubilized from mitochondria of rat liver with either 0.8 M-NaCl or 1 mM-EDTA (S. I. Chavin, unpublished).

Many membrane proteins are difficult to solubilize under these conditions (Jones & Kennedy 1969), others tend to aggregate irreversibly (Rosenberg & Guidotti 1969), and still others are unstable in aqueous solutions (Higashi *et al.* 1970). Furthermore, these difficulties may occur even in the presence of detergents and other denaturing reagents (Rosenberg & Guidotti 1969). Solubilization and fractionation may, however, be considerably facilitated by using reagents such as ionic detergents, urea, guanidine and organic solvents.

Rosenberg & Guidotti (1968, 1969) have accomplished the most successful resolution so far of the proteins of erythrocyte membranes by combining several existing methods (Marchesi & Steers 1968; Mitchell & Hanahan 1966), together with modifications of the conventional lipid extraction and detergent solubilization of membrane proteins. The composition of intact erythrocyte 'ghosts' is shown in Table 2 and the amino acid composition is shown in Table 3.

TABLE 2

Composition of intact and lipid-extracted ghosts of red blood cells

Component	*Intact ghost* %	*Lipid-extracted ghost* %
Protein	49.2	91.0
Lipid (total)	43.6	1.5
Phospholipid	32.5	1.5
Cholesterol	11.1	0
Carbohydrate (total)	7.2	7.5
Neutral sugars	4.0	2.5
Hexosamines	2.0	2.6
Sialic acids	1.2	2.4

TABLE 3

Amino acid composition of lipid-extracted membrane protein

Amino acid	*Residues per 100 residues*
Lysine	5.21
Histidine	2.44
Arginine	4.53
Aspartic acid[a]	8.49
Threonine	5.86
Serine	6.26
Glutamic acid[a]	12.15
Proline	4.26
Glycine	6.73
Alanine	8.15
Half-cystine	1.08
Valine	7.10
Methionine	2.02
Isoleucine	5.29
Leucine	11.34
Tyrosine	2.41
Phenylalanine	4.20
Tryptophan	2.49

[a] One mole of NH_3 was formed for every three aspartic and glutamic acids.

The procedure involves four steps: two aqueous extractions of the ghost, a lipid-extraction step, and solubilization in sodium dodecyl sulphate followed by fractionation on Sephadex gels. Eight fractions (I to VIII) of membrane protein were observed after these procedures. The extraction of the membrane with EDTA has a significant effect on the subsequent extraction with NaCl. The distribution of proteins in the various fractions is shown in Table 4.

TABLE 4

Distribution of proteins in the various fractions of the membrane proteins

Fraction	*Extraction procedure*	*% of total membrane protein*	
I	EDTA extraction	11	
II	NaCl extraction	41	
III	Lipid extraction	7	
IV-VIII	Residue after lipid extraction (Gel filtration in sodium dodecyl sulphate)	41	
IV			8
V			5
VI			13
VII			7
VIII			8
	Total	100	

Subsequent analysis using Sephadex chromatography, gel electrophoresis and amino acid analysis showed the fractions to be both chemically and physically distinct. In addition there was a non-uniform distribution of both neutral sugars and sialic acid among the subfractions. *N*-terminal analysis of each of the fractions indicated that there was a minimum of 12 different polypeptide chains in the membrane protein fractions. The average molecular weights of fractions IV to VIII ranged from 10 000 to 170 000, as determined by gel filtration on Sephadex.

Marchesi *et al.* (1970) have reported the resolution of a specific protein fraction from the erythrocyte membrane. The fraction is approximately 20% of the ghost protein and is reported to be free of lipid, neutral sugar and sialic acid. This fraction (known as spectrin) migrates as a single band on gel electrophoresis in the presence of sodium dodecyl sulphate, with an apparent molecular weight of 140 000. This value is corroborated by results obtained from sedimentation equilibrium and viscosity measurements. In addition, cleavage by cyanogen bromide and separation by gel electrophoresis produced the expected number (as determined by data on amino acid composition) of peptides for a protein molecule of this size.

The recent work of Strominger's group has indicated the possibility of extracting membrane proteins directly into organic solvents without denaturation (Higashi *et al.* 1970). In principle this is a very sensible way of handling such proteins, although it may not always be successful. The enzyme, an ATP-dependent C_{55} isoprenoid alcohol phosphokinase from the membrane of *Staphylococcus aureus*, has the following properties:

(*a*) it is extractable in active form into acidic butanol at room temperature, from which it precipitates reversibly at $-15\,°C$; (*b*) it is stable in organic solvents such as acetone, heptane, ether and benzene, but not in chloroform–methanol (1:1); (*c*) it is insoluble in water and relatively unstable in aqueous buffers; (*d*) it undergoes a 50% increase in activity when heated in butanol at 100 °C for 20 minutes; and (*e*) enzyme activity requires either phosphatidyl glycerol or cardiolipin or both, but this requirement may be partially satisfied by certain detergents (Higashi *et al.* 1970; Higashi & Strominger 1970).

THE INTERACTION FORCES INVOLVED IN PROTEIN–LIPID INTERACTIONS

Some of the possible ways in which lipids and proteins of cell membranes may interact are:

(*a*) Covalent bonding

(*b*) Electrostatic interaction

(*c*) Polarization interaction

(*d*) Dispersion interaction

(*e*) Hydrophobic interaction

(*f*) Other types of binding (e.g. bridging by metal ions). A detailed discussion of the importance of these different binding forces in protein–lipid interactions has appeared elsewhere (Chapman 1969*b*).

Briefly we can say that:

(*a*) *Covalent links* between phospholipids and amino acids are not frequent;

(*b*) *Electrostatic interaction* between lipid and protein has been postulated to occur in the model of the plasma membrane (Danielli & Davson 1935) and has been observed in certain model complexes;

(*c*) *Polarization interactions* are expected to be very small;

(*d*) *Dispersion interactions* are usually small and very sensitive to distance. With lipid chains in a parallel configuration these dispersion interactions add up and can be as much as 10–20 kcal/mole. They play an important role in determining lipid transition temperatures;

(*e*) *Hydrophobic interactions*: Kauzmann (1959) suggests that hydrophobic bonds are stabilized largely by entropic effects. He suggests that hydrocarbon groups in an aqueous environment increase the order or quasicrystalline order of water molecules in their immediate vicinity. When the non-polar groups leave the aqueous environment, a large gain in entropy occurs. The change in free energy that occurs on transfer of the non-polar group from the aqueous environment is estimated to be about 3000 to 5000 cal per non-polar amino acid side chain, or 1000 cal per CH_2 group.

Evidence for the importance of hydrophobic bonds has come from the theoretical work of Scheraga (1961) and Scheraga *et al.* (1962) as well as from experimental approaches emphasizing the effects of non-aqueous solvents (Singer 1962), hydrocarbons (Wishnia 1962; Wishnia & Pinder 1966) and fatty acids (Bull & Breese 1967) on proteins and the interactions of non-polar polypeptides (Lotan *et al.* 1966). The latter studies on model polypeptides with non-polar side chains indicate that, in aqueous solvents, the α-helix is a favoured conformation for hydrophobic polypeptides.

The configurations adopted in water by phospholipids and polar lipids forming micelles and liquid crystalline aggregates are also such that the molecules have their polar groups facing the aqueous environment and the hydrocarbon groups clustered together. Hydrophobic bonding is considered to take place in this situation as well. The idea that the hydrocarbon chains are squeezed out of solution, rather than attracted to each other, is a familiar concept in colloid chemistry (Hartley 1955).

The addition of electrolytes to the aqueous medium is considered to strengthen hydrophobic bonds, whereas electrolytes weaken electrostatic linkages. Non-polar substances, on the other hand, weaken hydrophobic bonds but strengthen electrostatic linkages.

It seems very reasonable that hydrophobic bonding may play an important role in many lipid-protein interactions.

(*f*) *Other types of interaction*: In addition to the processes just discussed, there are other ways in which lipids and proteins may interact. These include hydrogen bonding, entropy effects and bridging by metal ions. It is difficult to assess the importance of hydrogen bonding between the different groups in lipid-protein complexes. It is known that water competes for the hydrogen bonding sites (Kauzmann 1959).

Summarizing the general molecular situation of what may occur when lipids and protein are brought together in water, we can say that various consequences may follow, depending on the resultant free energy of the product and involving some, or all, of the types of interaction forces considered here. These are:

(1) No interaction and no formation of complexes. Both lipid and protein retain their original configuration;

(2) Interaction in which the lipid loses all, or part of, its configuration; i.e. there is a lipid phase change but the protein retains its configuration;

(3) Interaction in which the protein loses all, or part of, its configuration while the lipid retains its configuration;

(4) Interaction in which both lipid and protein lose all, or part of, their original configurations;

(5) Interaction in which both lipid and protein retain (largely) their original configurations.

As an example of (2), charged groups on the protein may be neutralized by a few lipid molecules, with the protein retaining (largely) its original configuration. As an example of (3), a protein may unfold its amino acids at the lipid surface whilst the lipid configuration remains largely unchanged. As an example of (4), the hydrocarbon chains of the lipid might be rearranged so that they lie within the hydrophobic central region of the protein associated with the non-polar amino acids, both polar groups of lipid and protein being then on the outside of the complex. An example of (5) would be a small interaction in which a sheet of lipid and a protein both largely retain their original configuration.

MODEL SYSTEMS

To provide some background to the ways in which proteins and lipids may interact in real membranes it is useful to summarize the behaviour of some model systems. We have chosen these systems to illustrate the effects of electrostatic interaction and also hydrophobic interaction in simple lipid–water systems.

(*a*) *Lipid–water systems*

From the many studies that show the basic phase behaviour of membrane lipid–water systems we now know that these lipids show an endothermic transition from gel to liquid crystal at a temperature dependent upon chain length and unsaturation, that the hydrocarbon chains of the lipid are in a mobile condition in the liquid-crystalline phase, and that other parts of the molecules are also mobile, for example the $N^+(CH_3)_3$ groups (Chapman 1968).

Various lyotropic liquid-crystalline phases can occur so that whereas lecithins exhibit a lamellar phase over a wide range of concentrations, other lipids can exhibit hexagonal and other phases (Luzzati 1968). (An excellent account of the structures of these various phases is given in the latter article.)

(*b*) *Lipid–metal-ion–water systems*

Electrostatic interactions of the polar head of the phospholipids affect their thermotropic mesomorphism. With J. Urbina I have investigated (unpublished studies), the interaction of different divalent cations on the endothermic phase transition of dipalmitoyl lecithin and ox-brain phosphatidylserine. From monolayer and liposome experiments, uranyl ions (UO_2^{2+}) are known to interact stoichiometrically with the polar groups of lecithin and to bind in the membrane surface of different biological membranes, presumably to phosphate groups. We have confirmed that the strong binding of this ion is specifically to the phosphate groups by means of infrared spectroscopy. It is also known that phosphatidylserine forms a stoichiometric complex with Ca^{2+} and other divalent ions (Hauser *et al.* 1969).

These interactions have very strong effects on the phase transition of both lipids. A titration of dipalmitoyl lecithin with UO_2^{2+} shows that the original transition temperature is 314.5 K (41.5°C); it is shifted, in the 1:1 complex,

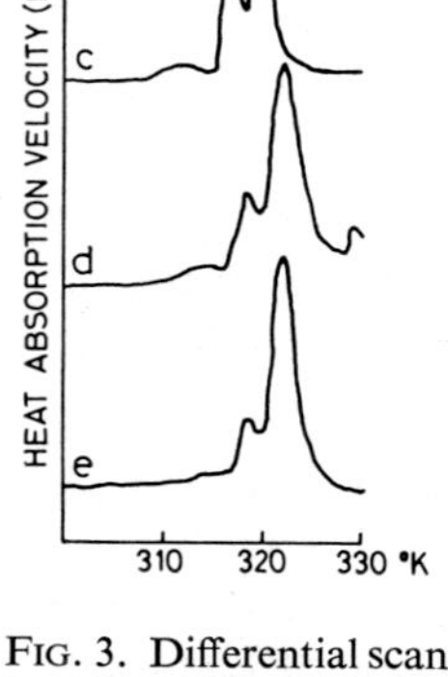

FIG. 3. Differential scanning calorimetry curves of dipalmitoyl lecithin–water (50–50 mixtures) in the presence of (from top to bottom) 0.0, 0.1, 0.2, 0.4 and 0.8 equivalent of uranyl ion (UO_2^{2+}) per equivalent of phosphorus (Urbina & Chapman, unpublished observations, 1971).

to 319 K (46 °C) (see Fig. 3). With phosphatidylserine containing a net negative charge, the binding of Ca^{2+} at 1:1 proportion shifts the transition from 290 K observed with free phosphatidylserine to 295 K for the complex.

(*c*) *Lipid–cholesterol–water systems*

These systems have been extensively studied (de Bernard 1958; Chapman & Penkett 1966; Lecuyer & Dervichian 1969). Cholesterol is insoluble in water but can be dispersed in lipid and is transported by it in an aqueous environment. The OH group is in the aqueous phase whereas the hydrophobic ring system is situated amongst the hydrocarbon chains of the lipid. At the temperature at which the lipid forms a gel phase, cholesterol prevents the lipid chains from crystallizing and the endothermic transition of the lipid is removed (Ladbrooke *et al.* 1968). In the liquid crystalline phase cholesterol inhibits some movement of the lipid chains (Williams & Chapman 1970).

(*d*) *Lipid–polypeptide–water systems*

Only a few studies have been made of lipid–polypeptide systems. Although they might be thought to be useful model systems for studying mainly electrostatic or mainly hydrophobic interactions, the studies that have been made have usually been associated with ion transport properties. Thus much work has been done on polypeptides and related molecules such as the antibiotics valinomycin, nonactin, gramicidin A and alamethicin.

In some cases, e.g. gramicidin A and alamethicin, it has been postulated that these molecules actually form channels bridging the width of the bilayer (Urry 1972; Mueller & Rudin 1968). In agreement with this idea is the finding that freezing the lipid below its transition temperature does not affect the ability of gramicidin A to facilitate ion transport across lipid bilayers (Krasne *et al.* 1971). In any case, there is good evidence that some of these molecules readily penetrate monolayers and bilayers and can modify the lipid. Thus alamethicin molecules are thought to be situated at a lipid–water interface, with the polar side chains in the aqueous phase and the non-polar groups among the hydrocarbon chains. (It is of interest that the amino acid composition of alamethicin is particularly hydrophobic.) The evidence for this orientation at the lipid–water interface comes from the high surface activity of alamethicin observed in monolayer studies and from microelectrophoresis measurements which show that lecithin particles have a net negative charge (Hauser *et al.* 1970).

The interaction of the molecule gramicidin S with lecithin is an interesting one in that it shows that the polypeptide solubilizes the lipid even when it is at the temperature corresponding to the gel phase (Pache *et al.* 1972).

(*e*) *Lipid–protein–water systems*

A number of model systems incorporating protein into lipid–water systems have been studied by physical techniques such as X-ray diffraction, n.m.r. and e.s.r. spectroscopy. Shipley *et al.* (1969) used X-ray diffraction to study the interaction of phospholipids with cytochrome *c* in the aqueous phase. The lipid was a mixture of lecithin and phosphatidylserine, with the ratio of phospholipid to cytochrome *c* being about 300:1. Cytochrome *c* modifies the swelling characteristics of the lipid dispersion and two lamellar (I and II) lipoprotein systems were produced which in excess water gave dimensions of 8.7 nm and 11.6 nm. These dimensions were consistent with the incorporation of one or two cytochrome *c* molecules (diameter 3.1 nm) between the lipid bilayers (Fig. 4). The presence of water was necessary for the structural integrity of these phases. When the overall ratio of protein to lipid was increased yet

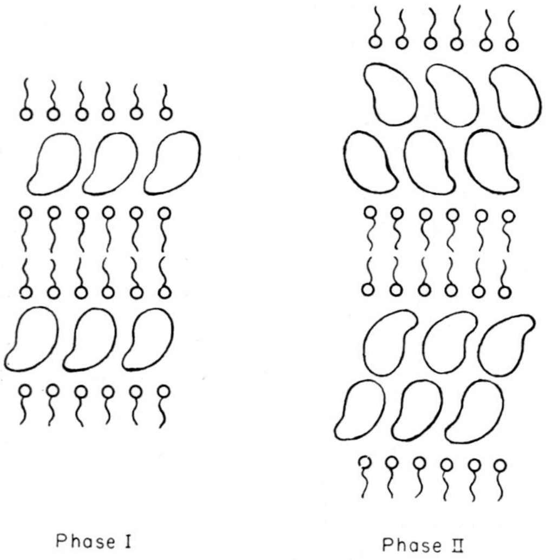

FIG. 4. A schematic diagram illustrating the way in which certain protein and lipid systems e.g. cytochrome *c* and phosphatidylserine–water systems, are organized (Shipley *et al.* 1969). A similar arrangement occurs with cytochrome *c* and phosphatidylinositol (Gulik-Krzywicki *et al.* 1969). Two lamellar phases (I and II) were observed with structures as shown. Tailed circles: lipid. The main interaction is considered to be electrostatic.

another lamellar phase (III) was observed. The dimensions of these model complexes were observed to be somewhat similar to those of myelin (a prominent second order reflexion at 8.6 nm). Similarly erythrocyte and mitochondrial membranes on centrifugation to a pellet give diffraction spacings of 11.5–12.0 nm, similar to the repeat pattern of lamellar system III.

Gulik-Krzywicki *et al.* (1969) used X-ray techniques to investigate complexes of ferricytochrome *c* with phosphatidylinositol, cardiolipins and mitochondrial lipids. They also studied mixtures of lysozyme with cardiolipin and with phosphatidylinositol. A variety of phases were observed. The ferricytochrome *c* complexes were lamellar and their structures were interpreted as indicating layers of protein–water between the lipid lamellae, some containing one protein layer and others two. However other systems, such as those containing lysozyme, gave a variety of phases, including hexagonal lattices, depending on the nature of the lipid. This work showed that the polymorphism of protein–lipid–water systems may be quite complex.

Rand (1971) has recently studied lecithin–cardiolipin mixtures with bovine serum albumin. He concludes that electrostatic interaction brings the components together so that shorter range polar and hydrophobic interactions can occur, causing subsequent gross conformational changes in the protein.

Bilayers have also been used to study protein–lipid complexes. Thus Steinemann & Lauger (1970) studied the interaction of cytochrome *c* with phosphatidylinositol in bilayer systems. At low ionic strength, about 10^{13} cytochrome *c* molecules/cm^2 are bound to the lipid surface and these authors concluded that the interaction was mainly electrostatic. The fast desorption of the protein after a rise in ionic strength showed that the protein did not penetrate into the lipid layer appreciably.

Clowes *et al.* (1972) studied the interaction of tetanus toxin with lecithin–ganglioside bilayers. In this case the toxin binds to the bilayer. The thickness of the lipid–protein complex increased from 7.9 ± 0.2 nm to 11.9 nm, which is consistent with the toxin maintaining its structure and not forming an extended layer over the surface, as was observed with erythrocyte protein (Cherry *et al.* 1971).

Cherry *et al.* (1971) studied the interaction of sialic acid-free erythrocyte apoprotein with erythrocyte lipids. They showed that the conductance increases up to three orders of magnitude and that the bilayer became electrically and physically unstable, leading to a rupture of the bilayer.

Studies with lipid vesicles have shown that soluble proteins such as lysozyme and cytochrome *c* cause a marked increase in permeability to ions. This has been correlated with the penetration of protein into the hydrocarbon region of the lipid membranes (Kimbelberg & Papahadjopolous 1971*a*, *b*). Recently a membrane protein, spectrin, isolated from erythrocyte ghosts (see p. 266 above), has been studied in these vesicles. When lipid vesicles are added to phosphatidylserine vesicles the ^{23}Na diffusion rate increases. Ca^{2+} and protein have a synergistic effect on ^{23}Na diffusion rate with phosphatidylserine and also phosphatidylserine–lecithin vesicles. In contrast to protein alone, the effect of calcium plus protein is seen at both neutral and acid pH. The nature of this interplay between protein and calcium is however unclear (Juliano *et al.* 1971).

Thus the importance of both ionic and hydrophobic associations has been pointed out in a variety of model systems. This has also been discussed by Green & Fleischer (1963) for mitochondrial membrane components, by Zwaal & van Deenen (1970) for the recombination of red blood cell lipids and apoproteins, and by Sweet & Zull (1969) for the interaction of albumin and lipid vesicles.

STUDIES OF CELL MEMBRANES

Let us now consider the information available on protein–lipid interaction in cell membranes. I shall briefly examine chemical evidence and also physical evidence.

Chemical evidence

The first point to note is that the ratio of protein to lipid in cell membranes varies from one membrane system to another. This is particularly notable when one compares the composition of myelin—20% protein and 80% lipid—to the composition of the erythrocyte membrane—49% protein and 43.6% total lipid.

The second point is illustrated by studies of the erythrocyte membrane. Thus the total lipid (weight per cent) is very similar in nearly all mammalian erythrocytes. The ratio of cholesterol to phospholipid is near 0.9 and the ratio of non-choline to choline-containing phospholipid is similar in human, bovine and porcine species. However, the bovine red cell contains little lecithin but has a high content of sphingomyelin, while the porcine red cell has a particularly large amount of glycolipid (12-14%) (Hanahan 1969). Thus there are differences in the classes of lipids found in various erythrocytes, i.e. in the same type of membrane system. There are also differences within individual classes; thus porcine erythrocytes contain more phosphatidylserine than the other erythrocytes examined.

We have also seen earlier (p. 266) that a variety of proteins is present in the erythrocyte membranes. The fraction of non-polar amino acids present in fractions IV, V and VI is 49 mol/100 mg and is greater than in the membrane protein as a whole.

Despite this, attempts have been made, using chemical methods, to determine how protein molecules are arranged in the erythrocyte membrane. Bretscher (1971) has shown, using radioactive labels that attach to the proteins of intact cells and cell ghosts, that a major protein (mol. wt. about 105 000 daltons) of the human erythrocyte membrane is in part on the external surface of the cell, and that different parts of the polypeptide are exposed on the two membrane surfaces. A further part of this polypeptide extends from one side of the membrane to the other. These conclusions are based on the assumption that the detailed molecular structure of the membrane is not substantially different in intact cells and ghosts. Bretscher (1971) suggested that this protein has a structural role.

Steck *et al.* (1971) studied vesicles of normal and inside-out orientation prepared from erythrocyte membranes. By treating each type of vesicle with proteolytic enzymes the two faces of the membrane could be selectively digested. The data suggested a highly asymmetrical arrangement of oriented proteins, some of which span the membrane (see Fig. 5).

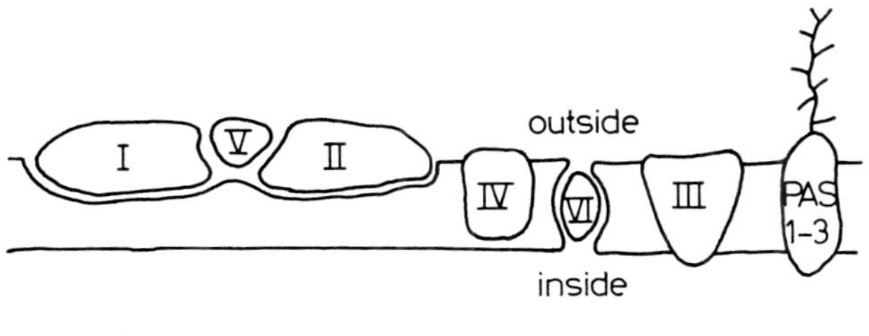

FIG. 5. A possible disposition for the various major protein and glycoprotein components. Components I, II, and V have been pictured as being rather superficial in location and physically associated *in situ*, while component VI has been assigned an indeterminate position. PAS 1–3 represents the sialo glycoprotein components. (After Steck *et al.* 1971.)

An important point which Steck *et al.* (1971) observed was a striking difference in the resistance to proteolysis of the surface of intact erythrocytes and of the isolated membranes. The lipids of intact membranes are much more resistant to degradation by phospholipase than those of the isolated membrane (Ibrahim & Thompson 1965). Zwaal *et al.* (1971) also observed that the phospholipids of intact erythrocytes are resistant to cleavage by highly purified phospholipase-C, whereas the isolated ghost showed cleavage of the lipids.

Thus isolation of the erythrocyte membrane increases its reactivity to a variety of agents and shows that caution is required in interpreting studies on isolated membranes.

Physical studies

Many physical techniques have been applied to cell membranes and it is convenient to summarize and integrate the conclusions on protein–lipid interactions by reference to a few particular membranes.

(1) *Outer segment membrane of the retinal rod.* A model of the outer segment has been developed, using chemical, electron microscope and X-ray data (Vanderkooi & Sundaralingam 1970). An important feature of this model is the globular nature of the membrane proteins. Fig. 6 shows the molecular

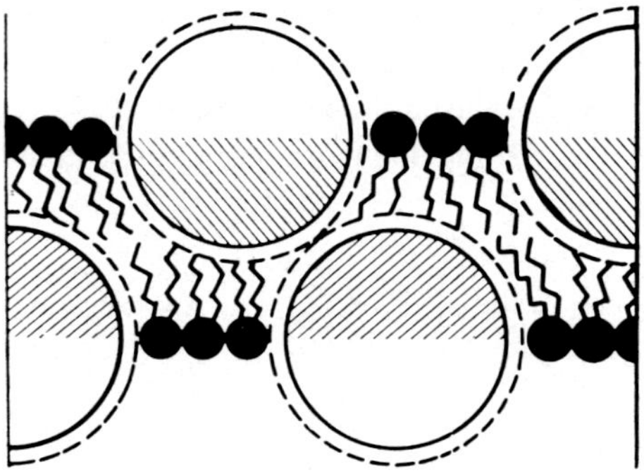

FIG. 6. A suggested arrangement for the membrane structure of the retinal rod. The filled circles are the polar heads of the lipids, the large circles are the proteins. (The solid line represents a 4.05 nm sphere and the dashed line a 4.6 nm sphere. The cross-hatched region is predominantly non-polar.) (After Vanderkooi & Sundaralingam 1970.)

arrangement. The authors suggest that there are protein–protein, protein–lipid and lipid–lipid interactions present but that these contacts will be constantly changing because of thermal motion. They argue that the lipids are hydrophobically bonded to the proteins, as shown by the difficulty in obtaining rhodopsin free of lipid in aqueous media.

(2) *Erythrocyte membranes.* Electron microscope studies using the freeze-etch method have been applied to a number of biological membranes.

The fracture faces of most biological membranes (not myelin) appear as smooth sheets interrupted by numerous particles (in erythrocyte membranes, about 8.5 nm in diameter, uncorrected for shadow thickness). When it became clear that the fracture face represents a plane through the interior of the membrane, it could be concluded that the particles represent localized regions where the bilayer membrane continuum is interrupted. In this context, it seemed legitimate to consider the membrane particles as the morphological representation of entities which are intercalated (but not necessarily sequestered) in the hydrophobic matrix of the bilayer membrane (Branton 1971). It is important to note that the particles observed in the erythrocyte membrane account for only about 30% of the total protein of the membrane (see comments by Singer 1971).

Freeze-etching has been used to follow the course of proteolytic digestion in red blood cell ghosts (Engstrom 1970). If the particles represent proteins within the membrane matrix, they should disappear when protein is removed from the ghosts before freeze-etching. Red blood cells were incubated in the proteolytic enzyme pronase for varying times. The course of protein hydrolysis was monitored by simultaneously measuring protein loss and particle loss. (Preliminary experiments indicated that pronase attacked all the cells in the preparation with equal vigour.) Loss of membrane protein initially caused extensive aggregation of the particles with little decrease in their number. The initial resistance of the particles to digestion by pronase is consistent with the view that the particles are buried within the hydrophobic region of the membrane matrix into which pronase diffusion would be greatly retarded. After more prolonged treatment with pronase, most of the particles were lost. This eventual removal of particles from extensively digested ghosts is consistent with the hypothesis that the particles are due to protein. These particles are considered to traverse the hydrophobic space of the bilayer and apparently can move when subjected to media of different pH. It is suggested that the plasma membrane is a planar fluid domain formed by a bilayer which is interrupted by localized yet mobile intercalations.

Nuclear magnetic resonance studies of erythrocyte membranes (Chapman *et al.* 1968; Kamat & Chapman 1968) have been compared with similar studies on serum lipoproteins (Chapman *et al.* 1969). The spectra of the serum lipoproteins show signals from the aromatic amino acids, and signals from the various molecular segments of the lipids which are present, e.g. the $[CH_2]_n$ group, the cholesterol ring methyl groups, the C=C groups, etc. The spectra of membranes in either the unsonicated or sonicated condition do not show the aromatic amino acid signals unless treated with detergent, whilst lipid signals only become more apparent after treatment with sodium deoxycholate.

The general conclusion is that a tighter, more compact structure of lipid and protein occurs in the membrane than occurs in the serum lipoproteins.

The circular dichroism spectra of erythrocyte ghosts in the region of 190–230 nm (as well as of some other membrane systems) show some of the features characteristic of proteins in a partially helical conformation with certain anomalous features such as low values of [θ] near 190 nm and a red shift above 220 nm. The red shift has been attributed to a number of different structural features of the membrane, including interactions between protein helices and an apolar environment for the protein helices. There is now convincing evidence that these anomalies are due to optical artifacts arising from the particulate, turbid nature of the samples (Urry 1972). Urry points out methods of correcting these distortions and suggests that about 50% or more of the red blood cell membrane protein is in an α-helical conformation. This has been interpreted to indicate that the proteins of the membrane are predominantly globular rather than spread out over the membrane surface. Models consistent with globular proteins intercalated in the membrane alternating with lipid bilayer structure have been suggested (Lenard & Singer 1966*a*, *b*; Wallach & Zahler 1966).

(3) *Sarcoplasmic reticulum membranes.* Sarcoplasmic membranes isolated from muscle tissues in the form of vesicular fragments which sustain Ca^{2+} uptake have been studied by n.m.r. spectroscopy (Davis & Inesi 1971). A reversible temperature-dependent structural transition was observed involving choline methyl groups of lecithin. The variation of the mobility of the methyl groups as a function of temperature correlated well with the temperature-dependent efflux of Ca^{2+}. Other agents which alter protein structure and simultaneously increase Ca^{2+} efflux also increase the fraction of mobile choline methyl groups. Irreversible heat denaturation of the membrane protein also alters the temperature dependence of the transition. Thus it was concluded that the observed thermal transition involves protein as well as lipid.

(4) *Acholeplasma laidlawii membranes.* Thermal (Steim *et al.* 1969) and X-ray studies (Engelman 1971) have shown that thermal phase transitions occur with these mycoplasma membranes. The temperature of the transition varies, depending on the nature of the fatty acids used to supplement the growth medium. These results are consistent with the idea that appreciable regions of the membrane contain lipid in a bilayer form. This membrane contains some 50% of protein in the membrane mass. The X-ray evidence (Engelman 1971) is interpreted to show that most of the protein is not in layers at the surface of the bilayer. We have also discussed this on the basis of calorimetric data (Chapman & Urbina 1971).

A possible model for the membrane protein structure which is consistent with the evidence is that the proteins are inserted almost entirely into the lipid bilayer, some perhaps penetrating it completely, some extending halfway through. The hydrophilic groups on the proteins would be in contact with the aqueous phase and the hydrophobic groups would be localized in the lipid hydrocarbon region. Engelman suggests that the central feature of membrane (as compared to non-membrane) proteins would then be that they can be substituted for groups of lipid molecules in the bilayer without greatly perturbing it.

Tourtellotte *et al.* (1970) have studied this membrane using a variety of techniques. The freeze-etch results show smooth areas and particulate components. The smooth area occupies more than 80% of the cell membrane while the particles represent 20% of the membrane.

(5) *The purple membrane of Halobacterium halobium.* Quite recently crystalline packing of visual pigment-like molecules has been observed in the 'purple membrane' fragment of the cell membrane of *Halobacterium halobium* (Blaurock & Stoeckenius 1971). The diffraction pattern of oriented sheets of purple membrane includes a number of sharp reflections centred on the axis (called the equator) in the plane of the membranes. The equatorial reflections are all accounted for by a regular, close packing of large numbers of hexagons in a plane; the centres of the hexagons are 6.3 nm apart. The scale of the array points to a crystalline arrangement of protein molecules. The extracted lipids, dispersed in water, show no sign of the equatorial reflections. The sharp reflections appear as lines extended at right angles to the equator, indicating that the protein molecules are arranged with their centres in planes parallel to the membrane surfaces. The variation of intensity along the line—reflections—suggests that the protein molecules are centred in two planes, one on each side of the membrane centre. The extensive equatorial pattern means that the protein molecules will be located in the membrane with unusual precision.

We have studied cell envelopes of *H. halobium* (D. Chapman & V. B. Kamat, unpublished). In high molarity NaCl solution the envelopes give broad featureless spectra, whereas in low molarity NaCl well-defined high resolution signals are observed, indicative of considerably increased segmental motion of amino acids of the protein.

(6) *Sindbis virus.* A number of animal viruses acquire a lipoprotein envelope by a process of 'budding' out through the surface of an infected cell. The Sindbis virus is a simple virus of this type, which Harrison *et al.* (1971) have studied using X-ray diffraction. The results were interpreted to indicate that the lipids are organized in a membrane around the spherical particle. The inner

polar groups of the bilayer are thought to interact with the core proteins and the outer polar groups with the external glycoprotein. In this system it is suggested that the protein does not form a bridge across the bilayer.

MEMBRANE BIOSYNTHESIS

Singer (1971) has suggested that in membrane biosynthesis a ribosome, or polyribosome, with bound messenger RNA, may become attached to a cell membrane. There has to be specificity to this attachment, so that the membrane proteins ultimately become associated with the right membrane. This might occur by specific interaction between a protein subunit in a membrane and its duplicate just synthesized on a polyribosome, or in part by a physical sequestration of specific messenger RNA molecules, such as occurs inside a mitochondrion (cf. Truman 1963; Beattie *et al.* 1967). Singer suggests that as the individual protein molecules are made, they would be inserted directly into the membrane without first being solubilized and not necessarily synchronously with lipid. By such a mechanism, clusters of identical protein molecules or clusters of several different protein molecules translated from the same polycistronic messenger RNA molecule (such as an enzyme complex) could arise in limited areas of a membrane, and different proteins could be incorporated into cell membranes at different stages of cell differentiation as the appropriate mRNA molecules became available (Dallner *et al.* 1966).

Experiments with *Escherichia coli* show that lipid synthesis is controlled by amino acid starvation in qualitatively the same way as RNA synthesis (Sokawa *et al.* 1968). Furthermore (Cronan & Vagelos 1972), the availability of auxotrophs has made possible the manipulation of the fatty acid composition of the phospholipids. Thus, the effects of blocking phospholipid synthesis either by deprivation of the required unsaturated fatty acid in the unsaturated fatty acid auxotroph or by deprivation of glycerol in the glycerol auxotroph have been studied in order to determine whether simultaneous phospholipid synthesis is required during the synthesis of the lactose transport system. Conflicting experimental results have been reported (Hsu & Fox 1970; Fox 1969; Overath *et al.* 1971). Wilson *et al.* (1970; Wilson & Fox 1971) extended the observation of Schairer & Overath (1969) indicating that the temperature profile of transport activity is specified by the fatty acid composition of the phospholipids. Wilson and co-workers found that properties of both the β-galactoside and the β-glucoside transport systems are characteristic of the phospholipid being synthesized at the time of transport system induction, rather than the predominant phospholipid of the membrane, and they concluded that newly synthesized proteins associate preferentially with newly synthesized lipids. However, similar experiments led Overath *et al.* (1971) to conclude that an induced transport system associates with the average lipids present at the time of induction rather than the lipids specifically synthesized at the time of induction.

Cronan & Vagelos (1972) have used a temperature-sensitive acyltransferase mutant to test the hypothesis that membrane lipid and membrane protein synthesis might be tightly coupled. They found no indication that a halt in membrane lipid synthesis produced any preferential effect on membrane protein synthesis.

A similar result was found during glycerol starvation of a glycerol auxotroph. Mindich (1971) showed with glycerol auxotrophs of *S. aureus* that a protein, the membrane-bound phosphotransferase, can be incorporated into membranes in the absence of phospholipid synthesis. Integration of the staphylococcal permease in the absence of lipid synthesis results in a transport system of low efficiency.

CONCLUSIONS

The variety of lipid structures in biological systems means that a whole variety of interactions with proteins can occur. Model experiments show that electrostatic, hydrophobic and metal-ion bridging mechanisms take place, sometimes separately and sometimes in combination. Protein separation techniques with membranes such as the erythrocyte membrane appear to confirm the importance of metal-ions for some of the protein–lipid structures.

Studies of the overall organization of cell membranes confirm the idea of a lipid bilayer being a dominant structure into which protein is intercalated. Evidence is accumulating to show that some of this protein can traverse the lipid bilayer.

In some membranes the protein appears to be particularly well ordered in the plane of the membrane. Some membrane structure may be heterogeneous as far as lipid is concerned, in the sense that some regions may be fluid at the growth temperature while other regions are in a gel condition (Oldfield *et al.* 1972). Experiments with antibiotics such as gramicidin A show that ion transport can still continue when the lipid is below its transition temperature because the antibiotic molecule forms a channel across the bilayer. Some proteins may be fixed in space in the plane of the membrane while others may be preferentially in a more fluid region and be more mobile.

Membrane biogenesis is still speculative, but interesting and important experiments on the lactose transport system may provide a clue to this process. Unfortunately, conflicting experimental results have been obtained on the question of whether newly synthesized protein associates preferentially with newly synthesized lipids.

References

BEATTIE, D. S., BASFORD, R. E. & KORITZ, S. B. (1967) *Biochemistry* **6**, 3099
BLAUROCK, A. E. & STOECKENIUS, W. (1971) *Nat. New Biol.* **233**, 152
BRANTON, D. (1971) *Philos. Trans. R. Soc. Lond. Ser. B.* **261**, 133
BRETSCHER, M. S. (1971) *J. Mol. Biol.* **59**, 351
BULL, H. B. & BREESE, K. (1967) *Arch. Biochem. Biophys.* **120**, 309
CHANGEUX, J. P., THIERY, J., TUNG, Y. & KITTEL, C. (1967) *Proc. Natl Acad. Sci. U.S.A.* **57**, 335
CHAPMAN, D. (1968) *Biological Membranes*, Academic Press, London
CHAPMAN, D. (1969*a*) *Introduction to Lipids*, McGraw-Hill, London
CHAPMAN, D. (1969*b*) in *Structure and Functional Aspects of Lipoproteins in Living Systems* (Tria, E. & Scanu, A., eds.), pp. 3–36, Academic Press, London
CHAPMAN, D. & PENKETT, S. A. (1966) *Nature (Lond.)* **211**, 1304
CHAPMAN, D. & URBINA, J. (1971) *FEBS (Fed. Eur. Biochem. Soc.) Lett.* **12**, 169
CHAPMAN, D., KAMAT, V. B., DE GIER, J. & PENKETT, S. A. (1968) *J. Mol. Biol.* **31**, 101
CHAPMAN, D., LESLIE, R. B., HIRZ, R. & SCANU, A. (1969) *Biochim. Biophys. Acta* **176**, 524
CHAVIN, S. I. (1971) *FEBS (Fed. Eur. Biochem. Soc.) Lett.* **14**, 269
CHERRY, R. J., BERGER, K. U. & CHAPMAN, D. (1971) *Biochem. Biophys. Res. Commun.* **44**, 644
CLOWES, A., CHERRY, R. J. & CHAPMAN, D. (1972) *J. Mol. Biol.* **67**, 49
CRONAN, J. E. & VAGELOS, P. R. (1972) *Rev. Biomembranes* **265**, 25
DALLNER, G., SIEKEVITZ, P. & PALADE, G. E. (1966) *J. Cell Biol.* **30**, 97
DANIELLI, J. F. & DAVSON, H. (1935) *J. Cell. Comp. Physiol.* **5**, 495
DAVIS, D. G. & INESI, G. (1971) *Biochim. Biophys. Acta* **241**, 1
DE BERNARD, L. (1958) *Bull. Soc. Chim. Biol.* **40**, 161
ENGELMAN, D. M. (1971) *J. Mol. Biol.* **58**, 153
ENGSTROM, L. H. (1970) Ph.D. dissertation, University of California, Berkeley
FOX, C. F. (1969) *Proc. Natl Acad. Sci. U.S.A.* **63**, 850
GREEN, D. E. & FLEISCHER, S. (1963) *Biochim. Biophys. Acta* **70**, 554
GULIK-KRZYWICKI, T., SHECTER, E., LUZZATI, V. & FAURE, M. (1969) *Nature (Lond.)* **223**, 1116
HANAHAN, D. J. (1969) in *The Red Blood Cell Membrane* (Jamieson, G. A. & Greenwalt, T. J., eds.), Lippincott, Philadelphia
HARRISON, S. C., DAVID, A., JUMBLATT, J. & DARNELL, J. E. (1971) *J. Mol. Biol.* **60**, 523
HARTLEY, B. S. (1955) in *Progress in the Chemistry of Fats and other Lipids* (Holman, R. T., Lundberg, W. O. & Malkin, T., eds.), vol. 3, p. 20, Pergamon, London
HAUSER, H., CHAPMAN, D. & DAWSON, R. M. C. (1969) *Biochim. Biophys. Acta* **183**, 320
HAUSER, H., FINER, E. G. & CHAPMAN, D. (1970) *J. Mol. Biol.* **53**, 419
HIGASHI, Y., SIEWERT, G. & STROMINGER, J. L. (1970) *J. Biol. Chem.* **245**, 3683
HIGASHI, Y. & STROMINGER, J. L. (1970) *J. Biol. Chem.* **245**, 3691
HSU, C. C. & FOX, C. F. (1970) *J. Bacteriol.* **103**, 410
IBRAHIM, S. A. & THOMPSON, R. H. S. (1965) *Biochim. Biophys. Acta* **99**, 331

JONES, T. H. D. & KENNEDY, E. P. (1969) *J. Biol. Chem.* **244**, 5981
JULIANO, R., KIMBELBERG, H. K. & PAPAHADJOPOLOUS, D. (1971) *Biochim. Biophys. Acta* **241**, 894
KAMAT, V. B. & CHAPMAN, D. (1968) *Biochim. Biophys. Acta* **163**, 411
KAUZMANN, W. (1959) *Adv. Protein Chem.* **238**, 3595
KIMBELBERG, H. K. & PAPAHADJOPOLOUS, D. (1971*a*) *J. Biol. Chem.* **246**, 1142
KIMBELBERG, H. K. & PAPAHADJOPOLOUS, D. (1971*b*) *Biochim. Biophys. Acta* **233**, 805
KRASNE, S., EISENMAN, G. & SZABO, G. (1971) *Science (Wash. D.C.)* **174**, 412
LADBROOKE, B. D., WILLIAMS, R. M. & CHAPMAN, D. (1968) *Biochim. Biophys. Acta* **150**, 333
LECUYER, H. & DERVICHIAN, D. G. (1969) *J. Mol. Biol.* **45**, 39
LENARD, J. & SINGER, S. J. (1966*a*) *Proc. Natl Acad. Sci. U.S.A.* **56**, 1828
LENARD, J. & SINGER, S. J. (1966*b*) *Science (Wash. D.C.)* **159**, 738
LOTAN, N., YARON, A. & BERGER, A. (1966) *Biopolymers* **4**, 365
LUZZATI, V. (1968) in *Biological Membranes* (Chapman, D., ed.), pp. 71–123, Academic Press, London
MARCHESI, V. T. & STEERS, E. (1968) *Science (Wash. D.C.)* **159**, 203
MARCHESI, S. L., STEERS, E., MARCHESI, V. T. & TILLACK, T. W. (1970) *Biochemistry* **9**, 50
MAZIA, D. & RUBY, A. (1968) *Proc. Natl Acad. Sci. U.S.A.* **61**, 1005
MINDICH, L. (1971) *Proc. Natl Acad. Sci. U.S.A.* **68**, 420
MITCHELL, C. D. & HANAHAN, D. J. (1966) *Biochemistry* **5**, 51
MUELLER, P. & RUDIN, D. O. (1968) *Nature (Lond.)* **217**, 713
O'BRIEN, J. S. (1965) *Science (Wash. D.C.)* **147**, 1099
OLDFIELD, E., CHAPMAN, D. & DERBYSHIRE, W. (1972) *Chem. Phys. Lipids* **9**, 69
OVERATH, P., HILL, F. F. & LAMNECK-HIRSCH, I. (1971) *Nat. New Biol.* **234**, 264
PACHE, W., CHAPMAN, D. & HILLABY, R. (1972) *Biochim. Biophys. Acta* **255**, 358
RAND, R. P. (1971) *Biochim. Biophys. Acta* **241**, 823
ROSENBERG, S. & GUIDOTTI, G. (1968) *J. Biol. Chem.* **243**, 1985
ROSENBERG, S. & GUIDOTTI, G. (1969) *J. Biol. Chem.* **244**, 5118
ROTHFIELD, L. & ROMEO, D. (1971) in *Structure and Function of Biological Membranes* (Rothfield, L., ed.), p. 251, Academic Press, New York
SCHAIRER, H. U. & OVERATH, P. (1969) *J. Mol. Biol.* **44**, 209
SCHERAGA, H. A. (1961) *J. Phys. Chem.* **65**, 1071
SCHERAGA, H. A., NEMETHY, G. & STEINBERG, J. Z. (1962) *J. Biol. Chem.* **237**, 2560
SHIPLEY, G. G., LESLIE, R. B. & CHAPMAN, D. (1969) *Nature (Lond.)* **222**, 561
SINGER, S. J. (1971) in *Structure and Function of Biological Membranes* (Rothfield, L., ed.), p. 145, Academic Press, New York
SINGER, S. Z. (1962) *Adv. Protein Chem.* **17**, 1
SOKAWA, Y., NAKUO, E. & KAZIRO, Y. (1968) *Biochem. Biophys. Res. Commun.* **33**, 1080
STECK, T. L., FAIRBANKS, G. & WALLACH, D. F. H. (1971) *Biochemistry* **10**, 2617
STEIM, J. M., TOURTELLOTTE, M. E., REINERT, J. C., MCELHANEY, R. N. & RADER, R. L. (1969) *Proc. Natl Acad. Sci. U.S.A.* **63**, 104
STEINEMANN, A. & LAUGER, P. (1970) *J. Membrane Biol.* **4**, 74
SWEET, C. & ZULL, J. E. (1969) *Biochim. Biophys. Acta* **173**, 94
TOURTELLOTTE, M. E., BRANTON, D. & KEITH, A. D. (1970) *Proc. Natl Acad. Sci. U.S.A.* **66**, 909
TRUMAN, D. E. S. (1963) *Exp. Cell Res.* **31**, 313
URRY, D. W. (1972) *Rev. Biomembranes* **265**, 115
VANDERKOOI, G. & SUNDARALINGAM, M. (1970) *Proc. Natl Acad. Sci. U.S.A.* **67**, 233
WALLACH, D. F. H. & ZAHLER, P. H. (1966) *Proc. Natl Acad. Sci. U.S.A.* **56**, 1552
WILLIAMS, R. M. & CHAPMAN, D. (1970) in *Progress in the Chemistry of Fats and other Lipids*, Vol. 2, (Holman, R. T., ed.), pp. 1–79, Pergamon Press, Oxford
WILSON, G. & FOX, C. F. (1971) *J. Mol. Biol.* **55**, 49
WILSON, G., ROSE, S. P. & FOX, C. F. (1970) *Biochem. Biophys. Res. Commun.* **38**, 617

WISHNIA, A. (1962) *Proc. Natl Acad. Sci. U.S.A.* **48**, 2201
WISHNIA, A. & PINDER, J. W. (1966) *Biochemistry* **5**, 1534
ZWAAL, R. F. A. & VAN DEENEN, L. L. M. (1970) *Chem. Phys. Lipids* **4**, 311
ZWAAL, R. F. A., ROELOFSEN, B., COMFURUS, P. & VAN DEENEN, L. L. M. (1971) *Biochim. Biophys. Acta* **233**, 474

Discussion

Luzzati: Some types of lipid–protein interactions have been fairly well defined in model systems similar to membranes:

(1) A hydrophobic type of interaction is characterized by a shrinking of the lipid bilayer in the presence of proteins, thus bringing some of the hydrocarbon chains in contact with the hydrophobic core of the protein. This type of interaction is confirmed by the effect of salts and by the fluorescent properties (Gulik-Krzywicki *et al.* 1969, 1970).

(2) A specific affinity of a basic protein of myelin for some lipids is observed, which leads to a doubling of the repeat distance of a lamellar phase, a fact indicating that the bilayer is asymmetric (L. Mateu *et al.*, unpublished findings, 1972).

(3) When one extracts the lipids from a biological membrane and studies them in the presence of water, one invariably finds that the lipids mix into a single phase. With serum lipoprotein, however, something different happens. These particles contain two kinds of lipids, phospholipids and cholesterol esters. In this case the two kinds of lipids, which stay together in the lipoprotein particles, come apart when they are extracted from the particles. One is thus led to postulate the presence of a protein specifically associated with the cholesterol esters (L. Mateu *et al.*, unpublished findings, 1972).

Chapman: The amount of protein is quite small compared with the amount of lipid in the myelin membranes.

Baddiley: Bretscher (1971) has proposed that in the erythrocyte membrane the composition of the phospholipids on the outer layer of the bilayer is different from that on the inner layer. I think he used erythrocyte ghosts in order to establish this. Could you comment on that, Professor Chapman?

Chapman: From the studies that have been made the protein appears to be asymmetrically organized in the membrane. Bretscher (1972) suggests that the lipids are asymmetric as well, i.e. that the lecithin and the sphingomyelin are all on the outside and the other types of phospholipid are on the inside. An easy way of testing this would be to put the membrane into an n.m.r. spectrometer and look at either the ^{1}H or ^{13}C resonances before and after adding manganese. If all the lecithin and sphingomyelin are on the outside then the manganese will broaden out all the $\overset{+}{N}(CH_3)_3$ signal. A quantitative estimate of total lecithin and sphingomyelin on each side of a natural membrane could be made by examination of this signal.

Changeux: You mentioned that there is some kind of local heterogeneity of motion within the bilayer structure of the membrane. How can you quantitate that? Is it possible to have more specific information about the proteins?

Chapman: I was talking about heterogeneity of the lipid. The calorimetric curves of, say, *Acholeplasma laidlawii* (Steim *et al.* 1969), or the X-ray information on the same system (Engelman 1971), or indeed recent X-ray studies on *E. coli* (Esfahani *et al.* 1971), show that not all the lipid has melted at the growth temperature. One has to go to a higher temperature for all the lipid to melt. What we suggest (Oldfield *et al.* 1972*a*) is that there may be regions where the lipid is in a rigid condition and other regions in a fluid condition. This raises a number of problems. For example when spin labels are used the spectrum shows that there is a highly fluid region in the membrane system. Now we find (Oldfield *et al.* 1972*b*) that the spin labels appear preferentially to go into the fluid region rather than into the rigid region. So if a cell membrane contains both these regions, one only detects the fluid region. (This is essentially a cautionary note about the use of the technique.) Some membranes are certainly going to be all fluid at the growth temperatures while other membranes are not. Why this should be the case I don't know.

Changeux: What fraction of the membrane is 'rigid', for instance in erythrocyte membrane?

Chapman: In *A. laidlawii* in some cases perhaps more than 80% of the membrane lipid might be in the rigid condition. Two other interesting facts related to this have not really been resolved. The lipid of myelin and at least some of the lipid of the erythrocyte membrane do not completely melt at the growth or body temperature (Jenkinson *et al.* 1969). If cholesterol is removed from myelin lipid a temperature of 55 °C is required to melt the lipid. Of course when cholesterol is present the lipid becomes liquid crystalline. Thus what I am saying is that in a number of membrane systems some lipid may be rigid at the growth temperature. In erythrocyte membrane because of the presence of cholesterol most of the lipid is in a liquid crystalline condition.

Changeux: Is it the fraction associated with proteins which is in a rigid condition?

Chapman: I am only saying there is a certain amount of lipid in a rigid form, but whether that is where the protein is situated I don't yet know.

Baddiley: Is there a strong correlation between the growth-temperature curve for a bacterial culture and the melting point of the lipids?

Chapman: Yes, and other poikilothermic organisms modify their fatty acids to correspond to their growth temperature. This does not mean that *all* the lipid needs to be fluid at the growth temperature.

Luzzati: Recent kinetic experiments on the melting of the hydrocarbon chains in lipids, lipoproteins and intact membranes have shown that in the presence of proteins and with heterogeneous lipids complex kinetic phenomena take place, which sometimes involve hysteresis effects. These seem to be related to a segregation of the different lipid molecules via slow diffusion in the plane of the bilayer.

Chapman: In lipid-water systems there is a reversible transition. On the other hand, with anhydrous lipids like the phosphatidylethanolamines hysteresis effects are observed (Chapman *et al.* 1966).

Silman: What are your views on the heterogeneity of the fatty acid chains of the phospholipids and on the fact that many biological membranes have so many different phospholipids, while *E. coli* can get along with just one type? To what extent is the heterogeneity an expression of the permeability properties of the membrane, and to what extent is it an expression of the interaction of the lipid with the protein to give a specific structure?

Chapman: My opinion is that the existence of different phospholipid classes is associated with particular interactions with protein. I also think that lecithin and sphingomyelin can substitute to some extent for each other, separating the regions where electrostatic interaction can occur between particular phospholipids and protein. An electrostatic interaction may be important in assisting hydrophobic effects. When lipids like gangliosides are present I think that they have specific interactions with protein. (We have put a ganglioside into a lecithin bilayer and added tetanus toxin. The tetanus toxin attaches itself to the ganglioside.) The fatty acids appear to produce the right kind of fluidity to satisfy the permeability processes for molecules coming into and out of the cell. But it isn't as simple as that. As I tried to point out, one finds heterogeneity even in fluidity within a membrane system.

Silman: What do you get out of the unsaturated fatty acids?

Chapman: It is too complicated to be definite yet about the role of the unsaturated fatty acids. We know that some membranes contain highly branched chains, while others contain highly unsaturated fatty acids. My own feeling is that broadly speaking it is again just a question of producing the right kind of mobility, or fluidity—taking into consideration what is available in that particular organism. Some detailed variations in the fatty acid structure may be particularly important for their involvement in hydrophobic interactions with protein. Other fatty acids are important as precursors of molecules such as the prostaglandins and it might be that prostaglandin production near particular membrane sites could be important.

Tanford: Most soluble proteins are known to be exquisitely specific. Is there any reason to believe that membrane proteins will be different? I think it is too naive to characterize interactions of membrane proteins by such general adjectives as 'hydrophobic' or 'electrostatic'. For example, a protein could be associated with the membrane by having specific binding sites for choline or other lipid head groups. Such an interaction would not be 'electrostatic' in the sense of non-specific attraction between opposite charges, and it would not be disrupted by addition of NaCl.

Chapman: I did not say that the proteins of the membrane have no specificity built into them. What I was suggesting was that perhaps the hydrophobic interactions between the protein and lipid may not need a marked specificity of fatty acids.

Tanford: I am only disagreeing with the idea that interactions with the surface have to be electrostatic.

Chapman: The interactions of protein and lipid in membranes may be electrostatic, hydrophobic, metal-ion bridging or a combination of all of these.

Elliott: The α-helical form and the random coil form are not identical but they are so close together that it is impossible to discriminate between them by the frequency of the amide-I bond.

Chapman: I agree. This is all based on your own work, Dr Elliott. To some extent it is supported by the CD spectra. I was pointing out that the evidence appears to support the idea that many membranes do not have appreciable quantities of β-structure. Despite all the anomalies the CD spectra appear to show the presence of marked amounts of α-helical structure.

Perlmann: I don't accept that interpretation. I had long arguments with J. Singer. All one can tell from CD spectra is that a conformational change has occurred. It is exceedingly dangerous to use these CD parameters to assign any conformation to a macromolecule.

Chapman: It was Urry (1972) who strongly pushed the idea of certain anomalies in CD spectra being due to scattering effects. Taking into account all the difficulties on the red blood cell membrane, he suggests that maybe 50% of protein is in the α-helical configuration.

Perlmann: There are cases where you get β-structure with similar band positions. With CD alone you can't tell which conformation is present.

Chapman: Urry points out these difficulties. Even taking model systems one can sometimes obtain the wrong answer. It looks α-helical even when it isn't.

Nevertheless, with that in mind, he says that about 50% of the erythrocyte protein is α-helical.

Perlmann: The information obtained from ORD and CD does not allow one to say there is so much of any given conformation present.

Baddiley: Some lipids such as glycolipids undoubtedly have very specific functions. For example, in the bacterial membrane we now know that diglycosyl diglycerides and phosphatidyldiglycosyldiglycerides in the bacterial membrane are anchors for the attachment of membrane teichoic acids (Toon *et al.* 1972). The poly(glycerolphosphate) chain of the teichoic acid is attached through a phosphodiester linkage to a hydroxy group, as yet unidentified, in the glycolipid. This is obviously a specific function of a lipid.

Chapman: I am sure that the phospholipid classes are important for specific interactions. The point I am uncertain about is whether all the fatty acids have to have a precisely defined structure for some particular hydrophobic interaction with the proteins of the membrane.

References

BRETSCHER, M. S. (1971) Major human erythrocyte glycoprotein spans the cell membrane. *Nat. New Biol.* **231**, 229–232

BRETSCHER, M. S. (1972) *Nat. New Biol.* **236**, 11–12

CHAPMAN, D., BYRNE, P. & SHIPLEY, G. (1966) *Proc. Roy. Soc. Lond.* **290**, 115

ENGELMAN, D. M. (1971) *J. Mol. Biol.* **58**, 153

ESFAHANI, M., LIMBRICK, A. R., KNUTTON, S., OKA, T. & WAKIL, S. J. (1971) *Proc. Natl Acad. Sci. U.S.A.* **68**, 3180

GULIK-KRZYWICKI, T., SHECTER, E., LUZZATI, V. & FAURE, M. (1969) Interactions of proteins and lipids: structure and polymorphism of protein-lipid-water phases. *Nature (Lond.)* **223**, 1116

GULIK-KRZYWICKI, T., SHECTER, E., IWATSUBO, M., RANCK, J. L. & LUZZATI, V. (1970) Correlations between structure and spectroscopic properties in membrane model systems. Tryptophan and ANS fluorescence in protein-lipid-water phases. *Biochim. Biophys. Acta* **219**, 1

JENKINSON, T. J., KAMAT, V. B. & CHAPMAN, D. (1969) *Biochim. Biophys. Acta* **183**, 427

OLDFIELD, E., CHAPMAN, D. & DERBYSHIRE, W. (1972*a*) *Chem. Phys. Lipids* **9**, 69

OLDFIELD, E., KEOUGH, K. M. & CHAPMAN, D. (1972*b*) *FEBS (Fed. Eur. Biochem. Soc.) Lett.* **20**, 344

STEIM, J. M., TOURTELLOTTE, M. E., REINERT, J. C., MCELHANEY, R. N. & RADER, R. L. (1969) *Proc. Natl Acad. Sci. U.S.A.* **63**, 104

TOON, P., BROWN, P. E. & BADDILEY, J. (1972) The lipid-teichoic acid complex in the cytoplasmic membrane of *Streptococcus faecalis* N.C.I.B. 8191. *Biochem. J.* **127**, 399–409

URRY, D. W. (1972) Reviews on biomembranes, vol. 1. *Biochim. Biophys. Acta* **265**, 115

General discussion

RECEPTOR PROTEINS

Changeux: It is dangerous to speculate about facts but it is even more dangerous to speculate about facts that are not well established. Nevertheless I would like to say a few words about some specific membrane proteins: the receptor proteins. At least three of them have been isolated: rhodopsin, the cholinergic receptor and the insulin receptor. Interestingly, these three proteins have a number of properties in common. (1) *The localization in the membrane*: they receive a signal from the outside which, at least in the case of the cholinergic receptor, controls the translocation of ions across the membranes. An important feature is that the permeant ions involved have to cross the hydrophobic phase of the membrane. (2) *The solubility:* the receptor proteins I mentioned are soluble in aqueous solvents only in the presence of detergents. Two of them, the cholinergic receptor and rhodopsin, are known to bind considerable amounts of detergents. About 20% of the apparent molecular weight of the cholinergic receptor protein solubilized by Triton X-100 comes from the detergent. (3) It is established for a number of membrane proteins, although this is not yet known for the cholinergic and insulin receptor, that the *amino acid composition*, although different from that of ordinary globular protein, still includes a large fraction of 'hydrophilic' amino acids. We are thus faced by the paradox of a protein with an amino acid composition close to that of a regular membrane protein but nevertheless embedded, at least in part, in a hydrophobic phase.

The model I wish to suggest is that a receptor protein presents at least two parts (globules or subunits). (1) One of these would recognize the regulatory signal, acetylcholine for instance, and would be some kind of standard globular protein facing the outside of the membrane. (2) The other, the 'ionophore', would bridge the two faces of the membrane and would be involved in the selective translocation of ions.

The suggestion I wish to make is that this second part of the protein might have an *inverted* structure, in the sense that the hydrophobic amino acids would face the hydrophobic phase of the membrane and the polar groups the centre of the molecule. These 'inside' polar groups would constitute some kind of hydrophilic channel through which the ions could cross the membrane.

This model implies that it might be possible to split the protein into two units: one, soluble in water, would bind the regulatory ligands; the other, insoluble in water, would bind the permeant. It might also be possible to test the prediction that the ions would cross the membrane in a hydrophilic phase but not in a hydrophobic one, etc.

Tanford: An alternative possibility, based on the same general assumption as you made, Dr Changeux, is that it is not the protein that is inverted, but the phospholipid bilayer. Specific sites built into the protein molecule could induce the hydrophilic head groups to turn inward, creating a hydrophilic passage through the membrane, the protein itself remaining in its normal type of conformation, with ionic amino acid side chains on the outside.

Katchalski: Dr Changeux, in cases in which the hydrophobic part of the protein is on the outside, is there a hole in the inner part through which the ions pass?

Changeux: According to the model the non-polar part of the 'ionophore' would interact with the hydrocarbon chains of the lipids but the polar groups of the lipids might very well interact with the exposed polar groups of the receptor moiety. The solubilization of the protein by detergents is viewed as a competition between lipid and detergent at the level of the exposed non-polar groups of the ionophore moiety.

The cholinergic receptor protein is known to be an oligomer in detergent solution but nobody knows whether this structure is still the same when the receptor is integrated in the membrane. The haemoglobin molecule has a single site for phosphoglyceric acid in the middle of the assembly of four subunits. Such a site might be a model for a hydrophilic 'channel'.

Chapman: The suggestion of haemoglobin analogues was also proposed by Wallach & Zahler (1966). Professor Tanford, why are you so reluctant to allow other conformations of protein to exist? I thought that in your paper you showed that there could be partial folding in unstable conditions, even with one solvent system. Why can't you have some kind of different folding of protein in a lipid solvent?

Tanford: I am not reluctant to admit any possibility, but I am reluctant to focus on one possibility to the exclusion of all others. As I said earlier (p. 103), there is no evidence to suggest that membrane proteins have dramatically fewer ionic amino acids than soluble proteins. Since ionic groups have strong affinity for water, I suspect that inverted protein conformations with ionic groups in the interior would have a prohibitively high free energy. I certainly don't visualize any of our kinetic intermediates as being of this type.

Changeux: That would mean that the interaction of these proteins with the membrane is essentially a polar one?

Tanford: No. I want to emphasize the possibility that individual membrane proteins may interact with phospholipids in specific ways, which may involve both polar and hydrophobic sites, and which may distort the normal bilayer structure that the phospholipids would adopt in the absence of the protein. Your model involves distortion of the protein but not of the bilayer. I want to distribute the burden more equally, so that a membrane protein would not need to be so drastically different from a soluble protein.

Changeux: I am not certain that such a large distortion is possible. In addition, the solubility properties of these proteins are very unusual.

Tanford: You don't need much hydrophobic surface to account for insolubility.

RECOGNITION BY MACROMOLECULES

Katchalski: Earlier we discussed how one macromolecule recognizes another. Dr Phillips, what hope is there of understanding what is going on in such recognition processes at the molecular level?

Phillips: By now we know a fair amount about interactions between rather small enzymes and various substrates. For example with lysozyme we see interactions with sugar moieties; with ribonuclease, we see how specificity towards pyrimidine bases arises; and, of course, there are many peptidases with which one can see interactions with amino acids. The problem in general is simply to get hold of the systems in which one can study the interactions of interest. It is rather remarkable that at this stage no one has succeeded in getting very good crystal specimens in which protein-nucleic acid interactions can be studied well, apart from the rather limited example of ribonuclease. On the other hand there is a fair amount of hope for development in this field. There is now progress with the immunoglobulins, and there seems to be some hope with Dr Gilbert's repressor protein.

Szwarc: I wish to give an example of recognition of a small molecule by a polymer. We prepared a polymer having the following structure:

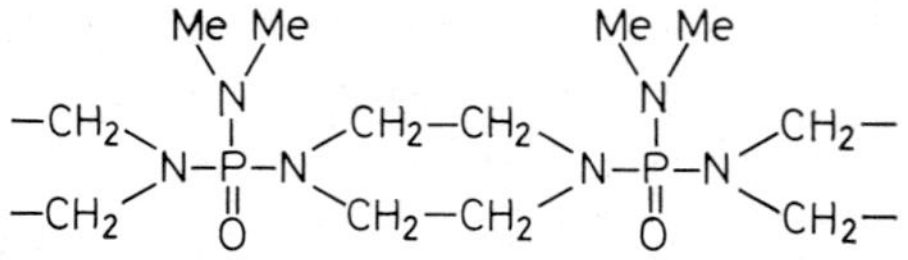

This is a polymer of relatively low molecular weight, completely miscible with water. It is prepared in dry acetone and precipitated with a hydrocarbon solvent, for example cyclohexane. It is then filtered and dried in a vacuum oven to constant weight, after which it is presumably pure. However, when such a dry polymer was dissolved in chloroform we found by n.m.r. that it contained 30% cyclohexane. This is remarkable because cyclohexane boils at 80 °C and the polymer was kept for many hours in a vacuum oven at 100 °C. When the polymer was precipitated by *n*-hexane, and then filtered and dried, no hydrocarbon was incorporated. When it was precipitated by a mixture of *n*-hexane and cyclohexane only cyclohexane was found in the polymer.

So, if you wish to use this language, the polymer recognizes cyclohexane. I believe the reason is very simple: piperazine units in the polymer backbone form chair-like structures, like cyclohexane itself, and when the polymer precipitates these chair-like rings stack one upon the other. If cyclohexane is present in the system, it becomes incorporated in the stack because it fits. Thus the occlusion takes place, although there is no hydrophobic or hydrophilic interaction or any hydrogen bonding; the geometry simply fits. Therefore the polymer recognizes cyclohexane. The melting point of the polymer is very high, over 200 °C, and the incorporated cyclohexane cannot get out.

Phillips: This is recognition by a specific structure in a synthetic polymer. The key thing is that the polymer itself is taking up a specific structure and then recognizing a complementary one.

Szwarc: Exactly; it is a key and lock mechanism.

Phillips: I go along with Dr Tanford in thinking of proteins as molecules that will recognize specific structures rather than behave in a random way. There is perhaps a gap in discussion between the biopolymer and synthetic polymer fields. In synthetic polymers one doesn't think so much about specific structures. In the synthetic polymer field people still talk very much about random coils rather than defined three-dimensional structures.

Flory: Surely it is well demonstrated that virtually all biopolymers occur as random coils under suitable conditions. Most synthetic polymers may occur in crystalline states of well-defined order under suitable conditions. It is a question of the temperature, chemical environment, etc.—a question of physical state. Biopolymers and synthetic polymers are not appropriately differentiated according to their physical states of order versus disorder. Attempts to differentiate the two categories (if indeed they need to be differentiated) on this basis call to mind vestiges of the vitalistic hypothesis. A number of ubiquitous polymers of biological origin—cellulose and rubber, for example—usually are relegated to the inferior category of the synthetics.

Phillips: I think I am judging whether the random coil is biologically important.

Flory: The random coil most certainly is important in elastic proteins such as elastin, which, it could be argued, is as essential as haemoglobin in serving the transport of oxygen. What is the helix coil transition if it does not involve the random coil?

Katchalski: It is quite obvious that biological specificity demands highly organized macromolecular structures. If one accepts, however, the supposition that the biosynthesis of biopolymers proceeds via initial formation of no apparent specific conformation, and that biopolymer degradation may well be preceded by denaturation, one has to accept some kind of random-to-ordered conformational transition as an integral part of the living process.

Phillips: I can accept that.

Flory: There are numerous examples of what we call random coil behaviour amongst native biopolymers: the elastic proteins are illustrative, as are shrinkage of collagen and contraction of muscle protein. To be sure, the specificity of an enzyme poses a sharper challenge. The random polymer in which there is no discernible order is a nondescript state with less appeal to those armed with sophisticated instruments. By the nature of currently fashionable methods, one's attention turns to structured constituents to the neglect of the non-structured or amorphous ones. The latter may escape observation but nevertheless be present. What we do not see is sometimes as important as what is readily sensed by physical methods. I seriously doubt that nature has been assembled with the objective of appealing to the aesthetic senses of man.

PREDICTING THREE-DIMENSIONAL PROTEIN STRUCTURE

Katchalski: The theoreticians have provided us with a good understanding from basic principles of the energetics of helical, β- and random coil structures, as well as of the helix-to-random coil transition. In working towards this understanding, the theoreticians have so far mainly utilized nearest-neighbour interactions. In order to cope with the structure of globular proteins one needs to take into consideration long-range interactions. Can you tell us, Dr Flory, whether there is some hope that the theoreticians will be able to achieve the above goal and predict the three-dimensional structures of globular proteins from their primary amino acid sequence?

Flory: The long-range interactions of a polymer chain with itself when packed in a dense conformation present formidable obstacles to an exact treatment. The refined calculations being carried out by Dr Levitt are certainly cast in the proper direction. Refining the crystallographic results will clarify the role of long-range interactions.

I don't want to deprecate the importance of the conformational calculations. On the other hand, I submit that the problem of starting from a given sequence of many peptide units, with full details of the chemical bonding, and deducing therefrom the most stable arrangement, especially if it is of the globular type, is beyond possibility of realization by calculations alone. It is a problem involving several hundred parameters and it does not factor into a product of functions with smaller sets of the variables occurring exclusively in each of the individual functions. The short-range problem will naturally resolve itself into functions of local variables, and even though these functions sometimes overlap, the overlapping is local. This latter problem can be solved by methods that are exact, but the total problem of deploying a macromolecule in space seems to me to be beyond any possibility of treatment, even with all the computers in the universe.

Lifson: Your point is well taken, but if we could calculate the equilibrium structure of a protein by many years of work with many computers, is that a worthwhile enterprise? The answer to this depends upon what we actually want to achieve by such a colossal effort. If it is only to prove that we can do it, then my answer would be in the negative. What we should really try to reach is a detailed quantitative understanding of the nature of those short-range and long-range molecular interactions and solvent effects which determine the protein structure. Such an understanding of molecular forces is a prerequisite for calculations of protein structure and can be derived only by study of molecular systems much simpler than proteins. If we could reach such a level of knowledge as to have confidence in our ability to find the true protein structure, at the price of years of effort, we might just as well spend such effort in asking many interesting questions about the structure and its relation to function in proteins and other biopolymers, to which there might be quicker answers.

Levitt: Certain situations in nature are very complex, like the arrangement of sand grains on a beach at a particular time; there seems to be no reason to attempt to predict such an arrangement in detail. These situations are dead phenomena. On the other hand, atoms in a protein conformation are ordered very precisely. Complex yet precise patterns characterize living systems, and attempts to predict such patterns will have to be made sooner or later. Protein folding is a particularly good example of the formation of a precisely ordered arrangement of atoms that is unique to living systems. But at present we do not even understand the nature of the forces that stabilize proteins, not to mention the interactions between proteins and water.

Huxley: But how will one know whether the principles are right unless one can apply them and calculate what the protein should fold into?

Lifson: We may apply them to simple systems where they can give relevant answers to relevant questions. We hope, for example, that gramicidin S, which is small, is no different in principle to a very large protein.

Levitt: I don't agree with you. In order to overcome the astronomical figure of 10^{100} possible conformations that Professor Flory is suggesting there has to be a special strategy in protein folding. That a protein ever folds up so precisely means that the actual folding process must have more to it than just searching for all possible arrangements and applying forces to get the most stable conformation. There just isn't enough time. You can only explore a few arrangements, say 10^{10} or 10^{20} different conformations.

Szwarc: You are not really talking about a problem of long-range forces but one of short-range forces. After contact due to a short-range force has been established, the next force comes into operation, and so on, because the pertinent groups find themselves close by.

Flory: I have not talked about long-range forces, only about long-range interactions, i.e. about interactions between units that are long range in the sequence of the polymer but happen to be at short range in space.

MECHANISMS OF LENGTH DETERMINATION IN PROTEIN ASSEMBLIES

Katchalski: Nature uses the template technique to synthesize macromolecules of a definite length and structure. In the aggregates that you have described, Dr Kellenberger, no templates seem to be involved. Would you like to discuss the manner in which the cell regulates the size and form of these and other aggregates? You have mentioned that specific genes are involved in such a regulation mechanism. What is the molecular mechanism by which they are expressed?

Kellenberger: Three mechanisms have been proposed but none of them are substantiated by experimental facts.

(1) *Template model*

The length of a tube is determined by a template, for example a given lenght of a nucleic acid or of a fibrous protein.

The only case in which RNA certainly plays a role in length determination is in the rod-like viruses. The length of infectious tobacco mosaic virus (TMV) particles is apparently at least approximately determined through the RNA. As far as I know, precise determinations have never been made, since it is nearly impossible to ensure that all particles observed in the electron microscope contain RNA and, when they do, whether it is of normal length (Francki 1966). In this case, since it is helically wound, the length of the RNA would be much longer than the length of the particle.

Phage tails could be determined by a certain length of DNA hanging out of the head. For phage λ an observation could be advanced in favour of this idea: Eiserling (quoted in Kellenberger & Edgar 1971) found that the lumen of the tail is negatively stained only when the phage head is also empty. When the phage is full there is no way of ascertaining that this lumen is occupied by DNA and not by another plugging substance.

The length of phage tails is about 0.1 μm—not much different from the average gene length (0.1–0.6 μm). One could also postulate the existence of a fibrous protein comparable to collagen and composed of α-helices. The peptide length would then correspond to about three times the length of the fibre, which would still be lower than the upper limit of estimated gene lengths.

This protein template model, as Dr Levitt has also suggested, is therefore not at all out of reach. For phage λ, one could tentatively predict that gene product U leads to such a fibrous protein; this could be tested experimentally.

(2) *The Vernier model*

This model has been proposed on several occasions, to my knowledge the most explicitly by Anderson & Stephens (1964). It is based on two concentric cylinders in close contact with each other and made by two types of subunits of different sizes. Growth of the double tube would be terminated at the given determined length when the two cylinders happened to be in register. As far as I know there is no example yet which would fit this attractive model.

(3) *The model of 'cumulated strain'*

This model (Kellenberger 1969) arose as a consequence of discussions I had with D. L. D. Caspar on the problems of quasi-equivalence. The principle is best explained by cylinders of the stacked disc type: the shape and bonding properties of the subunit are such that a disc is formed without constraint. In order to bind the discs together and to ensure a maximum of interacting surface contact, the subunits are slightly deformed. Each added disc needs, in consequence, increasingly stronger deformations to provide the maximum bonding surface. This continues until the required deformation is too strong and—by lack of affinity—a further disc is no longer bound. This situation is explained in Fig. 1.

a

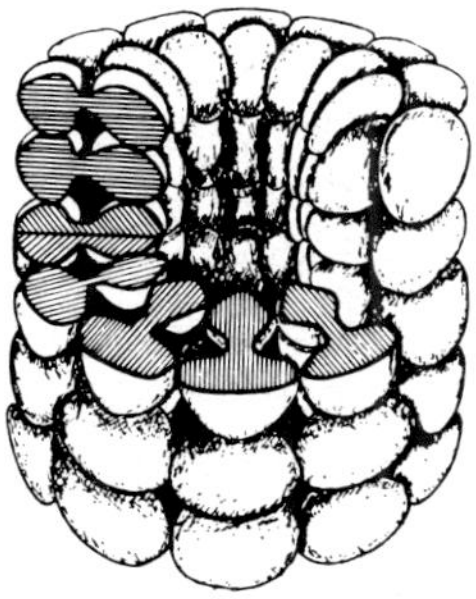

b

FIG. 1. (Kellenberger). Schematic explanation of the length determination by 'cumulated strain'. (*a*) The assembly of subunits whose geometry is such that they can provide a maximum interacting area only when they are deformed. From the centre to the left and right the deformation increases up to the point at which a further subunit can no longer interact. (*b*) The same model applied to a cylindrical structure. Deformation occurs only in the axial direction of assembly, but not for bonding within one disc.

In a T4 phage tail tube six subunits are thought to form a disc. Assuming a free energy for each bond of about 16.8 kJ (4 kcal) we would end with a total free energy of bonding between discs of about 100 kJ (24 kcal). Assuming for each deformed disc an investment of about 8.4 kJ (2 kcal) in deformation, then the 12th disc would contain 92.4 kJ (22 kcal) in deformation; a 13th disc would contain as much energy in deformation as is available through bonding. It is clear that this bond would no longer be achieved.

In this simple form, the model would predict that the outer discs can be 'melted away' much more easily than the middle ones. However, this is never observed. On the contrary it seems that the melting is an all-or-nothing process affecting the whole structure. This can only be explained by highly cooperative effects, by which the deformations are, at the end of the process of assembly, redistributed over all the subunits, only then reaching an equilibrium.

[*Note added in proof:* This more sophisticated model can be formulated as follows:

During the process of assembly the subunits induce conformational changes in each other (induced fit). These changes concern, however, not only the newly added subunit and the last assembled subunit but the already assembled part of the structure as a whole. In each step the energy stored in deformation is redistributed throughout the structure. Growth continues until a minimum energy state is reached, which would be characterized essentially by a minimum of hydrophobic regions staying exposed to the aqueous phase. Expressed in a simple-minded fashion, growth would be limited by the unavailability of hydrophobic regions for further bonding since the surface of the assembled structure would be entirely covered by the hydrophilic groups. This more sophisticated model of ('distributed') cumulated strain has the advantage of accounting for the observed sudden stabilization of the finished structure (non-reversibility), the all-or-nothing effect observed for its dissociation into subunits, and the very precisely determined number of subunits. The final state of equilibrium might be accessible to calculation; I doubt very much, however, that this is also so for the intermediate steps unless one knows much more about the extent and nature of the possible conformational changes which the subunit can undergo.

The model of cumulated strain finds an even more interesting application in the size determination of icosahedral virus shells. In the viruses with more than 60 subunits the hexon-covered faces would be limited in size by the cumulated strain.]

Levitt: Has anybody calculated the thermodynamics of the cumulative strain model? What sort of distribution of lengths do you get?

Kellenberger: The thermodynamic calculations have not been done yet.

The length distribution depends on how much one allows for conformational changes. Opinions vary, and facts are known for only very few proteins (haemoglobins). From electron microscopy observations I tend to think that proteins are much more flexible than is generally assumed. For example, empty capsids of large viruses are completely flattened by surface tension, but they are still not broken. Hence the quaternary structure of proteins must be very flexible and I think that not only is the bonding between subunits deformable but also—and mainly—the subunit itself. In the unsophisticated model at each step the strain is indeed increased by fixed amounts. (But see p. 297 for the situation in the 'sophisticated' model.)

Szwarc: Dr Flory, are the crystals of polyethylene under specific conditions always of the same size? That is, if you crystallize polyethylene you get single crystals, so if you keep the temperature constant are all the single crystals of the same size?

Flory: Crystallization is nucleation-controlled and, if the conditions are isothermal, which they usually are not, then the nucleation kinetics will control the size. Under these conditions the crystal size appears to be reproducible although statistically variable to some extent. The degree of regularity is not precise, as in a globular protein. It is generally less than commonly occurs in fibrous proteins. The level of order is much inferior to that occurring in biological structures.

Szwarc: In block polymers I showed structures that are regular within 10 or 20 nm and this is not a biological system.

Flory: I think the regularity of the globular protein is orders of magnitude beyond that of the block polymer. If you mean regularity of the bilayer I am not sure how to draw the comparison.

References

ANDERSON, T. F. & STEPHENS, R. (1964) Decomposition of T6 bacteriophage in alkaline solution. *Virology* **23**, 113–116

FRANCKI, R. I. B. (1966) Some factors affecting particle length distribution in TMV preparations. *Virology* **30**, 388–396

KELLENBERGER, E. (1969) Studies on the morphopoiesis of the head of phage T4. VII. Polymorphic assemblies of the same major virus protein subunit. In *Symmetry and Function of Biological Systems at the Macromolecular Level* (Engstrom, A. & Stranberg, B., eds.), pp. 349–366, Almquist & Wicksell, Stockholm

KELLENBERGER, E. & EDGAR, R. S. (1971) Structure and assembly of phage particles. In *The Bacteriophage Lambda* (Hershey, A. D., ed.), pp. 271–295, Cold Spring Harbor Laboratory, Cold Spring Harbor, N.Y.

WALLACH, D. F. H. & ZAHLER, P. H. (1966) *Proc. Natl Acad. Sci. U.S.A.* **56**, 1552

Chairman's Summary

E. KATCHALSKI

Our meeting started with Dr Szwarc describing how a synthetic polymer chemist tries to construct his polymers. Many of the polymerization techniques used by the chemist do not seem to appear in nature. There are, however, cases in which polymer biosynthesis proceeds via chemical reactions similar to those employed by the polymer chemist. For example, the mechanism of formation of many biopolymers *in vivo* closely resembles the synthesis of peptides by use of active esters as described by Dr Schwyzer.

The complex machinery involved in the biosynthesis of nucleic acids and proteins was lucidly described by Dr Zillig and Dr Revel. It is somewhat dismaying that the synthesis of a single protein molecule requires at least 100 different proteins, each with its own structure and role in the biosynthetic process. Dr Baddiley comforted us by explaining the 'relatively simple' machinery involved in the biosynthesis of the bacterial cell wall. However, a new type of component, the lipid component, was found to participate in the biosynthesis of the three-dimensional network which envelops the bacterium.

We then turned to a discussion of the three-dimensional structure of synthetic and native polymers. I should like to stress that the theoretical polymer chemist has gone a long way towards unravelling the intramolecular forces stabilizing simple regular macromolecular structures, as described by Dr Flory. Many of the biologists, however, are mainly interested in the structure and function of biopolymers possessing complicated conformations, such as those to be found in globular proteins. Unfortunately, the theoretician cannot as yet provide suitable tools for coping with such complicated structures.

Dr Tanford tried to persuade us that at least in the test tube one can get the 'wrong' kind of macromolecular conformation, which only slowly converts to the 'correct' one possessing the lowest free energy. Whether nature avoids such mistakes we still don't know. Dr Levitt has used all his theoretical skills to predict the secondary and tertiary structures of tRNA. He can relax for a while because the crystallographers are finding tRNA a tough nut to crack. The flexibility of segments of linear polymer chains even when located in a crystalline array was emphasized by Dr Corradini. His observation that flaws present in the polymer crystal can shift around quite easily offers a conceptual approach to explain local conformational changes in macromolecular structures and biological membranes.

Dr Flory's findings concerning phase separations *in vitro* leading to preferred macromolecular orientations might be of considerable importance in explaining the formation of transient structures within the protoplasm.

Dr Kellenberger and Dr Klug exposed us to the wonders of biology by showing us what nature can do when it wants to assemble subunits into higher structures, such as bacteriophages and TMV particles. Dr Susan Lowey and Dr Huxley introduced us into the muscle club by telling us about the structure of the myofibril and putting forward some of their ideas concerning the mechanism of muscle contraction. Highly specific macromolecular interactions were discussed by Dr Gilbert, who indicated that the binding constant governing the binding of the *lac* repressor to the *lac* operon is of the order of 10^{13}. In all the above assembly processes the interactions between the corresponding macromolecular units are highly specific.

Last but not least Dr Chapman introduced us to an even more complicated system, the biological membrane. In this system we are only beginning to understand what forces govern the interaction between its various components, and how these same components are arranged.

Before this symposium started we were worried, as is usual at meetings of people from different disciplines, about finding a common language. I think that we did indeed succeed in overcoming the language barrier. Moreover, this symposium again proved to me that the biologist is most certainly in need of the assistance of the physicist and chemist in order to understand thoroughly the complex systems that he handles. I hope that our polymer and physical chemists found such biological systems as enzymes, ribosomes, phages, muscles and membranes sufficiently fascinating to divert some of their energies from their basic work to assisting the biologists in their attempts to understand the living organism.

Index of contributors

Entries in **bold** *type refer to papers; other entries are contributions to discussions*

Indexes compiled by William Hill.

Subject index